THE
SKEPTIC

THE
SKEPTIC

A LIFE OF
H. L. MENCKEN

TERRY TEACHOUT

HarperCollins*Publishers*

THE SKEPTIC. Copyright © 2002 by Terry Teachout. All rights reserved. Printed in the United States of America. No part of this book may be used or reproduced in any manner whatsoever without written permission except in the case of brief quotations embodied in critical articles and reviews. For information, address HarperCollins Publishers Inc., 10 East 53rd Street, New York, NY 10022.

HarperCollins books may be purchased for educational, business, or sales promotional use. For information, please write: Special Markets Department, HarperCollins Publishers Inc., 10 East 53rd Street, New York, NY 10022.

FIRST EDITION

Printed on acid-free paper

Designed by Joseph Rutt

The text of this book is composed in Janson.

Library of Congress Cataloging-in-Publication Data

Teachout, Terry.
The skeptic : a life of H. L. Mencken / Terry Teachout.—1st ed.
p. cm.
Includes bibliographical references (p.) and index.
ISBN 0-06-050528-1
1. Mencken, H. L. (Henry Louis), 1880–1956. 2. Authors, American—20th century—Biography. 3. Journalists—United States—Biography. 4. Editors—United States—Biography. I. Title.

PS3525.E43 Z84 2002
818'.5209—dc21
[B]
2002024953

02 03 04 05 06 WBC/RRD 10 9 8 7 6 5 4 3 2 1

To Bill and Joe, gratefully

Grau, treuer Freund, ist alle Theorie,
Und grün des Lebens goldner Baum.

To me all men are equal: there are jackasses everywhere, and I have the same contempt for all. No petty prejudices!

Karl Kraus
Half-Truths and One-and-a-Half Truths

Refusal to face the verities, though not without immediate satisfactions, carries penalties. There's a Fool Killer, personifying the ancient principle; whom the gods would destroy, in this world; and he has a list; and that's a good way to put yourself on it. Then, the question's just one of time, of how soon he'll get around to you.

James Gould Cozzens
By Love Possessed

CONTENTS

Preface   *xi*

Prologue
"A PERMANENT OPPOSITION"   *1*

1 "I'D HAVE BUTCHERED BEAUTIFULLY"
   Birth of a Bourgeois, 1880–1899   *19*

2 "A SORT OF CELESTIAL CALL"
   Reporter and Editor, 1899–1906   *46*

3 "WIDER AND LUSHER FIELDS"
   Columnist and Critic, 1906–1914   *77*

4 "THE PURITANS HAVE ALL THE CARDS"
   At the Smart Set, *1915–1918*   *111*

5 "DON'T WAIT UNTIL THE FIREMEN ARRIVE"
   Becoming a Legend, 1918–1923   *143*

6 "I AM MY OWN PARTY"
   At the American Mercury, *1924–1928*   *187*

7 "I DISCERN NO TREMORS"
   Marriage and the Crash, 1929–1935   *240*

Contents

8 "WE'LL EVEN IT UP"
The Age of Roosevelt II, 1935–1941   *276*

9 "IT WILL BECOME INFAMOUS TO DOUBT"
Skeptic in Exile, 1942–1948   *298*

10 "UNTIL THOUGHT AND MEMORY ADJOURN"
Scourge at Bay, 1948–1956   *316*

Epilogue
"QUACKS AND SWINDLERS, FOOLS AND KNAVES"   *331*

Source Notes   *351*

Select Bibliography   *391*

Index   *399*

Illustrations follow page 142.

PREFACE

"The goods that a writer produces," H. L. Mencken wrote in *My Life as Author and Editor*, "can never be impersonal; his character gets into them as certainly as it gets into the work of any other creative artist, and he must be prepared to endure investigation of it, and speculation upon it, and even gossip about it." Dozens of writers have since taken him at his word. In addition to his own *Days* books, which amount to an informal (if highly selective) three-volume autobiography, he has been the subject of two full-length memoirs, Charles Angoff's *H. L. Mencken: A Portrait from Memory* (1956) and Sara Mayfield's *The Constant Circle: H. L. Mencken and His Friends* (1968); numerous monographs on particular aspects of his life and work; and several biographies, of which until recently the most up-to-date was Carl Bode's *Mencken* (1969), long the standard life of Mencken and on which I have gladly drawn.

But time and circumstances have caught up with Mencken's earlier biographers. In the three decades since Bode, several collections of Mencken letters have been published, including complete runs of his surviving correspondence with Theodore Dreiser, Marion Bloom, and Sara Haardt, and significant new research has been undertaken, much of it published in *Menckeniana*, the journal of Mencken studies. Most important, Mencken's private papers are now available to researchers. *The Diary of H. L. Mencken*, an abridged version of which was published in 1989, was only the tip of the iceberg. Opened along with the diary was a one-hundred-thousand-word manuscript of "Additions and Corrections" to the *Days* books. And on January 29, 1991, the thirty-fifth anniversary of Mencken's death, a four-volume manuscript called *My Life as Author and Editor* and a three-volume manuscript called *Thirty-five Years of Newspaper Work* were unsealed. These manuscripts, written during the forties and placed under time seal at Mencken's death, are not finished products; the diary makes it clear that he saw them merely as the raw material from which he would quarry polished memoirs of his journalistic career. Still, they are extraordinarily candid—even unguarded—and have added much to our understanding of his work and character. As he explained in his diary entry for December 29, 1941: "Up to this time, so far as I know, no really honest history of a newspaper, or of a magazine, has ever been written. . . . During the past 40 years I have seen both American newspapers and American magazines from the inside. If history is worth anything at all, then maybe this history that I have had a hand in will be worth setting down."

I AM, LIKE MENCKEN, A WORKING JOURNALIST, AND HAVE SPENT MOST of my adult life writing for and editing American magazines and newspapers. It stands to reason that the opening of Mencken's sealed memoirs of his professional life would have played a part in my decision to undertake a new biography of so celebrated an author and editor. Others have shared this impulse. When I went to work in the fall of 1991, another biography based on his private papers was already under way (it has since been published). But my own interest in writing about Mencken was also stimulated by the fact that the tone of American intellectual life has changed since 1969 in a way central to the reception of any new study of his life and thought. While the reputation of Mencken the stylist has remained high, critics from the early thirties on generally dismissed his claims to being taken seriously as a thinker. Realizing this, he placed his hopes of rehabilitation in the hands of future generations, arranging for the preservation of his papers in the hopes that his ideas might one day become more palatable. It took longer than he expected, but times have indeed changed: Mencken's social and political views, long thought irreversibly outdated, have become a resurgent strain in American thought. Like it or not, the Mencken *Weltanschauung* is once again a force to be reckoned with, and written about.

Unlike Mencken's previous biographers, I write, very broadly speaking, from his point of view, and I hope this has helped me to understand him more completely than either frankly hostile critics or those admirers who find his philosophy sympathetic but shrink (often quite properly) from its implications in the real world of action and ideas. My purpose, however, has not been to serve as an apologist, and I have not hesitated to be critical when I thought it appropriate, though I have also sought to place my criticisms—particularly as regards his alleged anti-Semitism—in an appropriate historical context. Perhaps the best way to describe this book is as what Henry James called a "partial portrait," an attempt to portray H. L. Mencken sympathetically but honestly, and to suggest something of how he stood in relation to his turbulent times.

This is *a* life of Mencken, not *the* life. I have made no attempt to be exhaustive, so as to avoid being exhausting. Accordingly I have said little or nothing about many of the people he knew well (such as the now-forgotten novelist Joseph Hergesheimer) and the events that took up much of his time (such as his work as a labor negotiator for the Baltimore *Sun*), usually because I felt that they were of little or no continuing significance in relation to the larger story of his life and work. Anyone who wishes to know more about him can easily consult more specialized sources, starting with his own memoirs. I have also been selective in discussing his writings, concentrating on those I believe to be most important and dealing sparingly or not at all with books of less intrinsic interest. This is a biography, not a work of literary criticism, and while I have tried to explain what it is about Mencken's style and ideas that continues to hold our attention, I have mostly used his work in order to illuminate his life.

FEW BIOGRAPHIES ARE COLLABORATIVE EFFORTS, BUT NONE IS WRITTEN without the aid of countless people. In my case the list of "collaborators" begins with the staff of the Enoch Pratt Free Library of Baltimore, Maryland, where most of Mencken's papers are now on deposit. "I have been a user of the Pratt Library since boyhood, and it gave me whatever education I may be said to have, so it seemed appropriate for it to receive my private records," he wrote in *My Life as Author and Editor*. He would have been pleased by the care with which the librarians of the Pratt watch over his literary remains, and the skill with which they guide visiting scholars through the maze of the Mencken Collection; in particular I must single out Vincent Fitzpatrick, curator of the collection and perhaps the most helpful colleague with whom I have ever had the pleasure of working.

I acknowledge with infinite gratitude the assistance provided by the Bradley Foundation and the Manhattan Institute for Policy Studies, which provided indispensable support during my final year of work. I also thank Tim Duggan, my ever-sensible editor; Sue Llewellyn, my erudite copy editor; Glen Hartley and Lynn Chu, my agents and protectors; Elaine Pfefferblit, formerly of Poseidon Press, who first suggested, then insisted, that I write a life of Mencken; Laura Demanski, Elaine's assistant and my friend in court throughout the early days of this long-lived project; Sydney Wolfe Cohen, who made a better index to the book than I could ever have made myself; Hilton Kramer and the late Samuel Lipman of the *New Criterion*, who commissioned the essay that inspired me to write about Mencken at greater length; Neal Kozodoy, the editor of *Commentary*, and Michael Pakenham, formerly the editorial-page editor of the New York *Daily News* and now the book editor of the Baltimore *Sun*, who helped make it possible for me to write about Mencken while earning a living; Richard Brookhiser, Andrew Ferguson, Leon Friedman, John Pancake, Richard Sudhalter, and

Jonathan Yardley, who said the right words at the right times; Jennifer Griffin, who introduced me to the marvels of 1524 Hollins Street; and Laura Lippman, who knows Baltimore well and helped me to know it better. Two particular debts are repaid (though only in part) on the dedication page of this book.

Last of all I want to thank my eighth-grade social studies teacher, Robert Nelson, who gave me my first book by H. L. Mencken. I'm not sure if the results were exactly what he had in mind, but I owe him anyway.

TERRY TEACHOUT

New York City
September 4, 2001

THE
SKEPTIC

"A PERMANENT OPPOSITION"

"I am going to Washington Saturday night to make a speech at the Gridiron Club dinner," H. L. Mencken wrote to a friend on December 7, 1934. "This is a dreadful ordeal for me, and I bespeak the prayers of all Christian people." Though the letter was both terse and ironic, as was Mencken's custom, it was impossible to misconstrue his meaning. Despite the fact that he thought "all save a small minority of the Washington correspondents" to be guilty of "intolerable incompetence and quackery," he was clearly looking forward to their semiannual stag dinner, for he had been invited to give the first speech of the evening, a face-to-face assault on Franklin Delano Roosevelt, president of the United States and the Gridiron Club's guest of honor. Nothing could have pleased Mencken more, or Roosevelt less.

The invitation might have surprised those who remembered Mencken solely as the reigning literary panjandrum of the twenties, a critic whose blistering attacks on the culture and customs of the Bible Belt (a term he coined) had made him the idol of every aspiring young writer in America. But he was also a hard-boiled newspaperman whose "Monday Articles," a fixture of the editorial page of the Baltimore *Evening Sun* since 1920, dealt as often as not with politics—and who, notwithstanding his reputation for radicalism, had only contempt for the orthodoxies of the age of Roosevelt. "All government, in its essence," he had written in 1922, "is a conspiracy against the superior man: its one permanent object is to oppress him and cripple him." Roosevelt's Herculean efforts to cure the Great Depression and give ordinary Americans a "new deal" by vastly expanding the size and scope of the welfare state had done nothing to change Mencken's opinion of big government, or of the voters who were allowing it to grow bigger.

Though he claimed that "in all my life I don't recall ever writing in praise of a sitting President," he had voted for Roosevelt in 1932, and subsequently gave him as much rope as he ever gave any elected official. While many of his Monday Articles in 1933 and 1934 were devoted to "the imbecility of the New Deal," he directed most of his fire at Roosevelt's brain trust of economic advisers. For the president himself he had kinder words: "We have had so many Presidents who were obvious numskulls that it pleases everyone to contemplate one with an active cortex. . . . I suspect that a large part of Dr. Roosevelt's hold on the plain people is that they recognize him to be what is called, for a lack of a better word, a gentleman."

Still, his appreciation of FDR's gentlemanly qualities stopped short of admiration. Roosevelt may have been a patrician, but he was also a politician, and Mencken considered himself innately superior to all such creatures, even highborn ones. The American politician, he wrote in 1926, is "a man who has lied and dissembled, and a man who has crawled. He knows the taste of the boot-polish. . . . He has taken orders from his superiors in knavery and he has wooed and flattered his inferiors in sense." And though his only previous visit to a Gridiron dinner had bored him, the bait dangled by club president James L. Wright was too tasty to pass up. Each banquet featured a "loyal opposition" speech by a critic of the current administration, to which the president replied at dinner's end. Instead of following the club's usual practice and inviting a politician to speak, Wright asked Mencken to do the honors, knowing the chore to be ideally suited to his well-known conception of the proper duties of a newspaperman:

"Finding virtues in successful politicians seemed to me the function of their swarms of willing pediculae; it was the business of a journalist, as I conceived it, to stand in a permanent Opposition."

The only hitch was that the invitation came at an awkward moment. Sara, Mencken's wife of four years, had just gotten out of the hospital, returning to the family apartment at 704 Cathedral Street under doctor's orders to stay in bed until further notice. Typically her husband made light of her illness in letters to friends: "The chiropractors want her to remain at rest as long as she can stand it. . . . My own belief is that they should liberate her at once, but I never argue with such fellows." But he knew better. Sara had a long history of acute respiratory infections, and Dr. Benjamin Baker, her longtime physician, had told Mencken that this one might kill her. The warning must have made it difficult for him to concentrate on routine chores, even one so agreeable as poking fun at the president of the United States before a room full of newspapermen.

Mencken's distraction became unmistakable on the day of the speech. Late that afternoon he and Paul Patterson, the publisher of the *Sunpapers* (as the morning and evening editions of the *Sun* were jointly known to Baltimoreans), drove down to Washington. On their arrival at the Willard Hotel, Mencken discovered that his garment bag contained nothing but soft shirts, which were unsuitable for a white-tie dinner. He hastily borrowed a spare shirt from Patterson and went off to shave and change. When he came back Patterson was shocked to see that the front of the shirt was spattered with blood. Preoccupied with his speech—and, presumably, with Sara's illness—Mencken failed to notice that he had cut himself shaving, and Patterson sent his chauffeur out to buy another dress shirt.

The two men then dropped in on a predinner cocktail party in the hotel, after which they descended to the anteroom of the banquet hall to mingle with the other guests. While looking for a men's room, Mencken heard a familiar voice calling his name. It was the president, who was sitting backstage, resting his withered legs in preparation for the long night to come. "I did not approach him,"

Mencken wrote in his diary the next day, "but he called to me and we had a pleasant meeting. He was extremely cordial, bathed me in his Christian Science smile and insisted on calling me by my first name." Mencken knew that by tradition, "reporters are not present" at Gridiron dinners, meaning that his speech, like FDR's reply, would not be quoted in the next day's papers. He could be as blunt as he pleased. Even so he had chosen to steer clear of outright abuse: "Inasmuch as [the speech] would embody a criticism of the President to his face, I decided to make it light and amiable in tone, avoiding altogether the harsh words that occasionally got into my *Evening Sun* articles." So he assured Roosevelt that his speech would be mild in tone, and FDR, "bursting with amiability," made a similar promise. He was lying. Mencken had no way of knowing that the thin-skinned president not only resented the columnist's earlier attacks but also held a grudge against the Gridiron Club for its acerbic anti–New Deal skits. Four hours later Roosevelt would have his revenge.

Mencken took his place on the dais at half past seven, seated next to Albert Ritchie, Maryland's outgoing governor and an old drinking companion. Well over four hundred guests were on hand, including Vice President John Nance Garner, Chief Justice Charles Evans Hughes, Gen. Douglas MacArthur, Joseph Kennedy, most of the cabinet, most of the leaders of the House and Senate, and all of Washington's top correspondents, as well as a few nonentities, among them the senator-elect from Missouri, an obscure machine politician named Harry Truman. Maryland terrapin was served, a tribute to the first speaker's birthplace and robust eating habits, and the skits began.

Just after nine o'clock Mencken stood up and unfolded his three-page typescript. Some of the members of the audience who were seeing him for the first time must have been startled to discover that he cut so undistinguished a figure. His 1942 draft card would sum him up matter-of-factly: "5' 8¼"; 185 pounds; race, white; eyes, blue; hair, black-gray; complexion, ruddy." An English journalist who met him three years later carried away a more colorful memory of what he looked like in his prime:

What I saw was a small man so short in the thighs that when he stood up he seemed smaller than when he was sitting down. He had a plum pudding of a body and a square head stuck on it with no intervening neck. His brown hair was parted exactly in the middle, and the two cowlicks touched his eyebrows. He had very light blue eyes small enough to show the whites above the irises, which gave him the earnestness of a gas jet when he talked, an air of resigned incredulity when he listened, and a merry acceptance of the human race and all its foibles when he grinned. He was dressed like the owner of a country hardware store. (On ceremonial occasions, I saw later, he dressed like a plumber got up for church.)

Mencken had no illusions about his lack of physical grace, but he wasn't self-conscious about it, or anything else. He knew his worth, and enjoyed his hard-won celebrity. (Among the papers he left behind at his death was a letter from Germany addressed, simply, "Herrn. Henry L. Mencken, U.S.A.") Every man in the banquet hall knew his name, and most could have rattled off his résumé from memory. For a decade he had divided his inexhaustible energies between the *Sunpapers*, which he served as columnist, consultant, and occasional reporter, and the *American Mercury*, the magazine that he cofounded in 1924 and edited until the end of 1933. His nineteen books and thousands of essays, articles, and reviews had won him the undying hatred of millions of devout believers in what he called "the whole Puritan scheme of things, with its gross and nauseating hypocrisies, its idiotic theologies, its moral obsessions." At the end of his life, one editorial writer would sum up a lifetime of middle-class outrage in a single sputtering sentence: "The wretched old crab did all he could with his vigorous pen to destroy the belief of men in their own souls." Now that pen was scratching away at Franklin Roosevelt and everything he stood for. Yet even those reporters who despised Mencken's politics knew him for what he was: America's greatest journalist.

"Mr. President, Mr. Wright, and fellow subjects of the Reich," he rasped, "I am put up here to speak a kindly word for the solvent and the damned, or, as the more advanced thinkers say, for the Rotten Rich." The raucous, gravelly voice was as homely as the man himself, the phrases coming in irregular, emphatic, almost explosive spurts, with every cigar-cured syllable bearing the stamp of "*Bawl*-mer" (as most native Baltimoreans, Mencken included, pronounced the name of their hometown). The speech, as promised, proved to be an even-tempered ribbing of the New Deal and its excesses: "Every day in this great country is April Fool Day. Even its wars usually produce quite as many comedians as heroes. Where on earth will you find a match for Congress, now that John Ringling has retired? Or for the NRA, now that cannibalism is prohibited by law?" It wasn't vintage Mencken, but it served its purpose. One hostile witness, the curmud-geonly Interior Secretary Harold Ickes, later admitted that his remarks "were cleverly cynical, as usual, but he wasn't particularly ill-natured."

After Mencken came two more hours of horseplay. Then, at last, it was FDR's turn. Roosevelt began by congratulating Mencken for "the temperateness of his remarks and criticisms. I had really expected more fireworks, in the inimitable Mencken style. When he deals so gently with the achievements and misachievements of the present administration I opine that we must be pretty good, after all." He noted that some of Mencken's "pungent strictures" were at times "irritating," but added that "my appetite is invariably whetted for more and his writings give me a chuckle after a hard day at the office"; he compared "the constructive character of his utterances this evening with the super-constructive character of his past writ-ings." He spoke always of "Henry," implying that he and Mencken, though they barely knew each other, were the best of friends.

Then FDR threw back his handsome head, turned off the Chris-tian Science smile, and unfurled a surprise that none of the reporters present would ever forget: He called them oafs and fools.

"There are managing editors in the United States," the president said, speaking in a slow, precise cadence as redolent of the ivy-covered halls of Groton and Harvard as Mencken's was of the red-brick row houses of Baltimore,

> and scores of them, who have never heard of Kant or Johannes Müller and never read the Constitution of the United States; there are city editors who do not know what a symphony is, or a streptococcus, or the Statute of Frauds; there are reporters by the thousand who could not pass the entrance examination for Harvard or Tuskegee, or even Yale. It is this vast and militant ignorance, this wide-spread and fathomless prejudice against intelligence, that makes American journalism so pathetically feeble and vulgar, and so generally disreputable.

A hush fell over the room. Only one man knew at once that FDR's remarks had been lifted verbatim from "Journalism in America," the first chapter of Mencken's own *Prejudices: Sixth Series.* Only one man grasped the president's purpose: to discredit his most famous critic by embarrassing him in front of his friends and colleagues. "I'll get the son of a bitch," Mencken whispered to Governor Ritchie, his reddening face a study in the limits of self-control. "I'll dig the skeletons out of his closet. I'll throw his 1932 campaign pledge speech right back at him." Aware that the Gridiron Club's rules gave Roosevelt the last word, Mencken nonetheless started drafting a reply, hoping vainly that an exception might be made. "Its chief point," he remembered later, "was the obvious one that whatever I wrote in 1927 I was still prepared to uphold in 1934, whereas there were other persons present who had made many categorical promises in 1932— for example, to balance the budget, to protect the dollar and to reduce the number of jobholders—and then repudiated them in 1933 and 1934."

As Mencken scribbled furiously, a few other people began to realize what Roosevelt was up to, and to chuckle at his audacity. Then, at the very end of the speech, FDR blandly announced that his closing remarks had been drawn in their entirety from the published works of "my old friend, Henry Mencken." The crowd exploded with laughter and cheers.

At dinner's end the president was wheeled past Mencken, who shook his hand and said, apparently to FDR's astonishment, "Fair shooting." Once Roosevelt was gone, though, several reporters came up to Mencken and complained about the president's unfair tactics. At least one cabinet member agreed with them. Writing in his diary the following week, Harold Ickes said that FDR "simply smeared [Mencken] all over." Roosevelt was unrepentant. "I did not really intend to be quite so rough on Henry Mencken," he wrote to Hearst editor Arthur Brisbane two weeks later, "but the old quotations which I dug up were too good to be true, and I felt in view of all the amusing but cynically rough things which Henry had said in print for twenty years, he was entitled to ten minutes of comeback."

"We'll even it up," Mencken later assured a colleague at the *Sun*. But what he said to Paul Patterson after the dinner, or to Sara on his return to Cathedral Street, went unrecorded, and no one else was allowed even a glimpse of his true feelings. Instead he tried to laugh the whole thing off. "I got in a bout with a High Personage at the dinner and was put to death with great barbarity," he wrote to a friend three days later. "Fortunately, I revived immediately, and am still full of sin." Yet there would be no more mention in the Monday Articles of the president's gentlemanly qualities. Sara died in May, and once Mencken returned to Baltimore from a European trip taken in a futile attempt to help him get over his grief, he opened fire at last: "Dr. Roosevelt is no longer the demigod that he was a year ago. He has lost his halo and his shining armor, and fast becomes, in the common sight, simply a politician scratching along." By year's end his rhetoric had grown harsher: "The greatest President since Hoover has carried on his job with an ingratiating grin upon his face, like that of a snake-oil vendor at a village carnival, and he has exhibited precisely the same sense of responsibility in morals and honor; no more."

What Mencken said about Roosevelt in public was mild compared to the verdicts he handed down to posterity in the privacy of his office, where he rapped out sulfuric diary entries and ferocious memoranda to himself on his Underwood typewriter. Certain that FDR was scheming to lure America into World War II, Mencken failed to grasp the full extent of Nazi Germany's depravity. Blinded partly by his hatred of Roosevelt and partly by his familial affection for German culture (Mencken was Saxon on his father's side, Bavarian on his mother's), he adopted an isolationist line that at its worst was rigid and callous beyond belief: "I find it difficult to work up any regret for the heroes butchered in World War II. Anyone silly enough to believe in such transparent quacks as Hitler, Mussolini, Stalin, Roosevelt and Churchill leaves the world little the loser by departing from it." When Roosevelt died his diary entry was just as spiteful: "He had every quality that morons esteem in their heroes."

Why did Mencken take it personally? Could he have found it galling that the middle-class son of a Baltimore tradesman had been bested before an audience of his peers by a rich man born to power and privilege? Whatever the reason, he never forgave, or forgot. Not long before a stroke forced him into retirement, he was asked by a student reporter, "Did you really dislike Franklin Roosevelt as much as you said you did, or was that just something to stir up the readers?" "Every bit of it," he replied. "In my book that man was an unmitigated S.O.B. He was an S.O.B. in his public life and an S.O.B. in his private life. Any other questions?"

By then he knew what he must already have suspected on that long-ago Saturday night in 1934: The winds of change had overtaken him. He himself had changed not at all. What he thought was no different in the twenties, when Walter Lippmann called him "the most powerful personal influence on this whole generation of educated people," than in the thirties, when the collectivist ideology he loathed had come to dominate the American political and intellectual scene. "All Government, in its essence, is organized exploitation," he wrote in the *Evening Sun* a year after the Gridiron dinner, "and in virtually all of its existing forms it is the implacable enemy of every industrious and well-disposed man." It was the same thing he had said in 1922, but nobody cared anymore. The Great Depression and the New Deal had intervened. In the end it is hard to blame H. L. Mencken for taking out his frustration on the gaudiest symbol of the whirlwind that had swept his power away.

ONCE THE POWER—LIKE THE FAME—HAD BEEN REAL. IN 1950 Edmund Wilson called Mencken "without question, since Poe, our greatest practicing literary journalist." If anything it was an understatement. As a literary critic, he wrote the first book ever published about George Bernard Shaw and was largely responsible for bringing Joseph Conrad, Theodore Dreiser, and Sinclair Lewis to the attention of American readers; as coeditor of the *Smart Set* and cofounder of the *Mercury*, he published the early work of Sherwood Anderson, Willa Cather, William Faulkner, F. Scott Fitzgerald, James Joyce, Eugene O'Neill, and Ezra Pound. In his spare time he produced *The American Language*, the pioneering study of the divergence of British and American English, which he saw through four editions and two supplementary volumes. He also launched three pulp magazines, wrote a bestselling autobiographical trilogy, edited a definitive anthology of his own writing, compiled a fat dictionary of quotations, translated Nietzsche's *Antichrist*, served as an unpaid literary consultant to Alfred Knopf, and dictated, by conservative estimate, a hundred thousand letters. Nor was his influence restricted to purely literary matters. As a working reporter and newspaper columnist, he covered everything from the Scopes evolution trial to the 1948 presidential conventions; as a popular philosopher, he produced a still-readable trilogy of "treatises" on democracy, comparative religion, and the history of ethics; as a social critic, he led the charge against the sterile pseudopuritanism of Prohibition-era American culture, an undertaking that his enemies on the left would always be quick to praise.

The hundreds of signed presentation copies on display in the Mencken Room of Baltimore's Enoch Pratt Free Library suggest how widely admired Mencken was by the best American writers of his day—especially those whose work he liked. Fitzgerald inscribed *Tales of the Jazz Age* to "the notorious H. L. Mencken, under whose apostolic blessing five of these things first saw the light"; Lewis inscribed *Main Street* to "H. L. Mencken—with apologies for making it *Henry* on page 263" (where the novelist had inserted a reference to "Henry Mencken" as one of the "subversive philosophers" whose books Carol Kennicott read wonderingly in Gopher Prairie, Minnesota). But even those authors for whom Mencken had little use found him difficult to ignore, as Ernest Hemingway grudgingly acknowledged in *The Sun Also Rises*: "I wondered where Cohn got that incapacity to enjoy Paris. Possibly from Mencken. Mencken hates Paris, I believe. So many young men get their likes and dislikes from Mencken."

To the public at large, he was best known for his knockabout assaults on the manners and morals of what he called, in his most celebrated coinage, the "booboisie," the Babbitts and Snopeses of Middle America and the South who were, after FDR, his favorite targets. He got as good as he gave. In 1926 some five hundred editorials about Mencken were published in American newspapers, at least four-fifths of which were unfavorable. Two years later Knopf brought out *Menckeniana: A Schimpflexicon*, an anthology devoted to scurrilous comments on his writing and person, all culled from his own bulging scrapbooks: "If Mencken only ran about on all fours, slavering his sort of hydrophobia, he would be shot by the first policeman as a public duty." "He seems to love the putrid, the sinful, the low, and uses any occasion to air his antipathy to the customs and beliefs of the average American citizen." "Mencken, with his filthy verbal hemorrhages, is so low down in the moral scale, so damnably dirty, so vile and degenerate, that when his time comes to die it will take a special dispensation from Heaven to get him into the bottommost pit of Hell." "Mencken is an outstanding, disgusting example of what constitutes a poor American."

His admiring contemporaries, by contrast, saw him as a real-life Fool Killer, wandering among the benighted with club in hand, ever ready to do battle with invincible ignorance. In the twenties liberal-minded Americans loved Mencken, just as they would later love FDR, for the enemies he made. Even his loathing of Roosevelt was initially seen by many older readers as another aspect of his crusty public persona, as E. B. White suggested in his parody "H. L. Mencken Meets a Poet in the West Side Y.M.C.A.":

> Poetry, religion, and Franklin D.
> The three abominations be.
> Why mince words? I do not feel
> Kindly toward the Nouveau Deal.
> Hopkins peddles quack elixir,
> Tugwell is a phony fixer.
>   Another lapse
>   For Homo saps.
> Yahweh!

By the mid-thirties, younger intellectuals, preoccupied with the search for a "usable past" and the rough beauties of proletarian culture, were dismissing Mencken as a reactionary period piece. Yet even in the early forties, when his reputation had reached its lowest ebb, one source of his declining fame would continue to win universal praise: his prose style. On the publication of *Happy Days*, the first volume of his memoirs, Otis Ferguson wrote in the *New Republic* that Mencken "wrote like a bat out of hell, and he picked his words like weapons"; on his death in 1956, Joseph Wood Krutch wrote in the *Nation* that "Mencken's was the best prose written in America during the twentieth century." When he died, one headline writer summed up the flashy surface of his career in five words: "Mencken, Critic Of All, Dies." Three decades earlier Walter Lippmann had captured something closer to the essence of the man: "He calls you a swine, and an imbecile, and he increases your will to live."

* * *

BUT IF MENCKEN IS NOW A CLASSIC, HE IS STILL AN IMMENSELY CONTRO-
versial one, and he has not yet lost his sharp corners and jagged
edges. The posthumous publication of his diary, with its caustic
remarks about blacks, Jews, and Roosevelt, triggered an avalanche
of criticism hardly less virulent than that collected in *Menckeniana:
A Schimpflexicon*. Many people who write about him, including
some of his most forgiving critics, have tended to portray him as a
one-dimensional ideologue, a self-caricature with stein and stogie.
That the "comic mask" of which Edmund Wilson spoke as early as
1921 has since become a permanent part of American literary history
is an eloquent tribute to his gifts as an essayist. Like all masks,
though, it must be removed in order to see the all-too-human face it
conceals.

Looking back on his career as a literary critic, Mencken told a correspondent, "Esthetic problems interest me only mildly. I am predominantly a reviewer of ideas and the more squarely those ideas are
based upon demonstrable facts the better I like them." He took his
own ideas with comparable seriousness: "Large parts of [my writ-
ings] have dealt with transient and forgotten themes, and are now
dead, but whenever I dip into them I am struck by the number of
ideas that still seem more or less valid and significant today. I have
thrown off, in my time, an enormous number of such ideas—possibly more than any other American writer of my time." Yet to examine these ideas even superficially is to be surprised by the extent to
which they contradict one another. Ambiguity is not a quality one
typically associates with Mencken. On the surface he was the most
confident of writers, a paragon of self-assurance; he was, as editorial
writers like to say of themselves, often wrong but never in doubt. But
his certainty was a theatrical illusion produced by the panache of his
prose. Though he posed as an apostle of common sense, the tug and
tang of ambiguity are part of what makes him so compelling, both as
author and as personality.

Mencken the passionate devotee of demonstrable fact, for example, was also Mencken the accomplished amateur pianist and some-time music critic. "When I think of anything properly describable as a beautiful idea," he wrote, "it is always in the form of music. . . . Nothing moves me so profoundly as the symphonies of Beethoven and Brahms. In fact, I get more out of them than I ever get out of books." And for all his rough-and-ready newspaper ways, the ex-poet who kept a complete edition of Oscar Wilde in his sitting room was an aesthete when it came to his own literary criticism: "The motive of the critic who is really worth reading . . . is not the motive of the pedagogue, but the motive of the artist. It is no more and no less than the simple desire to function freely and beautifully, to give outward and objective form to ideas that bubble inwardly and have a fascinating lure in them, to get rid of them dramatically and make an articulate noise in the world." As for the ideas of which he was so proud, anyone reading Mencken for the first time is likely to be struck by the conflict between the truculent pessimism of his philos-ophy and the infectious gusto of his temperament. "Here," he writes in *Notes on Democracy*, "the irony that lies under all human aspiration shows itself: the quest for happiness, as usual, brings only *un*happi-ness in the end." Yet there is a jolting quality to the relish with which he voices this grim sentiment, a quality that goes far beyond the sense of liberation with which the stoic accepts the inescapable facts of life. One might say of Mencken, as was once said of Paul de Man, the deconstructionist who in his youth published anti-Semitic arti-cles in pro-Nazi publications, that "he was the only man who ever looked into the abyss and came away smiling."

The delight Mencken took in staring into the abyss is the outward mark of a deeper rift in his thought. He was a lifelong skeptic who claimed to be "constitutionally unable to believe in anything absolutely," least of all anyone else's good intentions, and his scorn for "the uplift" is one of the most frequently sounded themes of his work. "I have little belief in human progress," he told a reporter in 1927. "The human race is incurably idiotic. It will never be happy." Yet he subscribed to the crudest possible version of the idea of progress: the belief that science might redeem the world. Mencken places the scientist at the apex of his personal pantheon, praising him as "one of the greatest and noblest of men. . . . he stands in the very front rank of the race." Much of his humor consists of loud jeering at those members of the booboisie who opted for faith over science: "If a man, being ill of a pus appendix, resorts to a shaved and fumigated longshoreman to have it disposed of, and submits willingly to a treatment involving balancing him on McBurney's spot and playing on his vertebrae as on a concertina, then I am willing, for one, to believe that he is badly wanted in Heaven." Even at the height of World War II, the ultimate refutation of the bland optimism of the nineteenth century, his correspondence still contained passages like this: "I am thus inclined to believe that the world is constantly growing better, at least on its more enlightened levels. . . . Every time the scientists take another fort from the theologians and politicians there is genuine human progress."

The strain of paradox that runs through Mencken's work can also be seen in his life. He is copiously praised in the memoirs of his contemporaries. "His readers," one friend wrote, "thought of him as bigoted, cantankerous, wrathful and rude. It was a case of mistaken identity. . . . I have never known a public figure who was so different from his reputation." Said another: "He believes and stoutly maintains that one strong enemy is more valuable than two mediocre friends, and then makes friends with the strong enemy soon after he shows up. He is in favor of a merciless autocracy and collects stamps for his brother's little child." Yet he was also a man who faithfully retreated to his office three times each day to read, write, and reflect on the shortcomings of his fellow men. He thought nothing of spending twelve-hour days at the typewriter, and though his friends adored him, he often seemed to view them as distractions from the current task at hand: "I get little pleasure out of meeting people, and avoid it as much as possible. . . . Nine-tenths of the people who call me by telephone I don't want to talk to, and three-fourths of the people I have to take to lunch I don't want to see." Those who bored, crossed, or otherwise irritated him were unceremoniously dropped, and sometimes dealt with brutally in the unpublished memoirs he left to the Pratt under time seal. Hamilton Owens, one of his closest associates at the *Sunpapers*, said of him, "Never was a man more gregarious, never one who strove more generously to keep his friendships green"; in his diary Mencken called Owens "a time-server with no more principle in him than a privy rat."

Mencken's peculiar notion of friendship paralleled his opinion of humanity as a whole: "The existence of most human beings is of absolutely no significance to history or to human progress. They live and die as anonymously and as nearly uselessly as so many bullfrogs or houseflies. They are, at best, undifferentiated slaves upon an endless assembly line, and at worst they are robots who leave their mark upon time only by occasionally falling into the machinery, and so incommoding their betters." His own sense of personal superiority was strengthened immeasurably by his early encounter with the writings of Nietzsche, the subject of his third book and the man whose thinking influenced him more than any other: "Like Nietzsche, I console myself with the hope that I am the man of the future, emancipated from the prevailing delusions and superstitions, and gone beyond nationalism."

But Mencken's prejudices against men in the mass were always open to revision in the particular. "The masses are animals, beyond salvation," he said. "But I'm out to collar the rest." Jews he professed to find "inherently incapable of civilization," yet his most intimate friendships were with George Jean Nathan and Alfred Knopf; blacks he regarded as congenitally inferior to whites, yet he gladly published in the *Mercury* the work of such writers as James Weldon Johnson and George Schuyler, and crusaded aggressively, even bravely, against lynching and for civil rights. For all his misanthropy, it cannot be said that there was no generosity in him, or love. The generosity was there in abundance, as his friends testified repeatedly. As for the love, one need only look at the diary entry in which he records the fifth anniversary of Sara's death: "It is amazing what a deep mark she left upon my life—and yet, after all, it is not amazing at all, for a happy marriage throws out numerous and powerful tentacles. . . . I'll have her in mind until thought and memory adjourn."

Much of Mencken's power as a writer lay in his ability to convey the complexities of his personality through the medium of his prose. One meets the man himself, incessantly skeptical, amused and amusing, in every sentence he wrote. His sheer vitality is as galvanizing today as it was to Joseph Conrad in 1922: "Mencken's vigour is astonishing. It is like an electric current. In all he writes there is a crackle of blue sparks like those one sees in a dynamo house amongst revolving masses of metal that give you a sense of enormous hidden power. For that is what he has. . . . [W]ho could quarrel with such generosity, such vibrating sympathy and with a mind so intensely alive?"

A further paradox is that the vividness of his personality may make it harder for latter-day readers fully to understand his work, which sounds far more modern than it is. (It had the same effect on younger readers in the twenties, which is why they were so disillusioned by what he wrote in the thirties.) In *Only Yesterday*, his influential popular history of America in the twenties, Frederick Lewis Allen placed the *Mercury* dead center in a list of such characteristic manifestations of the era as "the Dayton Trial, cross-word puzzles, bathing-beauty contests . . . radio stock, speakeasies, Al Capone, automatic traffic lights, [and] Charles A. Lindbergh." Yet those who knew its editor best called him a Victorian, and that, for better and worse, is what he was, in every sense of the word. He was born ten months before Billy the Kid was shot, and had finished his schooling, gone to work, and begun to publish before the close of the nineteenth century. "The only modern inventions that have been of any real use to me are the typewriter and the Pullman car," he told an interviewer in 1946. For all the vigor with which he blasted the booboisie, few American writers have been so completely accepting of their time and place. "If I had my life to live over again," he wrote in 1939,

> I don't think I'd change it in any particular of the slightest con-
> sequence. I'd choose the same parents, the same birthplace, the
> same education (with maybe a few improvements here, chiefly
> in the direction of foreign languages), the same trade, the same
> jobs, the same income, the same politics, the same metaphysic,
> the same wife, the same friends, and (even though it may sound
> like a mere effort to shock humanity) the same relatives to the
> last known degree of consanguinity, including those in-law.

Anyone seeking to understand H. L. Mencken should look with special care at the first three items on that list. A self-made man he may have been, but the materials out of which he made himself tell much about the singular man he became.

"I'D HAVE BUTCHERED BEAUTIFULLY"

Birth of a Bourgeois, 1880–1899

In 1883, when Henry Louis Mencken was nearly three years old, August, his father, bought a three-story row house that looked out on Union Square, a small park close to what was then the western edge of Baltimore. Except for the five years of his marriage and his first year as a widower, Mencken would live in that house until his death in 1956. Nothing about his life is as revealing as the fact that he spent so much of it in one place. Instead of immersing himself in the frenzied transience of modern-day America, he lived the settled life of a member of the European bourgeoisie, and liked it:

> The charm of getting home, as I see it, is the charm of getting back to what is inextricably my own—to things familiar and long loved, to things that belong to me alone and none other. I have lived in one house in Baltimore for nearly forty-five years. It has changed in that time, as I have—but somehow it still remains the same. No conceivable decorator's masterpiece could give me the same ease. It is as much a part of me as my two hands. If I had to leave it I'd be as certainly crippled as if I lost a leg.

Mencken would pay a price for the stability he loved so well. During the sixty-seven years he spent at 1524 Hollins Street, he watched his block, once a peaceful, tree-lined enclave, become a slum. Long after his death the streets facing Union Square would be partially (if temporarily) reclaimed, but the surrounding neighborhood continued to crumble. To the cop on the beat today, Southwest Baltimore is the innermost circle of urban hell, the dingy, drug-ravaged core of a blue-collar harbor town ringed by indifferent white-collar suburbs. Unemployment in Baltimore is high. So is the crime rate: Someone is murdered seven days out of ten, and most of those killings are drug-related (one out of ten Baltimoreans is a heroin addict). So are racial tensions, especially during the city's near-tropical summers. The city's population, eroded by white flight and the long decline of the steel and shipping industries, has been shrinking steadily ever since World War II, when it reached its peak; it was 703,057 in 1990, about 30,000 less than in 1920, when Mencken wrote his first Monday Article for the *Evening Sun*.

But Mencken's hometown has not changed quite so much since his death as the casual visitor might think. To go to an Orioles game at Camden Yards and watch the easygoing crowd root for the home team is to feel the pleasantness of life in a city slightly off the beaten path, an ingrown, insular, oddly comfortable place that time and prosperity have mostly passed by. The glossy riverside renovations of recent years have had little effect on the outward appearance of the rest of the city, many of whose neighborhoods, including Southwest Baltimore, still share the common architectural denominator of narrow streets lined with shabby old two- and three-story row houses, and a few Mencken-related landmarks have escaped the wrecker's ball. The family home survives in something quite close to its original condition; 704 Cathedral Street, the brownstone apartment house where Mencken lived with Sara, is still occupied, as is 811 West Lexington, his birthplace, though it now lies in the shadow of one of the city's most violent housing projects; Marconi's, the restaurant where the Menckens lunched together in the days of their long courtship, serves up crabcakes at the same address, just around the corner from the Enoch Pratt Free Library, to which Mencken left most of his private papers and three-quarters of the income from his copyrights.

Because so much of Baltimore looks much as it did in Mencken's day, and because he wrote about the city so memorably, the leap of imagination needed to conjure up the Baltimore of the 1880s, the place where the author of *Happy Days* spent his "fat, saucy and contented" childhood, is in certain ways a short one. Standing at the top of the six white marble front steps of 1524 Hollins Street and looking at the row houses that line Union Square, one sees what Mencken saw in 1927:

> The two-story houses that were put up in my boyhood, forty years ago, all had a kind of unity, and many of them were far from unbeautiful. Almost without exception, they were built of red brick, with white trim—the latter either of marble or of painted wood. The builders of the time were not given to useless ornamentation; their houses were plain in design, and restful to the eye. A long row of them, to be sure, was somewhat monotonous, but it at least escaped being trashy and annoying.

What Mencken was describing was a nineteenth-century middle-class urban neighborhood, a tightly knit community of proud home-owners who, like August Mencken, spent their days immersed in the intricacies of manufacturing or trade. "Such undertakings as these, however admirable in the great scheme of things, however productive of profit, drain off something from the men who direct them," Hamilton Owens wrote in *Baltimore on the Chesapeake*. "One doesn't ordinarily find . . . any great ebullience of spirit, any magnificent spilling over into colorful adventure." For all the pleasure Mencken took in it, the Baltimore of his youth was still a stodgy monument to the Protestant work ethic. Another of his *Sunpaper* colleagues called it "a slow, plodding, dull town." But the Baltimore of the 1880s was dull by choice as much as chance. Its citizens needed no reminding of the twin convulsions their parents and grandparents had survived:

the Know-Nothing riots and the Civil War. Antebellum Baltimore was called "Mobtown" (the term, according to Mencken, was still in use well into the 1880s) because of the frequency with which its residents rioted on election days. The nickname would acquire darker overtones. By 1850 nearly one out of every four Baltimoreans was foreign born, and the rapidly growing presence of German and Irish immigrants in Baltimore and elsewhere in the United States led to the founding of the Order of the Star Spangled Banner, a secret society of xenophobes whose members, sworn to vote only for native-born Protestants, answered the questions of outsiders with the phrase "I know nothing." Baltimore became a center of Know-Nothing activity in the fifties, and political clubs with names like the Black Snakes, the Rough Skins, the Red Necks, the Ranters, the Plug Uglies, and the Blood Tubs sprang up all over the city. When their members took to the streets, the police, bribed into passivity, looked the other way as rocks, muskets, and shoemakers' awls, the latter being the preferred Know-Nothing implement for hand-to-hand street fighting, were put into play. One riot left eight people dead and 250 wounded; in another, cannon fire was exchanged.

Know-Nothing violence was brought under control by 1860, but what followed, if less protracted, was even more divisive. Maryland, cut in two by the invisible line separating Northern wheatlands from Southern, slave-cultivated tobacco country, contained a large and vocal minority of secessionists enraged by the election of Abraham Lincoln. Tensions in Baltimore peaked on April 19, 1861, when the Sixth Massachusetts Regiment, marching through Baltimore on its way to Washington, D.C., in response to Lincoln's proclamation calling 75,000 militiamen into national service, was attacked by a secessionist mob. The resulting riot left four soldiers and twelve civilians dead, the first combat casualties of the Civil War. Maryland was placed under Union-supervised martial law; Baltimoreans fought on both sides in the war, returning at its end to find a city cloven by fear and distrust. To some degree that fissure still exists. Though comparatively few Marylanders think of themselves as full-

fledged Southerners, fewer still have completely forgotten the irredentist sentiments embodied in the official state song, "Maryland!
My Maryland!," composed by a Baltimorean, James R. Randall, in
the wake of the April 19 riot:

> The despot's heel is on thy shore,
>   Maryland!
> His torch is at thy temple door,
>   Maryland!
> Avenge the patriotic gore
> That flecked the streets of Baltimore,
> And be the battle-queen of yore,
>   Maryland! My Maryland!

The Civil War left Baltimore's principal markets in ruins, but
trade with the South soon resumed. Because Maryland had not
seceded from the Union, it escaped the Reconstruction laws the
South found so oppressive, and wartime government contracts had
put capital into the hands of a new generation of ambitious businessmen, among them Johns Hopkins and Enoch Pratt, whose names (if
nothing else about them) are known to every contemporary citizen
of Baltimore. Postwar Baltimore was "commission merchant" to the
South, a citywide warehouse supplying the burned-out ex-
Confederate states with the commodities they needed to rebuild, and
its newly prosperous citizens wanted, in the most literal sense, to
mind their own business. Too many of them remembered what life
had been like in the 1850s and 1860s to yearn for "colorful adventure"; too many immigrants had lived through the Know-Nothing
riots and told their children how they dodged the cannon fire and
shoemakers' awls. One Baltimore boy who heard the tales of
Mobtown days was Henry Mencken, whose maternal grandfather,
Carl Heinrich Abhau, was stoned more than once in the 1850s.
Grandfather Abhau's stories taught his grandson what an angry mob
can do to a helpless man. They were not far from Mencken's mind

when, in 1917 and again in 1941, he was forced into semiretirement because of his pro-German views.

Mencken's interest in his ancestors outlived his childhood. In later years he would devote a considerable amount of time and money to genealogical research. The first twenty-seven pages of the notes he dictated in 1925 for Isaac Goldberg, his first biographer, consist of a description of his family tree. A shorter account can be found in another set of notes he supplied eleven years later for his Associated Press obituary: "His father was August Mencken and his mother Anna Abhau, both born in Baltimore. His paternal grandfather, B. L. Mencken, was a native of Saxony who came to America in 1848, and married Harriet McClellan. These Menckens were in tobacco, but the traditional family trade was law, and various Menckens were lawyers, judges and professors of law in Leipzig during the 17th and 18th centuries. Others were professors of history, and one of them . . . established the first learned review in Germany. The Abhaus were from Hesse-Cassel, and through his maternal grandmother, Eva Gegner, Mencken had Bavarian blood." As this account suggests, Mencken thought well of his German forebears, particularly Johann Burkhard Mencke, rector of the University of Leipzig, who in 1715 wrote a pamphlet called *De Charlataneria Eruditorum* (The Charlatanry of the Learned). Two centuries later Mencken would describe this popular broadside as "a violent and highly effective lampoon upon the hollow pedantry of the day," subsequently commissioning an English translation that was published (at Mencken's expense) by Alfred Knopf.

For all his pride in his family's scholarly attainments, Mencken seems to have been no less impressed by the fact that they had dried up long before he came on the scene. "In Germany," he noted late in life, "the Menckenii have virtually disappeared, and at last accounts the official head of the clan was a country pastor. *Sic transit!*" The American branch of the family had little more to offer in the way of intellectual distinction. Mencken's great-grandfather ran a stock farm in Saxony and, later, kept an inn. Burkhardt Mencken, his grandfather, emigrated to Baltimore in 1848 to avoid being swept up in the Dresden revolution—"In his later life he used to hint that he had left Germany, not to embrace the boons of democracy in this great Republic, but to escape a threatened overdose at home"—setting up shop as a cigarmaker and thereafter going into business as a wholesale tobacco dealer. August and Charles, Burkhardt's sons, went into the tobacco trade on their own in 1873, and their firm, Aug. Mencken & Bro., was rated by Dun & Bradstreet at "more than $100,000, first credit" by the early 1880s. The firm's title was significant, for it was August who was mainly responsible for its day-to-day operations, just as he was responsible for the shaping of his oldest son's character.

BORN IN 1854, AUGUST MENCKEN LEFT SCHOOL IN HIS EARLY TEENS TO work in a Baltimore cigar factory, founding Aug. Mencken & Bro. at the age of nineteen. "In his later years," Mencken said, "my father was fond of boasting that he had but $23 of capital at the start and his brother but $21, but though this was literally true, they were backed by their father, certainly with a supply of tobacco and maybe also with a certain amount of cash." But August soon established himself as an energetic businessman in his own right, and his politics were as typical of his class as his bristling moustache and sideburns. He was a high-tariff Republican who ran a militantly nonunion shop and viewed the eight-hour day as "a project of foreign nihilists to undermine and wreck the American Republic." According to Mencken August's code of ethics "seems to have been predominantly Chinese. All mankind, in his sight, was divided into two great races: those who paid their bills, and those who didn't. The former were virtuous, despite any evidence that could be adduced to the contrary; the latter were unanimously and incurably scoundrels." An inveterate practical joker, he read Mark Twain with pleasure but was otherwise content to stick to the local newspapers; he went to church only when circumstances absolutely demanded it, and downed a generous hooker of rye whiskey before every meal, not excluding breakfast.

Self-made and successful, kindly but strong willed, and more than a bit of a philistine, August Mencken was in every way a man of his time and place, urban America in the late nineteenth century. "I never saw him in a place of disadvantage or embarrassment," H. L. Mencken remembered years later. "In the small world that he inhabited he always sat at the head of the table." Though Mencken would travel far beyond the boundaries of that world, he knew how much he owed to August. Four decades after August's death, he spoke of how his father had "instilled into me something of his general view of the cosmos. . . . There was never a time in my youth when I succumbed to the Socialist sentimentalities that so often fetch the young of the bourgeoisie. My attitude toward the world and its people is and has always been that of the self-sustaining and solvent class. It requires a conscious effort for me to pump up any genuine sympathy for the downtrodden, and in the end I usually conclude that they have their own follies and incapacities to thank for their troubles."

Six years after starting Aug. Mencken & Bro., August took a wife. Born in 1858, Anna Margaret Abhau was a small woman who, according to her oldest son, "looked, in her blue-eyed blondness, to be even younger than she actually was." Though he made no secret of his affection for Anna, this description is one of the few things Mencken wrote about her for public consumption. His only extended reminiscence of her is found in a private memorandum:

> My mother was a great worrier, and of a generally jittery disposition. She could think of more contingencies and catastrophes than anyone else I ever knew. . . . She lived long enough to see me more or less successful in journalism, and I think she took pride in my doings, but some of my early writings must have upset her, for she did not mention them. Her own reading was rather narrow in scope. . . . But she nevertheless managed to keep herself pretty well informed, at least for an aging woman of that era, and not infrequently she surprised me by the extent

of her knowledge. She had, in the formal sense, very little education, but she knew how to express her ideas in very clear and correct English, and it would have been ridiculous to call her ignorant. I do not recall ever opening with her any subject that she could not understand.

Anna and August being two conventional people, it should come as no surprise that their relationship led promptly to the customary denouement: Married in 1879, they became parents a year later. At the time they were living at what was then 380 Lexington Street (the street has since been renumbered), a two-story row house in a German neighborhood that August rented for sixteen dollars a month. It was in that house, on Sunday, September 12, 1880, at nine in the evening, that Henry Mencken, called "Harry" by his family, was born. Years later Anna would tell him that his birth had been hard, she being petite and he pudgy: "I was on the fattish side as an infant, with a scow-like beam and noticeable jowls. . . . If cannibalism had not been abolished in Maryland some years before my birth I'd have butchered beautifully." Difficult births were part of the normal course of life in the Baltimore of the 1880s, as were home deliveries. As Mencken pointed out, "It was rare for a solvent person to go to hospital for any purpose—even surgery was done in the house—and almost unheard of for a white woman above the level of a street-walker to go there for a confinement."

This last detail is telling, for it is possible to forget that we are as far removed in time from the world of Mencken's youth as he was from the world of George Washington. He was born just fifteen years after Lincoln was assassinated (and three years after the last Federal troops of occupation were withdrawn from the South), in the same year that William Ewart Gladstone succeeded Benjamin Disraeli as Queen Victoria's prime minister and Otto von Bismarck, a distant relative of August Mencken, was laying the foundations of the German welfare state. The thirty-eight United States of America (Colorado had been admitted four years earlier) contained

50,262,000 citizens (some 10 million of whom were illiterate), 54,319 telephones, 971 daily newspapers, one symphony orchestra, and 2 million acres of Indian territory. Census takers counted 1,131 "authors, lecturers and literary persons," 1,912 "hunters, trappers, guides and scouts," 12,308 journalists, 67,081 government officials, and 172,726 blacksmiths. Whatever their line of work, most people didn't live nearly as long as they do today (the average male life expectancy at birth was 40 years) or make as much money (the average annual family income was about $360, the equivalent of roughly $6,600 in 2001).

A year after Mencken was born, Wyatt Earp and Doc Holliday strode into the O.K. Corral and shot their way into the annals of popular culture (and onto page 2 of the *New York Times*). But while it would be another decade before the Census Bureau declared the frontier officially closed, the signs of its passing were already plain enough. A quarter of all Americans lived, like the Menckens, in urban areas, and the figure would rise to 40 percent by 1900. Four years had gone by since the Battle of Little Big Horn; in another three, Buffalo Bill Cody would take his Wild West Show on the road, and Harry Mencken would see it when, a few years later, it came to Baltimore ("We admired Buffalo Bill and shivered at the sight of his bloodthirsty Indians, but the general feeling was that the circus was better"). Though middle-class Baltimoreans still kept horses at the neighborhood livery stable, lit their homes with gas, and bought *McGuffey's Eclectic Readers* for their children, drastic changes in their daily lives were in the offing. It was in 1880 that Thomas Edison invented the first practical electric lights for home use. Factory-canned fruits, vegetables, and meats had just begun to appear in grocery stores; the first skyscraper was only three years away.

New cultural notions were in the air, too, but few of them had reached Baltimore, where the names of Henrik Ibsen and Friedrich Nietzsche, familiar in Europe, were virtually unknown. Johannes Brahms, Giuseppe Verdi, and Pyotr Ilich Tchaikovsky were the popular composers of the day, Anthony Trollope and Lew Wallace (formerly a Civil War general, now the author of *Ben-Hur*) the best-selling English-language novelists. Herman Melville, Ralph Waldo Emerson, and Henry Wadsworth Longfellow were all still alive. So were Charles Darwin and Karl Marx. Fyodor Dostoevsky had just published *The Brothers Karamazov;* Richard Wagner was at work on *Parsifal*, Pierre-Auguste Renoir on *The Luncheon of the Boating Party*, Mark Twain on *Huckleberry Finn*, Henry James on *The Portrait of a Lady*. Winston Churchill was five years old, Calvin Coolidge eight. Igor Stravinsky, Pablo Picasso, Louis Armstrong, Franklin Roosevelt, and Adolf Hitler had yet to be born.

BALTIMORE SEEMS TO HAVE BEEN FAIRLY QUIET ON SEPTEMBER 12, 1880. The lead story in the next day's *Sun* was "the burning of the steam planing mill in Uhlers alley, when two firemen were seriously and one slightly injured." Word of an uprising in Afghanistan led the front-page telegraphic summary of world news; the main story from Washington, D.C., was "Abuse of the Franking Privilege—Campaign Documents in the Mails." Further down the page was the news that "President Hayes has been made a vice-president of the American Bible Society." The second page led with this item: "THE SUN, as a family and business newspaper and advertising medium, is not surpassed by any journal in the country. It has a daily circulation of regular subscribers, besides its transient readers, of more than twenty-five thousand in excess of that of a paper published in Baltimore which meanly and fraudulently claims that it 'has more subscribers than any other paper published south of Mason and Dixon's line.'" That was the most colorful piece of writing in the whole paper. Neither the four-page main section nor the two-page "supplement" contained any illustrations, woodcuts, large-type headlines, signed columns, or traces of humor. The *Sunpapers* of Mencken's heyday, the lusty age of yellow journalism, were a long way off.

As yet unnoticed by the *Sun*'s editors was the extent to which Baltimore's primitive infrastructure was incapable of adequately serving a city whose population would balloon from 332,313 in 1880 to 508,957 in 1900. Open sewers running along the curbs of the cobblestone streets emptied directly into the back basin of the bay, and the city's water supply was unfiltered. "I can't recall ever hearing anyone complain of the fact that there was a great epidemic of typhoid fever every Summer," Mencken remarked in 1939, "and a wave of malaria every Autumn, and more than a scattering of smallpox, especially among the colored folk in the alleys, every Winter." Baltimore had the highest typhoid rate of any major American city of its time, and many families, the Menckens among them, slept under mosquito nets in warm weather. Few seemed to notice that much of the city smelled (in Mencken's words) like "a billion polecats," and fewer cared. Food, particularly fish scooped out of the Chesapeake Bay, was cheap and plentiful; first-class sipping whiskey, according to August Mencken's bill file, cost $4 a gallon in 1883. What was unavailable, be it high culture or modern sanitation, went unregretted. "The Baltimoreans of those days," Mencken remembered, "were complacent beyond the ordinary, and agreed with their envious visitors that life in their town was swell." It was a judgment in which their most famous son would happily concur: "This town is anything but perfection, but I know of no other more charming. What if it be ravaged by plagues, and blistered by a villainous climate, and sprawled over endless hills, and snouted and slobbered over by innumerable hordes of blue-nosed Puritans? Go to! There is yet its charm. We Baltimoreans like it, enjoy it, swear by it." For the rest of his life he would compare the world at large to his hometown, and find it wanting.

Mencken's lifelong love affair with the city of his birth was colored by the fact that his childhood was "placid, secure, uneventful and happy." *Happy Days*, the story of his first twelve years, is a book without shadows. Modern readers, their sensibilities formed in a harsher school of autobiography, often approach it with suspicion, assuming that its author must have had something to hide, but there is no reason to suppose that his memories of his days as "a larva of the comfortable and complacent bourgeoisie" are anything other than straightforward: "As Huck Finn said of 'Tom Sawyer,' there are no doubt some stretchers in this book, but mainly it is fact."

Interested readers of *Happy Days* can see the place where much of it is set. The Menckens moved to Union Square after a brief stay at a house on Russell Street, and the feature of their new home in which Harry took greatest delight remains intact: the fenced-in, hundred-foot-deep backyard. Together with his brother Charlie, twenty months his junior, Harry explored every inch of it. For him the Hollins Street yard was "a strange, wild land of endless discoveries and enchantments"; even now it is easy to see how a child would have found it vast and enchanting. It contained a peach tree, a cherry tree, a plum tree, a pear tree, and a grape arbor, from all of which the boys derived casual nourishment and more than a few bellyaches. "There was room in it," Mencken recalled,

for almost every imaginable boyish enterprise. Railroads ran from end to end of it, there was a telephone (made of cotton thread and two tin-cans) from the pony stable to the house, and at one time there was even a pond for boats. Always there was a dog, and usually a cat too. The trees were full of birds and my brother and I pursued them with air-rifles. . . . Even in the dead of Winter we were pastured in it almost daily, bundled up in the thick, scratchy coats, overcoats, mittens, leggings, caps, shirts, over-shirts and under-drawers that the young then wore.

When the resources of the backyard were temporarily exhausted, the outside world awaited:

The Hollins street neighborhood, in the eighties, was still almost rural, for there were plenty of vacant lots nearby, and the open country began only a few blocks away. Across the street from our house was the wide green of Union Square, with a fishpond, a cast-iron Greek temple housing a drinking-fountain, and a little brick office and tool-house for the square-keeper. . . . My grandfather's house was but three blocks away; my cousins lived next door; there were plenty of agreeable boys in the neighborhood. . . . Playing in the street was not dangerous, for there were no automobiles and no trolley-cars, and wagons had to go very slowly over the Baltimore cobble-stones. We played ball in the street. We spun tops on the cement walks of Union Square. . . . We walked the tops of the backyard fences, jumped out of stable hay-lofts, and pestered the neighborhood grocers by stealing their potatoes. These potatoes we would roast in the alleys. The alleys all seemed very romantic. Strange treasure was to be retrieved from their ash-cans.

Mencken felt his upbringing to have been permissive, at least by the standards of the day: "We were paddled freely for misdemeanors, but always gently and never after the age of six or seven. Most of the other punishments that I heard of from playmates were unknown. We were never sent to bed, or locked in a dark room, or deprived of food." This laissez-faire approach to discipline presumably had something to do with the fact that the four Mencken children (Harry and Charlie would soon be followed by Gertrude and August *fils*) were all well-behaved and precocious. Harry was poor at sports, but he could read by the age of three and write by the time he was five. In September 1886, he entered F. Knapp's Institute, a private school across from City Hall in downtown Baltimore, some two miles from Hollins Street. (He rode there by streetcar every day.) Not long after he started school, his baby fat melted away, leaving him "dreadfully round-shouldered." Too clumsy to shine on the playground, he concentrated on his studies, and his first-term report card, dated December 24, 1886, shows the results: "Deportment: Excellent. Industry: Praiseworthy. Advancement: Very satisfactory. Cleanliness: Very neat."

The ponderous teaching methods of Friedrich Knapp and his staff appear to have had little impact on Harry, who later claimed that "the professor and his goons certainly never taught me to speak German, or even to read it with any ease." Many years would pass before Mencken, having developed a serious interest in the culture of his ancestors, scraped together enough German to translate Nietzsche, conduct correspondence, and quote Goethe in the original on appropriate occasions. (August Mencken spoke almost no German, and his open hostility toward the beery *Gemütlichkeit* of Baltimore's German-American community left its mark on his oldest son.) Harry did his work well, quickly coming to be seen as "a sort of star pupil," but F. Knapp's Institute was in decline by 1886, and the formative intellectual and aesthetic experiences of his childhood took place elsewhere.

The earliest of these dates from 1888, when August purchased an old-fashioned square piano for four hundred dollars from Charles M. Stieff, Baltimore's very own piano manufacturer, and persuaded Harry to learn how to play it. The boy became a competent sight reader, a skill that served him well in later life, though he spent so much time playing piano at adolescent parties that he never learned to dance. As an adult he occasionally played the parlor piano in whorehouses while randier friends visited the girls upstairs. None of the "preposterous lady Leschetizkys" with whom he studied bothered to introduce him to the rudiments of harmony, but that did not stop him from composing "fifty or sixty pounds" of short piano pieces (a pound or so of which still survives), all in the style of the nineteenth-century salon music that remains to this day the staple diet of many small-town piano students. In spite of his teachers, Mencken developed a permanent taste for Austro-German classical music, and had he fallen into the hands of a first-rate teacher, his life might well have taken another course:

My early impulse to compose was no transient storm of puberty, explicable on purely endocrine grounds. It stuck to me through the years of maturity, and is still far from dead as I slide into the serenity of senility. . . . I have written and printed probably 10,000,000 words of English, and continue to this day to pour out more and more. It has wrung from others, some of them my superiors, probably a million words of notice, part of it pro but most of it con. In brief, my booth has been set up on a favorable pitch, and I have never lacked hearers for my bally-hoo. But all the same I shall die an inarticulate man, for my best ideas have beset me in a language I know only vaguely and speak only as a child.

After music came the printed word. In the summer of 1888 Harry read his first full-length books, a juvenile adventure story called *The Moose Hunters* and a copy of *Grimm's Fairy Tales*, given to him that June as a school prize. August and Anna soon noticed his growing interest in reading, and he found a Baltimore No. 10 Self-Inker Printing Press under the tree that Christmas. In the course of teaching him how to use it, August accidentally smashed all the lowercase *r*s in the Black Letter font that came with the press. Once Harry learned how to run it on his own, he decided to set up shop as a card printer and promptly went to work designing a business card for himself. He wanted to set his name in Black Letter, but was unable to spell out "Henry L." or "Harry" with the type that had survived August's heavy-handed tutelage, leaving him no alternative but to use, for the first time, his initials:

$$\mathfrak{H. L. Mencken}$$
Card Printer
1524 HOLLINS ST.
BALTIMORE, MD.

The joy Harry Mencken took in his Christmas present may well have paved the way for his first great literary discovery. The Hollins Street sitting room contained an old-fashioned secretary full of books that August had acquired seemingly at random, ranging from Ignatius Donnelly's *Atlantis* to a five-volume *History of Freemasonry in Maryland*. Among them was a complete set of Mark Twain, from which Harry, looking for something more stimulating than *Grimm's Fairy Tales*, plucked *Adventures of Huckleberry Finn*, then just four years old. "I had not gone further than the first incomparable chapter," he later wrote, "before I realized, child though I was, that I had entered a domain of new and gorgeous wonders, and thereafter I pressed on steadily to the last word."

As an adult Mencken would call his childhood encounter with Twain "probably the most stupendous event of my whole life." The author of *Happy Days* was partial to hyperbole, but this is no stretcher. Mark Twain gave him the beginnings of a style and a philosophy, and pointed him toward his life's work. At the age of eight he had found the perfect model—perhaps the only possible model—for the racy prose with which he would make his name. Mencken's youthful fascination with *Huckleberry Finn* foreshadowed many of his other mature tastes, from his interest in the subtleties of American dialect ("The shadings have not been done in a hap-hazard fashion, or by guess-work, but pains-takingly, and with the trustworthy guidance and support of personal familiarity with these several forms of speech") to his dislike of the sententious moralizing of so many nineteenth-century American authors ("Persons attempting to find a motive in this narrative will be prosecuted; persons attempting to find a moral in it will be banished; persons attempting to find a plot in it will be shot"). Time and again he would praise Twain in terms that made clear just how powerful an influence the older writer had on the younger: "How he stood above and apart from the world, like Rabelais come to life again, observing the human comedy, chuckling over the eternal fraudulence of man! What a sharp eye he had for the bogus, in religion, politics, art, literature, patriotism, virtue! What contempt he emptied upon shams of all sorts—and what pity!"

If Harry was lucky to have discovered Huck Finn at such an impressionable age, he was luckier still to have had an approving father. Many men in Victorian America might well have regarded Mark Twain as unsuitable fare for an eight-year-old child. (In 1876 *The Adventures of Tom Sawyer* was banned from the public libraries of

Brooklyn and Denver.) But August, a freethinker of long standing, merely said "Well, I'll be durned!," shyly confessed his own love of Twain's writings, and gave his son free access to the secretary and its contents. Harry read his way straight through the collected works and then tore into the rest of his father's library, paying special attention to a three-volume set of Dickens and a five-volume *Chambers Encyclopedia*. In due course he acquired his first library card and began "an almost daily harrying of the virgins at the delivery desk" of the Hollins Street branch of the Enoch Pratt Free Library. It was not until after he left F. Knapp's Institute that he would begin to explore English literature in a more orderly way, but even as a small child, books were at the center of his life: "I began to inhabit a world that was two-thirds letterpress and only one-third trees, fields, streets and people."

That summer the Menckens rented a country house in Ellicott City, ten miles from Baltimore. Harry played in the woods and swam in the river, and his daily wanderings took him to the printing office of the *Ellicott City Times*, a weekly paper that went to press on Thursdays. For the rest of the summer, he faithfully returned to the *Times* every Thursday. "I was captivated not only by the miracle of printing," he remembered, "but also by the high might and consequence of the young man in charge of the press. . . . He might give me a blank glance when he halted the press to take a chew of tobacco, but that was all. He became to me a living symbol of the power and dignity of the press—a walking proof of its romantic puissance." What the Baltimore No. 10 Self-Inker Printing Press had begun was finished by the ancient Washington handpress of the *Ellicott City Times*. From the summer of 1889 onward, the smell of printer's ink would be to Harry Mencken what the smell of greasepaint is to a stagestruck young man. Later Christmas presents (a microscope, a camera, a box of watercolors) would lay passing claim to his affections, but it was the printing press "that left its marks, not only upon my hands, face and clothing, but also on my psyche. They are still there, though more than fifty years have come and gone."

Harry was so wrapped up in printing that he failed to realize that "there was just as much delight to be got out of putting words together as out of setting them up in type. I recall vaguely that the compositions I was put to writing at F. Knapp's Institute irked me at the start, and that it was a red-haired girl named Mary Schilling, an orphan and a boarder at the school, who first taught me that concocting them might be stimulating. But whether she achieved this by precept and example, or simply by arousing my rivalry, I simply don't remember." By the end of his six-year stay at the school, he had made considerable progress as a writer, as his surviving composition books show. Yet in 1892 August Mencken sent him to the Baltimore Polytechnic School, an institution intended not for literary-minded bookworms but for budding engineers, chemists, and mechanics. It was a decision the boy would never understand:

> I had, in those days as in these, very little interest in mechanics, and next to no capacity for using my hands. I was eager to escape from F. Knapp's Institute, which was fast deteriorating, but the only thing taught at the Polytechnic which really interested me was chemistry. Perhaps the notion of sending me there was suggested to my father by the fact that two of the Lürrsen boys, our neighbors . . . had been students at the school. . . . I can think of no other motive in my father, for he certainly did not plan to train me for any mechanical pursuit. It was his hope that I would come into his business, and succeed him as its head.

A half century later he brought *Happy Days* to an abrupt close early in 1892. "I was then," he wrote, "at the brink of the terrible teens, and existence began inevitably to take on a new and more sinister aspect." For Harry Mencken, the idyll was ending.

NOT THAT LIFE AT THIRTEEN WAS ALL THAT DIFFERENT FROM LIFE AT twelve. The Baltimore Polytechnic, like F. Knapp's Institute, had its share of inept teachers, as well as a few good ones, among them a pair of drunkards who taught English *con amore* and introduced Harry to "not only Thackeray, Addison and Shakespeare, but also Tennyson, Howells, Swinburne and Kipling. . . . I had read almost the whole canon of the English classics down to the beginning of the Nineteenth Century before I was fifteen years old and to the classics I had added scores of books of second, third and fourth importance. I knew such things as Pepys' Diary, Boswell's 'Life of Johnson,' and the *Spectator* backward and forward." It was at the Polytechnic, too, that he became obsessed with Rudyard Kipling's *Plain Tales from the Hills* ("The chief decoration of my bedroom in Hollins street was a group of four or five portraits of him, brought together in one frame") and, a few years later, began to write Kiplingesque verse of his own.

Nor did Hollins Street suddenly become a cave of sorrows because Harry Mencken was a year older. The family's daily routine remained the same. August awoke early each day, ate a large breakfast, and rode the streetcar downtown to Aug. Mencken & Bro., where he answered the morning mail, inspected the previous day's new cigars, and otherwise occupied himself until noon. He returned just as the Mencken boys were coming home from school, quaffing his second rye whiskey of the day and sitting down to a lunch whose calorie content "probably ran to 3000," after which he retired to his swivel chair, kicked off his shoes, put his feet up on the sofa, opened the paper, and immediately dropped off to sleep for a half hour or so before going back to the office. (As an adult H. L. Mencken would also take a daily nap, though his was just before supper.) Anna ran the house with a firm hand, shopping at the neighborhood market on Saturday mornings and meticulously tending the family garden, from which she pulled, according to *Happy Days*, at least a million weeds.

Still, Harry's life was changing slowly but irreversibly, and not for the better. In a set of autobiographical notes prepared for the use of scholars and put under time seal, Mencken left no doubt as to his bleak view of adolescence: "The teens are at once too grotesque and too pathetic to be dealt with in the mood of the three 'Days' books.

They belong, intrinsically, to pathology, and it is no wonder that they offer a happy hunting ground to quack psychologists. The individual passing through them has lost the artlessness of childhood but is still far from the rationality of maturity." It is far from surprising that he chose not to write about his own adolescence in detail, but it is not very hard to imagine what it must have been like.

The main theme of his awkward age was the joys of reading (he finished an average of three books a week between 1893 and 1898), which soon led him to think of becoming a professional writer. Ironically it was his Christmas present for 1892, a camera, that seems to have pointed him in the direction of journalism. Like countless boys of the nineties, Harry did his own developing and printing; unlike them he wrote a scientific paper called "A New Platinum Toning Bath for Silver Prints," which he unsuccessfully submitted to several photographic magazines in the summer of 1894. It was his first attempt at an extended piece of writing.

The tempo of his life then accelerated sharply. In the course of his fourteenth year, he began to read the Romantic poets and was seduced by the daughter of a neighbor, who gave him a mild case of urethritis and an itch to write bad poetry of his own. In the spring of 1896 he tried his hand at a short story, an effort called "Idyl" which, years later, he curtly dismissed as *Ladies' Home Journal* fare (Anna Mencken was a charter subscriber to that magazine). Here is how it begins: "The gossips of the little village of Beaufort had long ago whispered that Mr. George Hudson and Miss Maude Van Horn would shortly become engaged to be married." Around the same time he discovered Thomas Henry Huxley, the apostle of Darwin who in 1869 coined the term "agnosticism," thereby supplying Mencken with the first pillar of what would become his intellectual credo. As an adult he would call Huxley "the greatest Englishman of the Nineteenth Century—perhaps the greatest Englishman of all time." As a teenager he responded to the clarity and simplicity of Huxley's prose, and happily dispensed with whatever shards of religious belief had survived his father's uncomplicated skepticism.

Harry graduated from the Polytechnic in June 1896. He was both valedictorian (an honor for which he collected one hundred dollars from August) and, at fifteen, the youngest boy in his class. "All I learned at the Polytechnic," he later claimed, "was forgotten within a year after my graduation. I can't imagine a more useless education than that I received there. . . . Most of the boys of my class have got nowhere in this life. One of them is a prosperous lawyer; the others are unheard of." He knew the world of business was not for him, but August was determined that his oldest son should enter and, in time, run the family firm. It was suggested at one point that Harry attend college before going to work, but he dismissed the idea after the fact in terms that suggest not so much a desire to plunge into the mainstream of life as a futile attempt to accept the inescapable: "It seemed to me nonsensical if I was to go into business that I should waste any time at college. I had then and still have a very low opinion of the advantages to be obtained from a college training. I had already discovered by experience that I could learn much more by my own efforts than I could get out of teachers."

What made Harry's situation so uncomfortable was that he knew what he wanted to be: a newspaper reporter. It was a peculiar ambition for a teenager who, as the adult Mencken pointed out, "grew up in an amazingly unliterary environment. There had been no author in my family for a century, and my immediate relatives, though very far from illiterate, showed no interest in books." Yet journalism was a calling that in other ways seems logical for a boy already in love with the smell of printer's ink and the prose of Mark Twain, one of the first nineteenth-century American writers to romanticize the reporter's life, writing about newspapermen as if they were cowboys: "Jones will be here at 3—cowhide him. Gillespie will call earlier, perhaps—throw him out of the window. Ferguson will be along about 4—kill him. That is all for to-day, I believe. If you have any odd time, you may write a blistering article on the police." He was also familiar with other writers of this school, including Richard Harding Davis, the dashing foreign correspondent whose *Gallegher and Other Stories*, published when Harry was eleven, tells the archetypical tale of the office boy at a big-city newspaper who plays detective, brings in a scoop, and wins a place on the staff.

Wherever he got the idea of becoming a reporter, it must have taken hold quickly. Given his provincial background, newspaper journalism would have been the only logical outlet for his budding literary impulses. It was also one of the few unimpeachably masculine professions open to a young writer, an important consideration for a round-shouldered scholar who played piano at school dances instead of flirting with the girls. (As an adult he would often be reluctant to acknowledge his aesthetic streak.) But Harry Mencken was no rebel. The only bombs he ever threw, then or later, were made of paper and ink. He knew what was expected of him, and a few months before his sixteenth birthday, he went to work for Aug. Mencken & Bro.

Harry liked cigar making well enough, but his heart sank when he was promoted from the factory floor to the front office, and a stint as salesman proved even worse. Eventually August returned him to the drudgery of the office, after which the boy confessed to Anna that he wanted to leave the family business and become a newspaperman. She gave him her blessing, and he told August at some point during the summer of 1898. "He was naturally pretty well dashed," Mencken remembered, "but he did not protest with any rancor, and it was understood between us that we were to resume the discussion in a year or so." In the course of that talk, August confessed that he had dreamed of becoming an engineer as a young man, but resigned himself to his fate and started Aug. Mencken & Bro. upon discovering that he lacked the education necessary to pursue his goal. August told Harry the story "as if to console me," but it seems more likely that it strengthened his resolve to break away.

For the next three years most of his excess energies went into writing. Shortly after graduating from the Polytechnic, he had broken into print for the first time with a mock-lugubrious "Ode to the

Pennant on the Centerfield Pole" ("O sea-sick flag!/O aged rag!/Though high thy drooping head may raise/We know thou hast seen better days") published in the Baltimore *American*, and a year later he turned the trick again with a baldly Kiplingesque poem about the storming of the Dargai Heights by the British army ("The big guns roar a chorus to the song,/Sung by the tuneless, hurtling musket balls"). None of his other poems appealed to the editors of the *American*, though it wasn't for want of trying. At one point, he was turning out a poem a day.

In May 1898 he enrolled in a free correspondence school for writers run by a magazine called the *Cosmopolitan*, and though his painstakingly drafted assignments bear little trace of the author-to-be—even the handwriting is unrecognizable—something about them appealed to Harry's "professor." "You see the interesting side of things," he wrote to his earnest pupil, "and present it attractively. One serious fault that you have, however, is a tendency to use long and pompous words in funny passages. This sort of humor is done to death by newspapers and had better be avoided. . . . Make it your aim to use simple and direct language."

The following year Harry found a new school, one that proposed to teach him how to be a journalist by mail, and the application he filled out eventually found its way into his scrapbooks:

> *Age?* 19.
> *Occupation?* Clerical work in factory.
> *Time to devote to literary work?* 4 hours a day. . . .
> *Extent of your reading heretofore?* A systematic study of English literature.
> *Kind of reading you prefer?* The works of modern writers of fiction: Thackeray, Kipling, [Richard Harding] Davis (when at his best), and Tennyson's poems.
> *What experience in writing?* Have written short stories and some doggerel on topics of the day, but have made little effort to sell.

Anything published? The doggerel aforesaid—in newspapers.
Do you wish to become an all-around journalist? Yes.
Do you write easily? Moderately.
Do you expect or hope to make journalism or literary work your means of livelihood, or to make it a means of general culture? Expect to begin as a reporter & after that trust to hard work & luck for something better.

One wonders what the faculty of the Associated Newspaper Bureau School of Journalism would have made of this matter-of-fact document had they seen it. But Harry Mencken never sent it in. Within weeks he would be enrolled in a different kind of school.

ON THE EVENING OF DECEMBER 31, 1898, SHORTLY AFTER SUPPER, August Mencken had a convulsion and fainted. Harry was in bed recovering from the flu, but he ran downstairs at his mother's call, saw at once that his father was gravely ill, and started out on foot to fetch the family doctor, Dr. Z. K. Wiley, a Confederate battle sur-geon whose office was down the street. He would never forget the emotions he felt during that eleven-block dash: "I remember well how, as I was trotting to Wiley's house on that first night, I kept saying to myself that if my father died I'd be free at last." Wiley was out drink-ing but returned an hour or so later, tipsy but functioning, to spend the rest of the night at Hollins Street. He diagnosed acute nephritis and applied leeches over the kidneys. August soon lapsed into repeated con-vulsions: "Charlie and I held him in bed during one of them, listening to his ravings. They were in large part incoherent. This nursing wore us out, and as the end of the second week approached the whole family was more or less disabled. I was asleep in the next room when he died."

August died on Friday, January 13, 1899, five months short of his forty-fifth birthday, and was buried on Sunday. "On the Monday evening immediately following," Mencken wrote in *Newspaper Days*, "having shaved with care and put on my best suit of clothes, I pre-sented myself in the city-room of the old Baltimore *Morning Herald*, and applied to Max Ways, the city editor, for a job on his staff." It probably didn't happen as neatly, or as quickly, as that—he told Isaac Goldberg that two weeks lapsed between his father's death and his first visit to the *Herald*—but it must have happened as fast as he could manage it. He understood that his father's death, however tragic, made it possible for him to go into journalism; he knew, too, that August had left behind an estate large enough to take care of Anna for the rest of her days. One of Harry's aunts thoughtfully warned his mother that most newspapermen died drunk, but Anna, despite her worrying disposition, made no objection to his decision to try his hand at a new career.

Long afterward H. L. Mencken would make light of his three unhappy years in the tobacco business, dismissing them with the cool confidence of a successful man:

> I ceased to be a child when I was graduated from the Baltimore Polytechnic in June, 1896, and I began to be a man when I went to work for the Morning Herald in June, 1899. . . . There was a time, probably at seventeen or thereabout, when I played with the notion of suicide, but this was only the green sickness of youth, and it soon passed. I never entertained it seriously, for I never believed seriously that I'd stick to business. When my father died I was a little more than eighteen. Within the week following his death I applied for a job on a newspaper, and within six weeks I had it.

Could the eighteen-year-old Harry Mencken possibly have been that cocksure, or cold-blooded? It seems more likely that he believed himself, at least for a year or two, to be caught in a trap of his father's making; that his decision not to go to college was motivated not by cold reason but by a sense of growing desperation; that the relief which swept over him when he realized that August might die was quickly followed by shame and guilt. (It is worth noting that his eleven-block run to fetch the drunken Dr. Wiley turned his flu into a serious cold, one that left him with "a cough that persisted for years." This may well have been the first appearance of the hypochondria that would dog Mencken for the rest of his life.) No matter: The long idyll of childhood and the brief sorrows of adolescence were at an end. A newspaperman was ready to be born.

"A SORT OF CELESTIAL CALL"

Reporter and Editor, 1899–1906

Happy Days, the first volume of Mencken's autobiographical trilogy, breaks off in 1892, when its author was twelve. *Newspaper Days* picks up the story with the indelible image of Harry Mencken solemnly marching into the city room of the *Morning Herald* in search of a job. It was a scene straight out of Norman Rockwell: "I was eighteen years, four months and four days old, wore my hair longish and parted in the middle, had on a high stiff collar and an Ascot cravat, and weighed something on the minus side of 120 pounds."

This six-year gap, while frustrating to the biographer, makes artistic sense. Instead of chronicling the "green sickness" of adolescence at depressing length, Mencken preferred to skip from one happy interlude to the next. Hence the title of the first chapter of *Newspaper Days*, "Allegro Con Brio." The phrase, an Italian musical term meaning "fast, with spirit," is found at the heads of the first movements of Beethoven's Third and Fifth Symphonies, two of Mencken's favorite pieces, and is used no less appropriately in *Newspaper Days*, a book that, like its predecessor, is free of angst: "In the present book my only purpose is to try to recreate for myself, and for any one who may care to follow me, the gaudy life that young newspaper reporters led in the major American cities at the turn of the century. I believed then, and still believe today, that it was the maddest, gladdest, damndest existence ever enjoyed by mortal youth. . . . Life was arduous, but it was gay and carefree. The days chased one another like kittens chasing their tails."

Mencken deliberately omitted from *Newspaper Days* one key detail of how he happened to go to work for the *Herald*: Arthur Hawks, a friend from the Baltimore Polytechnic, was already working there, and "testified that I was a worthy fellow with some talent for writing." But little else in his rosy description of the newspaperman's life is blown out of proportion—at least as far as he was concerned. As soon as he got a toehold, he climbed straight to the top of the heap. "I was a success from the start, and at twenty-four was a city editor, and in command of men twice my age," he wrote in 1941. "At twenty-five I was editor-in-chief of a daily paper in a city of 500,000 population." No wonder he remembered his first newspaper days with nostalgia and delight: They were an unbroken string of triumphs.

His rapid rise was made possible not only by his talent but because there was more room at the top of the newspaper business than ever before. By 1899 new technologies had transformed American journalism almost beyond recognition. Web-perfecting and stereotyping presses made possible the production of cheap, large-circulation dailies aimed not at the well-to-do but the newly literate working classes, creating a mass market for popular journalism and jobs galore for bright young men capable of churning it out. The term "yellow journalism" might not yet have been coined, but the genre was already familiar to millions of big-city readers. One historian of American newspapers culled the following headlines from the front page of an 1884 issue of the *New York World*: A BROTHER ON THE WARPATH; HE ATTACKS HIS SISTER'S DENTIST AND THEN TRIES TO SHOOT HIM; ANOTHER MURDERER TO HANG; LOVE AND CIGARETTES CRAZED HIM; HE PAWNED THE DIAMONDS: HOW AN ENTERPRISING BROKER MANAGED TO FAIL FOR $50,000; DID SHE TAKE THE DIAMONDS? A HOTEL MAID ACCUSED OF STEALING JEWELS WHICH WERE MYSTERIOUSLY RETURNED. The same historian cites the first chapter of *Newspaper Days* as the archetypal story of how a reporter was born:

> Just as some boys of Mark Twain's generation had dreamed of
> standing in the pilot house of the "Natchez" or the "Eclipse,"
> so adventurous youngsters of the nineties rolled the magic
> names of the Sun, the Herald, the Record, and the Star on their
> tongues and imagined themselves interviewing prime minis-
> ters, exposing grafters, trapping murderers, and rescuing kid-
> napped heiresses, all in the role of "star reporter."

Such was Harry Mencken's dream, and although he never rescued
any kidnapped heiresses, much of the rest of it—and more—came
true.

Not, though, at first. Whatever Max Ways's private thoughts as
he gazed upon this well-scrubbed upstart from Hollins Street, he
managed to keep a straight face and ask the necessary questions:
"Had I any newspaper experience? The reply, alas, had to be no.
What was my education? I was a graduate of the Baltimore Polytechnic. What considerations had turned my fancy toward the newspaper business? All that I could say was that it seemed to be a sort of
celestial call: I was busting with literary ardors and had been writing
furiously for what, at eighteen, was almost an age—maybe four, or
even five years." Given the answers, the results were predictable.
Polite though Mencken's reception was, the bottom line consisted of
"the reply that city editors had been making to young aspirants since
the days of the first Washington hand-press": There were no jobs
just now. Ways told Mencken to come back another night, doubtless
never expecting to see him again.

According to *Newspaper Days* he came back the next day, and
every succeeding one for four straight weeks. According to the autobiographical notes he prepared in 1925, he waited until the end of
February. Either way the result was the same: Ways agreed to let
Mencken try his hand at a few assignments, and the young reporter
duly submitted his first two "stories" on the evening of February 23.
The first (about an unsolved stable robbery) was forty-nine words
long, the second (about "an exhibition of war scenes by the cineo-

graph") thirty-three. "I was up with the milkman the next morning to search the paper," he wrote in *Newspaper Days*, "and when I found both of my pieces, exactly as written, there ran such thrills through my system as a barrel of brandy and 100,000 volts of electricity could not have matched." H. L. Mencken lived to exaggerate, but this particular stretcher must have come within an arm's length of the truth. Though the files of the *Herald* long ago crumbled to dust, his own copies of those two forgettable clippings survive, pasted on the first page of the first volume of his personal collection of the thousands of newspaper and magazine stories he would write between 1899 and 1948.

Mencken was not yet a newspaperman, only another cub reporter trying out for a job, and for all his copious reading and high-flown ambitions, he was naive in the ways of journalism. His first visit to the city room of the *Herald* was also the first time he met a professional writer. Nor had he become one simply by virtue of publishing two paragraphs in a newspaper. He remained in the employ of Aug. Mencken & Bro., where he reported to work at eight o'clock each morning and left at five-thirty, giving him an hour or so to go home, change clothes, cram down his dinner, and make his way back downtown to the city room of the *Herald* to do whatever Max Ways told him to do, gratis, between seven and eleven P.M. Not until he secured a paid position at the *Herald* would he quit the cigar business for good.

It took him four months to earn his stripes, and at first he spent them riding trolley cars to and from the suburbs of Baltimore, shelling out as much as sixty cents a night in unreimbursed fares for the privilege of writing up oyster suppers and Sunday sermons. Within a few days, though, Ways discovered that his new cub reporter knew a thing or two about music, and assigned him to cover second-string concerts for the *Herald*. In April he was entrusted with a performance by the world-famous concert band of John Philip Sousa: "He seemed to lead in a more natural and unaffected way than is his habit, and the grotesque, though graceful, mannerisms, which were formerly such a characteristic feature of his conducting, were noticeably absent."

By then he was being given "city assignments of the kind that no one else wanted—installations of new evangelical pastors, meetings of wheelmen, interviews with bores just back from Europe, the Klondike or Oklahoma, orgies of one sort or another at the Y.M.C.A., minor political rallies, concerts, funerals, and so on." Yet some of the stories he wrote, however unpromising their subject matter, hint at the beginnings of a style. The first "big" story preserved in his scrapbooks, a quarter-page feature (with three photographs) on "Baltimore's New and Modern Dog Shelter," already sounds a soon-to-be-familiar note: "The fugitive four-footed hoboes are pursued by uniformed officials, caught without pain to either party, conducted to a handsome wagon which cost as much as a landau, taken to an elegant whitewashed dog hotel with all modern conveniences, and at last, if influential friends do not intercede, are wafted to the Great Beyond by a method which is certain, cleanly, painless, swift and scientific."

It's easy to see from pieces like these why Ways gave Mencken a break, and Mencken in turn was fortunate to have found so sympathetic a teacher: "He didn't bawl out errant young reporters; he summoned them to the desk and lectured them upon the nature and causes of their errors. His judgment of news was sound and he had a good eye for effective writing. . . . He worked me, of course, very hard. At least half of his other reporters were drunkards or incompetents and so the few good men, chiefly young ones, got little time to rest and meditate." Though only eighteen, he clearly had the potential to become one of the *Herald*'s few good men, and his diligence was rewarded. After a month or so of unpaid labor, he was given an expense account, and on July 2 he went to work as a staff reporter for the *Herald* at a salary of eight dollars a week, a dollar more than he was getting at Aug. Mencken & Bro.

Even by comparison with the old-fashioned newsrooms immortalized by Hollywood, the *Herald*'s fifth-floor city room would strike a modern-day visitor as primitive. It contained two telephones, no teletypes, and few typewriters—much of the paper's copy was still written by hand—though each desk had its own spittoon, a modern convenience of which Mencken made regular use, then and later. (He would keep one on the floor of his Hollins Street office to the end of his life.) The paper had no morgue, employed no rewrite men, and published no early editions; the staff came to work at half past noon and went home at eleven. Still, the pleasantly raffish atmosphere must have left the *Herald*'s newest staffer feeling like an ecstatic initiate of the most exclusive club in town.

Newspapermen in 1899 thought of themselves not as paid-up members of the middle class, but as a class apart. It was more than just a matter of their hard-drinking, largely nocturnal lifestyles. From his earliest days at the paper, Mencken's professional idol was Charles Dana, publisher of the *New York Sun*, who cherished his independence not merely from party politics but from all organizations, including labor unions and foreign governments. Writing about the *Sun* a quarter century later, Mencken recalled that it taught a generation of reporters "to see and savor the life that swarmed under their noses, and to depict it vividly and with good humor. . . . Dana saw through all the Roosevelts, Wilsons and Coolidges of his time; they never deceived him for an instant. But neither did they outrage him and set him to spluttering; he had at them, not with the crude clubs and cleavers of his fellows, but with the rapier of wit and the bladder of humor." Like so much of what he wrote in praise of others, this tribute has the ring of a credo.

If Mencken was shaped by the *Sun*, it was partly because he no longer had time to read anything but other newspapers. Instead "the heavy reading of my teens had been abandoned in favor of life itself . . . if I neglected the humanities I was meanwhile laying in all the worldly wisdom of a police lieutenant, a bartender, a shyster lawyer, or a midwife." Max Ways threw him in at the deep end of the trough, assigning him to the Southern Police District of Baltimore, where he saw his first hanging, in which "no less than four poor blackamoors were stretched at once." Mencken tells the story in *Newspaper Days*, adding a drier, more revealing footnote in an unpublished volume of "additions, corrections and explanatory notes":

> My first hanging made very little impression on me. Several of the older reporters were upset, and one fainted, but I was unperturbed. I recall clearly that I came much closer to being overcome when I saw my first dead workman, killed on the job. . . . As I saw the poor man lying in the dirt of the cellar, in his shabby clothes and scuffed shoes, I was really greatly touched, and I have never forgotten him. The men I saw hanged had all done something to deserve it, but here was an honest and laborious fellow who had been struck down in the midst of his work. Later on I saw many other such workmen lying in their blood, and the spectacle became less poignant, but it always moved me more or less. It seemed so dreadful to see an honest man done for, with the marks of his toil upon him. I couldn't help thinking of the cruel disaster that his death may have been to his wife and children. . . . [T]here was no Workingmen's Compensation Act, and more often than not the widow got nothing. Seeing such things made me a strong advocate of that act. I believe that every industry should take care of its own dead and wounded.

To read this sober passage is to be reminded of the danger of taking the *Days* books at face value. Nowhere in *Newspaper Days* proper does Mencken give any indication of the impact made on his younger self by the daily barrage of violent crime, summary punishment, and sudden death. But whatever impression life in the raw made on him, it never succeeded in turning him into a moral crusader. Some turn-of-the-century journalists became muckrakers, passionately committed to the cause of reform, but most, Mencken among them, were disillusioned by their first deep drafts of human nature: "I made up my mind at once that my true and natural allegiance was to the Devil's party, and it has been my firm belief ever since that all persons who devote themselves to forcing virtue on their fellow men deserve nothing better than kicks in the pants." However genuine the compassion he felt at the sight of honest workingmen struck down in the midst of their labors, he was both a bourgeois by birth and a skeptic by conviction, and nothing he saw during his newspaper days shook him from those twin affiliations, the cornerstones of his mature work and thought.

We know this because he tells us so, but it must be remembered that he is our only witness to his earliest newspaper days. None of his colleagues at the *Herald* left behind memoirs, and next to none of his correspondence from this period survives. Though we know something about the newspapers of other cities, the long-lost world of the Baltimore *Herald* exists only in the scrapbooks of Mencken's youth and the pages of *Newspaper Days*, and the second of these records, if for the most part factually reliable, is still not so much a personal history as an impressionistic entertainment—and one, moreover, with a theme. The carefully worked-up anecdotes in *Newspaper Days* are intended to prove the superiority of "the old-time journalist's concept of himself as a free spirit and darling of the gods, licensed by his high merits to ride and deride the visible universe" to the emerging notion of the newspaperman as pillar of the establishment. They are also meant to serve as a cautionary tale for insufficiently suspicious readers. One chapter, "The Synthesis of News," expands at length on Mencken's wry assertion that "journalism is not an exact science" by describing various fake stories he published in the *Herald*, often in collaboration with "enemy reporters" from the *Sun* and the *American*.

Because *Newspaper Days* was written to amuse as much as to inform, casual readers may come away from it supposing that its author spent the better part of his youth hobnobbing with cops, madams, bartenders, and crooked pols. In fact he was a police reporter for less than two years, and no sooner did he get the hang of his new job than he started moonlighting. In the fall of 1899, he sold a cheeky poem about Kipling to the *Bookman* advising his favorite poet to "forget politics and go back to Mandalay." The following April he sold the first of a half dozen stories to the magazine *Short Stories*, bringing in fifteen dollars for his troubles. Mencken would spend the next six years writing stories in his spare time, but there was little in them to suggest that he would become more than an accomplished hack.* When he stopped writing fiction, it was because he had decided he was no good at it: "Those [stories] that I wrote sold readily enough but they were hollow things, imitative and feeble."

More important than the quality of his stories was the effect their publication had on his standing at the *Herald*, many of whose brighter staffers were trying in vain to establish themselves as more than mere newspapermen. Early in 1900 another paper offered him a job, and the *Herald* responded by giving him a raise, to ten dollars a week. Successive job offers kept his salary growing by increments—to eighteen dollars by the beginning of 1901. By that time he had been shifted twice, first to the Northwest and Western Police Districts, then to the Central District (which then included Police Headquarters). Then he became the *Herald*'s man at City Hall, the paper's plum assignment, making him its highest-paid younger reporter. He would cover City Hall until he became an editor himself.

Long before he gave up beat reporting, the *Herald*'s editors, noting his success in magazine work, started giving him a different sort of assignment. When the chronic bronchitis he had contracted around the time of his father's death grew more serious, his doctor ordered him to quit work for a few weeks and go to Jamaica for a

---

*Here is the opening of "The Cook's Victory," his first sale to *Short Stories:* "Captain Hiram Johnson, of the oyster pungy *Sally Jones*, thought that buckwheat cakes reached the maximum of deliciousness when they were a light, unbroken brown. Consequently, when Windmill, the colored cook of the *Sally*, placed before him a dozen which ranged in hue from a dirty, speckled russet to a charred and lustrous black, he was very angry indeed."

vacation. He turned it into a busman's holiday, filing a story on August 26, 1900, called "At the Edge of the Spanish Main: Observations of a Baltimorean in Banana-Land." Not long afterward he launched his first editorial-page column (signed "H. L. Mencken," the first time that byline appeared in print). He wrote his columns in between other assignments, receiving no extra pay for his efforts; at various times they were called "Rhyme and Reason," "Knocks and Jollies," "Terse and Terrible Texts," "Untold Tales," and, finally, "Baltimore and the Rest of the World."

The modern American newspaper column was still in its infancy in 1900, and Mencken's way with the genre—a miscellany of prose and light verse, shoveled together without regard for continuity— was typical of the times. Not for a decade would he start to produce the compact, coherent first-person essays that won him local fame and left a permanent mark on the columnist's craft. Instead he used "Rhyme and Reason" as a vehicle for the poems he had been writing since his Polytechnic days. Several are "imitations" of such well-known poets as Robert Herrick and Algernon Swinburne, all evincing the thoroughness with which the schoolboy had studied his models; the rest vary widely in quality but are almost without exception similarly derivative, usually of Kipling. The sentimental ones are banal ("Sweethearts often quarrel upon/The slightest provocation,/Because they love/The pleasure of/The reconciliation"), the humorous ones fresher and more personal.*

---

*Here is the first stanza of the occasionally reprinted "Song of the Slap Stick":

> Why is a hen? (Kerflop!) Haw, haw!
> Toot, goes the slide trombone;
> Why is a hen? (And a swat in the jaw!)
> And the ushers laugh alone.
> Why is a—(Bang!)—is a—(Biff!) Ho, ho!
> Boom! Goes the sad French horn;
> Why is a hen? (Kerflop!) Do you know!—
> And the paid admissions mourn!

As Mencken gradually emptied his desk drawer, he switched from verse to prose. He tried his hand at aphorisms, most bearing the stamp of Mark Twain and Ambrose Bierce and a few impressively sharp ("It is a wise candidate that knows his own platform"), respectful imitations of the then-popular *Fables in Slang* of humorist George Ade (one of his most important early influences), and snippets of literary criticism. "Some day," opined the apprentice columnist, "the critics will awaken to the fact that . . . 'Huckleberry Finn' is the greatest novel yet produced by an American writer." A quarter-century later, he remembered this prophecy with understandable satisfaction: "Not bad criticism certainly for a boy of twenty."

The most promising of his early columns was a series called "Untold Tales," a semifictional chronicle of life at City Hall consisting of thinly veiled abuse of local politicians, all of whom were given triple-barreled pseudonyms such as J. Socrates Lithium and Q. Erat Demonstrandum. The column is mostly unintelligible today, but Mencken's cheerfully debunking tone, the product of close observation not only of Baltimore's local politicians but of the reporters who covered them, comes through loud and clear: " 'The word reporter,' he would say, 'is a synonym for "grafter." Some reporters can be bought for a schooner of half and half. Others demand a 15-cent drink. Still others want cold denarii, and a few hold out for a pretty good-sized pot. But the majority of them are cheap, and all of them are purchasable.' "

Not only did the young Mencken fumble with verse and short stories, he even started a novel, though it never got off the ground. "The fashion at the time," he later wrote, "was for romantic fiction with characters out of history, and it occurred to me to use Shakespeare. I knew next to nothing about his life, and by the time I got fifty pages of the story on paper that fact began to oppress me sorely. So I abandoned my first and last novel, and no one save its author has ever seen the following fragment of it to this day [August 1937]." Judging by the surviving pages, he was smart to give it up: "Now though I was full joyous ay the glorious outcome of my first day's hunting, yet I was already man enough to knoe that most of my glory was due to good fortune."

For all the uncertainty of his attempts to break free of the confines of daily reporting, his star continued to rise at the *Herald*, if only because of the lack of competition. It was his good luck to have gone to work for a second-rate paper. The *Sunpaper* may have been dull, but it was professional; the hit-or-miss quality of the *Herald*'s staff, by contrast, allowed him to expand into a vacuum. His writing, if uneven, was always publishable, and his skills as a reporter were growing fast, not least because he worked so hard, on occasion turning out a dozen stories in a single day. By 1901 he had run through the reporting assignments available to him without showing any sign of having outstripped his abilities. In September he was pulled off the City Hall beat and made dramatic editor of the *Herald*, writing the daily theatrical column and occasional reviews. (Robert I. Carter, then the *Herald*'s managing editor, gave him a piece of advice he never forgot: "The first job of a reviewer is to write a good story—to produce something that people will enjoy reading. If he has nothing to say he simply can't do it. If he has, then it doesn't make much difference whether what he says is fundamentally sound or not.") A month later he was promoted to Sunday editor, a job that sounded more impressive than it was—much of his energy went into supervising the production of the funny pages—leaving him with free time for outside writing.

Around this time, he sold the first of several stories to Ellery Sedgwick, editor of *Frank Leslie's Popular Monthly*, who praised him lavishly, paid him fifty dollars (his biggest fee to date) for "The Flight of the Victor," and became one of his best customers. The praise was as welcome as the money: Sedgwick, who would later become editor of the *Atlantic Monthly*, the Boston-based magazine closely identified with the New England–bred strain of taste and thought soon to be tagged "the genteel tradition," was already, in Mencken's words, "a Brahmin of the Brahmins." His dealings with Sedgwick were his first direct contact with the larger world of American letters, and with a species of American literati he would spend his adult life ridiculing. Soon he would flaunt his newspaper background and lack of academic credentials as badges of superiority, but in 1901 he was still too young and too obscure to dismiss the genteel tradition out of hand. He was pleased to have caught the eye of so noted an editor, and it seems likely that his colleagues were all the more impressed with him as a result.

As his responsibilities increased, Mencken had no time for editorial-page fooleries, and so "Baltimore and the Rest of the World," the final incarnation of his column, ended its run. It would be another eight years before he returned to an editorial page with a signed column, by which time his voice was almost fully formed. But his style had become pithier in the past year and a half, during which "Baltimore and the Rest of the World," in his words, "gradually lost its humorous character and became a sort of private, individual column." The last column, published on February 23, 1902, shows what he meant. Already under fire are several of his all-time favorite targets, including "Col. William Jennings Bryan"; J. Gordon Coogler, a maladroit southern versifier whose ineptitude would become a byword with Mencken; and the ever-popular Church of Christ, Scientist. (Mary Baker Eddy's faith-healing Christian Scientists shared a rung with Holy Rollers and chiropractors in Mencken's hierarchy of idiocy.) Even his interest in what he would later dub "the American language" is foreshadowed in a discussion of a glossary of circus words: "A proposed victim is known as a 'sucker,' to the confidence men who follow the circus, and 'fanning a guy' is to make sure he has no weapons on him before they proceed to get his money." But it is the remainder of the column that shows us most clearly where he was headed. Reporters cover the news, expressing their opinions about it only by implication. At twenty-one Mencken was already different: His opinions, not the news, were what interested him most. Instead of merely looking for trouble, he would make his own.

It is touching to turn from this pugnacious column to the photograph that serves as the frontispiece to *Newspaper Days*. Taken two years and three months after the demise of "Baltimore and the Rest of the World," it shows a smiling, cigar-chewing stripling doing his best to look like a middle-aged newspaperman, and failing. The boy in the picture may have been a prodigy of sorts, wise beyond his years in the ways of the world, but to call him a man would have been to stretch the point until it snapped. He was still "a young reporter to whom all the major catastrophes and imbecilities of mankind were still more or less novel, and hence delightful," and he was having, as the picture shows, the time of his life.

Hindsight is always right, and always wrong. We see the man of letters in the boy wonder, and assume the one to have been the necessary product of the other. Nothing could be less true. Popular journalism is a black hole capable of swallowing a lifetime's ambition. The city rooms of newspapers the world over are littered with middle-aged wheelhorses who once had great careers ahead of them. What set Mencken apart was not just his talent, remarkable though it was, but his certainty that there was more to life than a *Herald* byline. "My dislike—perhaps I had better say disdain—of neighborhood celebrity was developed in my earliest newspaper days, and has served me well all my life," he recalled. "I struck out from the start for wider and lusher fields than Baltimore could offer, and I was always fearfully aware of the possibility that I might become a mere local worthy."

The nineteen-year-old Mencken had taken his first step into the lush fields of "lovely letters" (to use his own tart phrase) when he started publishing in the pulps. It was a wrong turn, and he soon sensed its futility. But it is hard for a young writer to walk away from his early work—to accept that the poems and stories and articles over which he sweated and slaved are of no more lasting value than the term papers he wrote in school. Out of this misguided but forgivable loyalty to one's youth, many first books are written and published, and it was as much by chance as premeditation that Mencken's was a collection of poems. He had been approached as early as September of 1901 by Richard G. Badger, a Boston vanity publisher who offered to bring out a collection of his stories for a fee of three hundred dollars. Whether the young author had the sense to realize this was a mistake is neither clear nor relevant. Not having three hundred dollars to spare, he had no choice but to put temptation behind him. But the proposition put an idea in his head, for the following November, he submitted a collection of eleven short stories to F.A. Stokes & Company of New York. Though his proposal was rejected, the longing to produce a book remained, and another opportunity soon presented itself.

The title page of *Ventures into Verse*, published in the summer of 1903, reads as follows:

Ventures into Verse, Being Various BALLADS, BALLADES, RONDEAUX, TRIOLETS, SONGS, QUATRAINS, ODES and ROUNDELS/All Rescued from the POTTERS' FIELD of Old Files and here Given DECENT BURIAL [Peace to Their Ashes]/BY Henry Louis Mencken/WITH ILLUSTRA-TIONS & OTHER THINGS/By CHARLES S. GORDON & JOHN SIEGEL.

The volume is identified as a "FIRST (and Last) EDITION." The next page contains a "warning" from the author: "Some [of the poems] are imitations—necessarily weak—of the verse of several men in whose writings he has found a good deal of innocent pleasure. The others, he fears, are more or less original." The next page contains a quatrain titled "Preliminary Rebuke/Don't shoot the pianist; he's doing his best." If all this seems like a top-heavy prolegomenon to a forty-six-page pamphlet containing forty poems and issued in a hand-set edition of one hundred copies (mainly, Mencken admitted long after the fact, as a vehicle for the publishers, Marshall, Beek & Gordon, to show off their typographical craftsmanship), one may blame the times. *Ventures into Verse* was in every way except its belated date of publication a product of the precious nineties. The poems, all previously published in the *Herald, Leslie's Weekly, Life,* the *New England Magazine,* and other publications, were mostly written between 1896 and 1900 and dug up by Mencken to pad out his earliest editorial-page columns.

Years later he would recount the brief history of *Ventures into Verse* with the slightly embarrassed brusqueness of a man who had long since gone on to bigger and better things:

> The edition was limited to 100 copies—mainly bound in folded brown paper with red and white labels, in the arty manner of the time, but with a few done in bare binder's boards with red cloth backs. I took half and the publishers took half. I say in "Newspaper Days" that "I sent ten of my copies to the principal critical organs of the time," but that is an underestimate; I must have sent out at least 25. As incredible as it may seem, it got a number of friendly notices. . . . But other critics fell upon the book with a considerable ferocity, and as the reviews drifted in I began to realize, too late, its glaring deficiencies. . . . I made a resolution to write no more verse, and have kept it pretty well to this day, though with a few backslidings. So far as I can recall, only three orders for the book ever came in.

Mencken came to see his early poetic efforts as an important part of his literary apprenticeship: "I am convinced that writing verse is the best of all preparations for writing prose. It makes the neophyte look sharply to his words, and improves that sense of rhythm and tone-color—in brief, that sense of music—which is at the bottom of all sound prose, just as it is at the bottom of all sound verse." He never claimed to have written any good poems, and rightly so. The adolescent Mencken had caught Kipling's galloping rhythms, as the volume's envoi, "To R. K.," makes clear: "You have taught the song that the bullet sings—/The knell and the crowning ode of kings/The ne'er denied appeal!" But his "serious" poems lack intensity, and those written in dialect are hopelessly gauche, particularly "The Orf'cer Boy," which Theodore Dreiser later circulated among Mencken's friends as a prank ("Oh, 'e aren't much on whiskers an' 'e aren't much on 'eight,/An' a year or two ago 'e was a-learnin' for to write"). The young author's sheer technical skill is not to be despised—he could doubtless have written Max Beerbohmesque parodies had the genre interested him—but the *daemon* that drove Kipling is nowhere to be found in the pages of *Ventures into Verse*, and it is barely believable that the author of these pastiches would go on to forge the first great vernacular American prose style of the new century.

However embarrassed Mencken may have been by *Ventures into Verse* (he devotes only one paragraph to it in *Newspaper Days*), he took it seriously enough at the time to subscribe to a clipping service and paste the reviews he received into the first of more than two hundred scrapbooks now owned by the Enoch Pratt Free Library. From it we learn that *Ventures into Verse* was reviewed by newspapers and magazines across the country, and that most of the notices were favorable. Two are of more than passing interest. The first, appearing in the *Chicago Record-Herald*, was as prescient as it was condescending: "These verses . . . are salable rather than sincere. Nowhere are they in any abiding and profound sense the call of the spirit. Yet nowhere do they not show the art and the power to do things, eminently worth while. . . . In the hope of prodding him up to an awakening to his higher capabilities we must say, in parting, that Mr. Mencken belongs to that class which is cursed with a fear of being thought serious." And "The First Nighter," drama critic of the *Cleveland Leader*, reveals in his review that Mencken's writing had already made an impression on at least one out-of-towner: "Mr. Mencken is a dramatic critic of discrimination as well as a poet and humorist, and I always read his opinions of actors and plays in the full belief of his capacity and honesty."

However modest a claim *Ventures into Verse* has on our attention today, its reception shows that Mencken, young as he was, had outgrown the *Herald*. He was looking for intellectual stimulation outside the paper, both through people he met in his capacity as drama critic and in a very different setting. Starting in 1902 he played chamber music regularly with a small group of congenial friends, thus laying the foundation for what later came to be known as the Saturday Night Club, the center of his adult social life. It was no surprise that he had come to view the *Herald* as a center of "sterile Philistinism." The only question was how long the paper would be able to hold onto its most talented staffer, and there was no question about what his next professional step would be. Five months after the publication of *Ventures into Verse*, he was promoted for the first time to a managerial position.

It is an iron law of newspaper life that every really good reporter finds himself pressured to become an editor. In January of 1903, Lynn Roby Meekins, the first managing editor of the *Saturday Evening Post*, was hired as managing editor of the *Herald*. Max Ways had left to go into politics, and Meekins found himself running a paper in decline; after looking over the staff he came to the predictable conclusion that Mencken was his most valuable asset. In November the twenty-three-year-old dramatic editor and Sunday editor of the *Herald* was promoted to city editor, Ways's old job. Mencken subsequently gave up his theatrical column (though he continued to do occasional drama reviews and write articles for the *Sunday Herald*). He had to: He was too busy to keep it going. "I began business," he later wrote, "by discharging eight or ten reporters. What followed was a dreadful spell of hard work. The remaining reporters, with few exceptions, were either drunkards or idiots. I read copy, and often rewrote one-third of it. Gradually I accumulated a better staff and by the beginning of 1904 the *Herald* was a well-run and amusing paper. . . . To this day, in fact, no American provincial paper that I can recall has ever been so briskly written. We were often beaten on news by the Baltimore *Sunpaper*, but our bright young men wrote rings around it every day."

Mencken would spend comparatively little time supervising the careers of other people, and the last sentence of *Newspaper Days* reveals the distaste he felt for exercising managerial authority: "Since

1910, save for a brief and unhappy interlude in 1938, I have never had a newspaper job which involved the control of other men's work, or any responsibility for it." But like it or not, it was through the exercise of power that he came to understand which side he was on. Life as a reporter had taught him to be skeptical of idealism; life as an editor led him to reject it totally. "All the ideas that are thought of when I am thought of," he wrote in 1925,

> are the products either of heredity or of environment. I inher-ited a bias against the rabble. I come of a family that has thought very well of itself for 300 years, and with some reason. My father, when I first became aware of him, was a capitalist engaged in an active struggle with labor. It was thus only natural that I should grow up full of suspicions of democratic sentimen-tality. As I came to manhood and began to deal with men myself, I noticed quickly that the failures were all incompetents—that God had marked them for the ditch, not man. When I was 23 years old I was the city editor of a big city daily, and the boss of 30 or 40 reporters, not to mention many printers and other such fellows. All this naturally confirmed my bias. I had to struggle daily against the incompetence of men full of gabble about their sacred rights. They were stupid and lazy. I was a better man than any of them, and on all counts. The worst were those who believed that there were conspiracies against them. They were Socialists, though most of them didn't know it.

Experience and predisposition were crystallizing into ideology: What began as a "bias" evolved into a simple but comprehensive worldview from which he would never again deviate. His new way of looking at life was so all-encompassing that it invaded even his now-infrequent ventures into short-story writing. "The Bend in the Tube," written in the summer of 1904 and published in *Red Book* that December, reveals not only a distinctive literary voice but a point of view to go with it. Boggs, the antihero of "The Bend in the Tube," is an embodiment of everything Mencken found foolish and distasteful about his fellow men:

> First of all, Boggs was a lunatic, which was nobody's fault. Secondly, he had a new theory concerning the redemption of silver certificates and the circulation of national banks, which was his own fault. In the third place, he was an anarchist, which was the fault of a good many people, individually and collectively. . . . Beginning manhood as a divinity student, he had left college under a cloud of heresy, and then, dropping a step further down the scale of brute creation, he had become a reporter on the *Tribune*, a third-class morning paper in a provincial city.

In due course he falls in with a group of woolly-minded barroom radicals who adhere to the doctrine that "whatever is shouldn't be," and finds himself wrestling with "problems that philosophers gave up as insoluble two thousand years before he was born." The effort drives him mad with envy and leads him to attempt the murder of a colleague.

The editor of the *American Mercury* would never have chosen a short story as the vehicle for his ideas, but he would have found no fault whatsoever with the ideas dramatized in "The Bend in the Tube." Everything about Mencken—his politics, his religious beliefs, his personal relationships, his taste in literature—was conditioned by his certainty that life was a struggle for survival to which some men were ill-suited by the accident of birth. His philosophy would become more elaborate, especially after he began to read Nietzsche, but it never strayed far in spirit from the social Darwinism in shirtsleeves that he acquired as city editor of the *Herald*, listening to drunken reporters gripe about the unfairness of life.

SOME OF THE HARSHNESS OF "THE BEND IN THE TUBE" MAY ALSO BE due to the fact that it was written not long after the experience that turned Mencken from a cocky young city editor into "a settled and indeed almost a middle-aged man, spavined by responsibility and aching in every sinew." A fire broke out in downtown Baltimore on February 7, 1904, and Mencken, who had been out drinking until three-thirty that morning, rolled out of bed at half past eleven and headed for the offices of the *Herald* to direct the paper's coverage of what promised to be "swell pickings for a dull Winter Sunday." He stayed on the job for the next sixty-four hours. When the fire was put out ten days later, one square mile of the city—the whole of Baltimore's downtown business district, including all but two or three of its office buildings—had been reduced to smoke and ashes.

It is impossible to improve on Mencken's own account of how he covered the great Baltimore fire of 1904, originally published in the *New Yorker* and later reprinted as the penultimate chapter of *Newspaper Days*. There is no better portrait of a turn-of-the-century American journalist at work. The bare facts are spectacular enough. The *Herald* building caught fire at nine P.M. and was promptly abandoned. (Weeks later, Mencken would visit the burned-out shell of the building and retrieve his old copy hook, "twisted as if it had died in agony." For years it reposed on the desk of his second-floor office at Hollins Street, one of the few souvenirs of his newspaper days to survive the fire.) Mencken, Meekins, and a skeleton staff walked a mile to Camden Station, caught the next train to Washington, set up shop in the composing room of the *Washington Post*, and put out a four-page special edition that hit the streets of Baltimore at six-thirty Monday morning. The eight-column front-page headline was written by Mencken:

HEART OF BALTIMORE WRECKED BY
GREATEST FIRE IN CITY'S HISTORY.

The next week, he wrote in *Newspaper Days*, was "brain-fagging and back-breaking, but it was grand beyond compare—an adventure of the first chop, a razzle-dazzle superb and elegant, a circus in forty rings." The *Post* was already committed to providing another Baltimore paper with emergency facilities, so the *Herald* was forced to improvise. Tuesday's paper was published in Washington, Wednesday's at the plant of a Catholic weekly in Baltimore. Then Mencken and his colleagues went to Philadelphia, having struck a deal with the owner of the *Evening Telegraph*, and put out the *Herald* from there for the next five weeks. The papers were loaded onto a special train at two A.M. each morning and hauled back to Baltimore.

Mencken's role in this journalistic tour de force did not go unnoticed by his peers. A 1906 profile in *Editor & Publisher*, then as now the house organ of the newspaper business, noted that his work during the fire "was brilliant and attracted special attention," but his brilliance could not conceal the fact that the *Herald* was in trouble, undermined by inept management and a general trend away from morning papers. Mencken had already told Ellery Sedgwick of his growing doubts about the *Herald*, and Sedgwick wrote him on February 9, two days after the fire, with a tentative job offer. Once things had settled down in Baltimore, the two men met in New York, and Mencken was formally offered the post of assistant editor of *Leslie's*. "He proposed to pay me $40 a week," Mencken told Isaac Goldberg, "and to get me a pass once a month to visit my mother in Baltimore. This was more than twice the salary I was getting on the Baltimore *Herald*, but I couldn't leave my mother in Baltimore and so refused."

Whether Mencken took the offer seriously is an open question—his dislike of New York was as sincere as his attachment to Baltimore—but the part about his mother rings true. As unlikely as it may sound, the sharpest, cruelest, most self-assured wit in the history of American letters, the fearless scourge of puritanism in all its forms, was a mama's boy. He lived at home until Anna died in 1925, and did not move out until his marriage five years later. "I owed to her, and to her alone, the fact that I had a comfortable home throughout my youth and early manhood," he wrote long after her death, and nowhere in his published or unpublished writings does he admit to having found anything peculiar about this arrangement. The same man who wrote in *Newspaper Days* of his friendship with one of Baltimore's leading madams (lauding her as "a woman who, though she lacked the polish of Vassar, had sound sense, a pawky humor, and progressive ideas") made a point of recording in a private memo the following detail of life at Hollins Street: "Whenever I got home, whether late at night or early in the morning, I found a plate of sandwiches waiting for me, made by [Anna's] own hand."

For whatever reason, Anna's oldest son chose to stick with a sinking ship, and found himself bailing faster than ever. In the summer of 1904 he covered both presidential conventions for the *Herald*, his first encounter with what would become one of his best subjects. The time would come when his convention dispatches crackled with disdain, but for now he contented himself with straightforward reportage, enlivened by glints of wit:

> With perspiration standing out in beads upon its forehead and its collar wilted and its spirits depressed, the republican party, by its representatives in convention assembled, today nominated Theodore Roosevelt, of New York, to succeed himself as President of these United States and of the territories thereof, and sovereign lord of the dominions beyond the seas. . . . 'Tom' Platt, who four years ago tried to give Mr. Roosevelt his final quietus, bared his neck to the yoke of the inevitable by announcing that New York stood by the Rough Rider to a man.

Six weeks after he came home from the Democratic convention, the *Herald* started an evening edition on the cheap, using the staff of the morning paper. As city editor of both editions, Mencken came to work at eight A.M. and worked until midnight, sleeping on a couch in his office. Even bright young men can work such hours for only so long; by week's end, the *Morning Herald* was shut down and the *Evening Herald* got underway in earnest, with Mencken as its city editor and part-time dramatic critic. In due course Lynn Meekins became editor in chief and Mencken was promoted to managing editor, a step whose significance was not lost on him. Most of the bright young men he knew who had gone into the newspaper business with him in 1899 were still reporters, but he was running a paper—indeed, his name was on the letterhead of the *Evening Herald*, and the office boys now called him "the Old Man." Moreover, it was as managing editor that he found his calling, writing editorials and unsigned columns that need no byline to be immediately identifiable:

> Today [grafting] is just winning recognition as an art. A generation hence it may be one with astronomy, sculpture, high jumping and the art tonsorial. . . . There are many things in the German Emperor's make-up that no good American admires, but his most hostile critics must give him credit for virility, industry and honesty, which are virtues every man likes to fancy he possesses. . . . A gentleman of bulging brow arose at the meeting of the National Educational Association the other day and undertook to lay the bastinado of intellectual excommunication upon the popular novel. . . . The best way to make men better is not to throttle ambition and competition, but to let generation after generation fight it out, with the unfit perishing and the fit surviving. . . . Beside [Mark Twain] the writers of best-selling novels cut ludicrous figures and immortals such as Poe seem hopelessly puny and tame.

Perhaps the funniest piece Mencken wrote for the editorial page was his response to a critique of American journalism by no less a light than Henry James. The expatriate novelist, returning to the land of his birth for the first time in twenty years, delivered an address at Bryn Mawr College in which he dismissed the "newspapers and common schools" of America as "influences which keep our speech crude, untidy and careless." Mencken, who at no time in his life found James's prose to be anything other than indigestible, fired back with the brassy self-confidence of a self-made man whose style owed nothing to *The Wings of the Dove* and everything to *Huckleberry Finn:* "The average newspaper reporter writes better English than Henry, if good English means clear, comprehensible English. . . . Take any considerable sentence from any of his novels and examine its architecture. Isn't it wobbly with qualifying clauses and subassistant phrases? Doesn't it wriggle and stumble and stagger and flounder? Isn't it 'crude, untidy, careless,' bedraggled, loose, frowsy, disorderly, unkempt, uncombed, uncurried, unbrushed, unscrubbed? Doesn't it begin in the middle and work away from both ends? Doesn't it often bounce along for a while and then, of a sudden, roll up its eyes and go out of business entirely?"

In articles like this you can hear the *click* of a style snapping into place. It made sense that their anonymous author should now have wanted to try his hand at a real book, and the subject was conveniently close to hand. Harrison Hale Schaff, head of the Boston publishing house of John W. Luce, invited Mencken to submit a proposal for a book about George Bernard Shaw, whose plays the former dramatic editor of the *Herald* had been seeing, reading, and reviewing since 1901. Three weeks later Mencken sent him the introduction and a sample chapter of what was to become the first book ever written about Shaw. Schaff liked what he read, and soon the finished manuscript of *George Bernard Shaw: His Plays* was in the mail. The proofs arrived at Hollins Street around the time of Mencken's twenty-fifth birthday, and there followed a scene (lovingly described in *Newspaper Days*) sure to warm the heart of anyone who has ever published a book:

> I was so enchanted that I could not resist taking the proofs to the office and showing them to Meekins—on the pretense, as I recall, of consulting him about a doubtful passage. He seemed almost as happy about it as I was. "If you live to be two hundred years old," he said, "you will never forget this day. It is one of the great days of your life, and maybe the greatest. You will write other books, but none of them will ever give you half the

thrill of this one. Go to your office, lock the door, and sit down to read your proofs. Nothing going on in the office can be as important. Take the whole day off, and enjoy yourself." I naturally protested, saying that this or that had to be looked to. "Nonsense!" replied Meekins. "Let all those things take care of themselves. I order you to do nothing whatsoever until you have finished with the proofs. If anything pops up I'll have it sent to me." So I locked myself in as he commanded, and had a shining day indeed, and I can still remember its unparalleled glow after all these years.

Alas, the glowing young author had produced not a masterpiece but another student effort, wholly well-meaning and hopelessly dull. *George Bernard Shaw: His Plays*, he admits at the outset, is "a little handbook for the reading tables of Americans interested enough in the drama of the day to have some curiosity regarding the plays of George Bernard Shaw, but too busy to give them careful personal study or to read the vast mass of reviews, magazine articles, letters to the editor, newspaper paragraphs and reports of debates that deal with them." Most of the 107-page book is given over to stilted plot summaries of Shaw's plays (only fifteen of which had been published), and the staid rhythms of the introduction betray no trace of the youngster who had wrestled Henry James to a draw: "In all the history of the English stage, no man has exceeded him in technical resources nor in nimbleness of wit. Some of his scenes are fairly irresistible, and throughout his plays his avoidance of the old-fashioned machinery of the drama gives even his wildest extravagances an air of reality."

It's tempting to suggest that Mencken was inhibited by his lack of literary credentials. He had not yet realized that his homemade prose style was as well-suited to serious criticism as it was to yellow journalism, a discovery that would prove to be the main source of his power and influence as a critic. Even more surprisingly, he failed to come to grips with Shaw's didacticism. The "Shaw" of *George Bernard Shaw: His Plays* is a creature of his admiring imagination, an artist for art's sake whose passionate political convictions, all of which Mencken found repugnant, are somehow irrelevant to his theatrical art: "His private opinions, very naturally, greatly color his plays, but his true purpose, like that of every dramatist worth while, is to give a more or less accurate and unbiased picture of some phase of human life, that persons observing it may be led to speculate and meditate upon it. . . . A preacher necessarily endeavors to make all his hearers think exactly as he does. A dramatist merely tries to make them think." Within a few years he would change his mind about Shaw, dismissing him as an "Ulster Polonius" who bent his plays out of shape in order to sermonize on the issues of the day, but in 1905 Mencken was too dazzled by the Irishman's verbal flair (and, one suspects, too impressed by his unabashed lack of gentility) to see him as he was.

Today *George Bernard Shaw: His Plays* is remembered only as a curiosity—it receives only a single-sentence mention in Michael Holroyd's three-volume Shaw biography—and that judgment, though harsh, is not unjust. In retrospect the only striking thing about Mencken's first real book is its choice of subject. It is little short of amazing that a twenty-four-year-old editor at a backwater newspaper in a medium-size American city should have produced the very first full-length study of a world-famous European playwright. Moreover, the book, partly because it was first out of the gate and partly because Shaw was still controversial in the United States in 1905, was widely reviewed. Some Shaw-hating critics included Mencken in their raking fire ("Circus criticism, written by clowns for clowns, is popular in this country, and it just fits the case of G. B. Shaw," spat the dramatic critic of the *Chicago Record-Herald*), and a few prigs found his prose dangerously daring (the *Nation* condemned his English as "rather too colloquial for elegance"), but most of the reviews were respectful. The *New York Post* called *George Bernard Shaw: His Plays* "interesting and clever"; the *New York Times* called it "well written and informing"; the *Boston Transcript* called it "sensible and sound."

Mencken was still no celebrity. The *San Francisco Bulletin* identified him in a headline as "H. L. Meneker," while the *Chicago Tribune* described him as "an author of whom the reviewer is lamentably ignorant." Nor did his Opus 1 bring him much in the way of profit: His first and last royalty check came to two hundred dollars. And though he sent a copy to Shaw, he heard nothing back (though he learned, years later, that Shaw had liked the book). But he was pleased with the fruits of his labors, and remained so, if only for sentimental reasons, to the end of his life: "The book, on re-reading, turns out to be very fair work for a man of twenty-four. I'd change parts of it if I had it to do again today, but certainly not all of it. It contains some acute criticism and some very amusing writing. . . . My style had not yet formed, but the rudiments of it are visible."

George Bernard Shaw: His Plays served as a punctuation mark in Mencken's career: After it he wrote no more fiction. Three more stories would appear under his name in *Red Book* and the *Monthly Story Magazine*, but they had all been written prior to the completion of the Shaw book, the response to which, coupled with the continuing encouragement of Ellery Sedgwick, persuaded him that criticism was his *métier*. Schaff had already made another proposal, this one even better suited to Mencken's interests and inclinations: a study of Friedrich Nietzsche. At the time Mencken knew little about the German philosopher, but he started borrowing books from the Pratt and the Library of Congress—all in the original German, for whose intricacies his studies at F. Knapp's Institute had left him ill prepared—and by the end of 1906, he had cut a deal with Schaff.

MUCH WOULD HAPPEN BEFORE HE SIGNED THE CONTRACT FOR HIS FIRST major book. Four months and one week after his twenty-fifth birthday, he was appointed "Secretary and Editor" of the *Herald*, an appointment that led to the first two important journalistic stories about him to be published. The first, which appeared in *Editor & Publisher* on February 3, 1906, told the tale of the rise of a prodigy in terms that must have piqued the interest of talent-hungry editors across the country: "He became a reporter on the old *Morning Herald* in July,

1899. His ability was quickly recognized and his work was generally copied in the press of the country." Three months later, he achieved the even greater distinction of being interviewed by a New York newspaper. On May 4, the *New York Herald* ran a piece headed BLACK HEADLINES HERE TO STAY/MR. MENCKEN, OF THE BALTIMORE HERALD,/ SAYS PUBLIC DEMANDS DISPLAY/OF IMPORTANT NEWS. The quotes are too stilted to be authentic ("Big black headlines were devised by the first yellow journalist to fill an obvious want, and they will continue to fill this want until the last press falls to pieces. . . . But the sensational display of exaggerated or silly stories meets no sane want at all, and already the vice is being laughed out of existence"), but that he was thought worth quoting at all was a landmark.

Whatever pleasure Mencken may have taken in his modest notoriety was tempered by the continuing need to breathe life into a dying paper. Managerial problems ate up his time (the last clipping in his *Herald* scrapbooks is dated February 8). Then, on June 17, the *Herald* published the following notice on its editorial page: "Tomorrow the property of the *Herald* Publishing Company will pass into new hands, and there will be no further publication of the *Sunday Herald*, the *Evening Herald*, or the *Weekly Herald*." No one was surprised or worried, and most of the *Herald* staffers found jobs quickly. Mencken himself went to work for the *Evening News* the very next day as news editor, having been offered employment by every other paper in town, but the pay was inadequate—forty dollars a week— and the circumstances unsatisfactory, for Charles Grasty, publisher of the *News*, went in for crusades at the expense of hard news coverage. Within a week or two he knew he had come to the wrong place, and soon had the opportunity to correct his mistake when the *Sun* offered him the job of Sunday editor. The pay was the same, but with the promise of a three-dollar raise.

The last chapter of *Newspaper Days* is called "Sold Down the River," but it is as good-humored as the rest of the book, with good cause: Mencken had climbed out of the wreckage of the *Herald* without a scratch. Seven years of hackwork had taught him the basics of his craft, and the presidential conventions of 1904 offered him his first chance to write about the national scene. He had seen nine men hanged, written hundreds of news stories, and published two books. He had developed a philosophy of life; he had tasted fame, and knew he wanted more of it. Just as important, he knew what he did not want to do. Poetry and fiction had been abandoned, management tried and found wanting. A few more years would pass before he cut himself loose from irrelevant responsibilities and began to concentrate solely on achieving his professional goals, but the man himself was, to a degree remarkable for one so young, a finished product.

Finished but not seasoned: Mencken's inability to find anything but hot air in Henry James indicates the limits of his education as exactly as it does the breadth of his ambition. Despite his inexperience, the twenty-five-year-old critic already felt secure enough to write off one of America's most celebrated novelists, an opinion from which he never wavered. Such self-assurance was impressive in so young a man, and the time was coming when it would serve him well in his battles against the cultural establishment. Where a lesser man would have folded his hand and slunk away, Mencken time and again played it for all it was worth, sure that his compass was pointed in the right direction. He *knew* Theodore Dreiser was a genius and William Jennings Bryan a dolt, and that was that. In the case of James his certainty would have been still more impressive had it not been so wrong-headed. Instead, it smacked of the philistinism of his father, the bourgeois burgher from Southwest Baltimore who knew what he liked and liked what he knew.

This aspect of Mencken's character, which will surface repeatedly in the course of his life, can be explained in part by the fact that he was for all intents and purposes an autodidact. Perhaps he was right to assume he would have gotten nothing out of college but the "balderdash" of "chalky pedagogues"; certainly it would have done little to prepare him for his chosen career as a newspaperman. But he was destined to be more than a newspaperman, and the deeper waters in which he had started to swim were over his head. The intensive reading of his adolescence had been carried out in isolation, and however much "worldly wisdom" he acquired at the *Herald*, there was no one on the staff who carried enough intellectual guns to challenge his unformed ideas, no mentor prepared to suggest there were more good books than the ones he had already read.

He knew enough to know how isolated he was. What he could not yet have known, or even suspected, was that the confident world of his youth was crumbling under his feet. Between 1899, when the eager teenager from Baltimore Polytechnic asked Max Ways for a job, and 1906, when the smooth-faced editor of the Baltimore *Herald* was sold down the river, Queen Victoria died and was succeeded by King Edward VII; William McKinley was assassinated and succeeded by Theodore Roosevelt; Sigmund Freud published *The Interpretation of Dreams*; Albert Einstein formulated the Special Theory of Relativity; Richard Strauss composed *Salome*; Pablo Picasso completed his Blue and Rose Periods; *The Great Train Robbery* was released; Wilbur and Orville Wright flew for the first time; Enrico Caruso made his first phonograph recordings; Guglielmo Marconi transmitted radio signals across the Atlantic Ocean; and Henry Ford founded the Ford Motor Company. All these things happened while Henry Mencken was wrapped up in the life and politics of a provincial city, busily learning his trade. Now he would spend the rest of his life practicing it, a nineteenth-century Baltimorean at large in a world turned inside out by modernity.

"WIDER AND LUSHER FIELDS"

Columnist and Critic, 1906–1914

The *Sun* has moved three times since H. L. Mencken went to work there in 1906. Inside its present headquarters, a giant shoebox of red brick and glass, the raucous clacking of typewriters and teletype machines has given way to the muted patter of plastic computer keyboards. Much else has changed at Baltimore's sole surviving daily paper. The twin *Sunpapers* of Mencken's heyday are now one—the long-ailing *Evening Sun*, at whose creation he was present and on whose editorial page he ran rampant for two decades, was shut down in 1995—and the morning edition that remains is owned by a media conglomerate, its destiny controlled by executives from elsewhere. The faces faintly mirrored in the city room's video screens are as much of a break with tradition as the screens themselves. Like all urban newspapers, the *Sun* now goes out of its way to hire minorities and women, and in the brave new world of identity politics, the author of *Prejudices* has become a sometimes uncomfortable memory, a relic of the unlovely past, though his words are proudly displayed in the paper's newly remodeled lobby: "I find myself more and more convinced that I had more fun doing news reporting than in any other enterprise. It is really the life of kings." This pride of place is fitting, for he was the first member of the paper's staff to become nationally known, and a half century after he filed his last

column, his is still the first name that comes to mind whenever anyone outside Maryland has occasion to speak of the *Sun*.

Founded in 1837, the *Sun* had become the local "paper of record"
decades before Harry Mencken first set foot in a city room. A century
later he would help write its official history, and his description of the
paper's coverage of the great fire of 1904 nicely catches its tone:
"THE SUN, in those days, was not a paper to get excited and
descend to indecorum, even in the presence of a fire that was destroy-
ing the heart of Baltimore. It got out no extra, but devoted itself to
preparing a full report for early next morning, with all names spelled
precisely right and no essential fact forgotten. The other papers published extras, but not the imperturbable *Sunpaper*." By the time he
came aboard, the *Sun*'s only serious rival was the *Evening News*, which
was livelier and more politically independent than its "somewhat
stodgy" competitor. Baltimore's paper of record gradually responded
to the pressure of competition, as well as to the revival of civic vigor
that followed the fire. In April the *Sun* hired its first columnist; in July
it hired Mencken; in November it moved to new quarters at Baltimore and Charles Streets, in the heart of the city. Yet the atmosphere
remained that of "a good club rather than a great industrial plant."
Staffers were for all intents and purposes hired for life, "or at least
until and if [they] committed some grave and unforgivable offense
against the honor and dignity of the paper."

Except for the hiatuses of the world wars, Mencken would spend
the rest of his working life writing for the *Sunpapers*, and during those
exiles he remained in close touch with the paper and its staff. Even
amid the spreading disillusion of old age, one finds in his private reminiscences of the paper a note of grumpy affection. "The attempt to
make a great newspaper of the *Sun* was an adventure that I enjoyed
thoroughly," he wrote in 1942, shortly after he unceremoniously
removed himself from its pages for the next-to-last time, "and there
were times when it almost took on to me the proportions, considering
my generally unsentimental view of things, of a great romance." But
his love was far from blind, and his long-standing relationship with

the paper was a marriage of convenience. It was, as he said, a good club, populated by the cheerful cynics whose company he liked best and in whose admiration he basked. He tried out in its accommodating pages most of the ideas with which he would later wow a national audience. Most of all the *Sun* afforded him a measure of security and stability that few freelancers ever know—a consideration of prime importance to a man who, for all the abandon of his public persona, played it safe in his private life.

It defies easy explanation that so aggressive a writer should have been so personally cautious. Had Harry Mencken been more deeply scarred by his father's early death than he realized—or cared to admit? Did the relief that swept through him when August died, and the shame that surely followed it, make him cleave too closely to his mother out of a sense of guilt? Or was there some other trauma in his young life that he succeeded in hiding from the eyes of prying biographers? He was never given to soul-searching, and nothing he wrote about himself answers any of these questions. All we have to go on are the facts. By 1908 he was doubling his paycheck through outside writing, and he could have tripled it by moving to New York. Instead he stayed put, secure in the knowledge that whatever lunacies might beset him during the day, he could come home each night to Hollins Street and a plate of his mother's sandwiches. For all the pleasure he took in a schooner of beer, there was no chance in the world that he would die drunk.

MENCKEN'S OMNICOMPETENCE WAS AS APPARENT TO HIS NEW EMPLOYERS as it had been to his old ones, and no sooner did he arrive at the *Sun* than he started to take on extra duties. His first unsigned editorial ran on his second day at the paper, and within two months he was doubling as drama critic. Later on he decided that his coming to the *Sun* marked the beginning of his maturity as a writer: "Any literate boy of nineteen can write passable lyrics, but I doubt that any man ever had a sound prose style before thirty. My own began to take form after the suspension of the *Herald* in 1906; before that I had written sound jour-

nalese, but it was without any character. In 1906 I suddenly developed
a style of my own and it was in full flower by the end of that year." In
fact his *Sun* editorials marched in lockstep with the pieces he had written for the *Herald*, in subject matter as well as style. Old friends such as
William Jennings Bryan and Kaiser Wilhelm II peer out from among
a bumper crop of "light leaders" about baseball, whiskers, sauerkraut,
and mothers-in-law. Even his reviews have a familiar ring: "Poor old
Nietzsche, the mad German, is dead these 16 years, but his soul goes
marching on," he wrote by way of introduction to a now-forgotten
play by the now-forgotten playwright Henry Arthur Jones.

Though he was disposed to be tactful, the *Sun*'s new part-time
drama critic chafed at the banalities to which he was subjected in the
line of duty, and he grew tired of breaking his lance on the well-
made plays of nonentities: "The audience was not large and at no
time did its applause reveal enthusiasm. The play, indeed, is rather
vague and puzzling, without coherence or dramatic directness. . . .
The chief trouble with the play is that the central incidents are often
incredible." Such plain speaking "grossly offended" the local theater
managers, who made their displeasure known. The owners of the
Sun backed their man to the hilt, but Mencken stopped reviewing
plays in 1909, having concluded that "the managers were right . . . it
was a bit unjust after all for me to treat them so hardly. They had to
take whatever plays the theatrical syndicate sent them." Besides, it
was no great loss. He had tastier fish to fry.

IT IS HARD TO IMAGINE THAT THE NAME AND WORK OF FRIEDRICH NIETZ-
sche could ever have been unknown to educated Americans. Yet such
was the case when the twenty-six-year-old Sunday editor of the *Sunpaper* accepted Harrison Hale Schaff's invitation to write an introductory study of the author of *Thus Spake Zarathustra*, about whom he
knew "next to nothing." Only five volumes' worth of Nietzsche's collected works had been translated into English, and there was little in
the way of serious criticism to consult. (When Nietzsche died in 1900,
the *New York Times* described him in a spectacularly uncomprehending

four-paragraph obituary as "the apostle of extreme modern rationalism and one of the founders of the socialistic school, whose ideas have
had such a profound influence on the growth of political and social life
throughout the civilized world.") Unfazed by the lack of readily available translations or the sketchiness of his German, Mencken went to
the Pratt, unearthed a complete set of Nietzsche in the original, and
started slogging away. He delivered the completed manuscript just
seven months later, and by the middle of 1908 his clipping service was
forwarding reviews. The one he liked best was by Nicholas Murray
Butler, the eminent president of Columbia University: "Unfortunately we do not know who Mr. Mencken is. Neither his title pages,
his introduction, nor yet *Who's Who*, gives any trace of him. Nevertheless, he has written one of the most interesting and instructive books
that has come from the American press in many a long day."

The Philosophy of Friedrich Nietzsche, like *Ventures into Verse* and
George Bernard Shaw: His Plays, is a young man's book, but it does
more than hint at things to come: The great stylist has at last got the
bit between his teeth and started to gallop. No one else would have
proclaimed that belief in the gospels "is [now] confined to ecclesias-
tical reactionaries, pious old ladies and men about to be hanged," or
praised *The Antichrist* in the manner of a brass band with bass-drum
obbligato: "Beginning *allegro*, it proceeds from *forte*, by an uninter-
rupted *crescendo* to *allegro con moltissimo molto fortissimo*. The sentences run into mazes of italics, dashes and asterisks. It is German
that one cannot read aloud without roaring and waving one's arm."
That is the language of journalism, not scholarship, and it helps
explain why Mencken's pioneering contribution to the now-vast
Nietzsche literature is unknown save to his own devotees. Rarely do
academics make reference to it, even in popular treatments. Their
silence is forgivable: Mencken was no systematic philosopher, and
The Philosophy of Friedrich Nietzsche tells us rather more about its
ebullient author than its enigmatic subject, the prophet of post-
modernity who saw the godless future unfolding in his mind's eye,
briefly intuiting its hopes and fears before vanishing into the black

hole of madness. Mencken was too commonsensical, too solidly grounded in the reporter's world of things as they are, to readily grasp the essence of so tortured a soul. His Nietzsche is an all-American type, a world-improving, can-do go-getter delighted to have seen through the fraud of Christianity and gone beyond good and evil: "He applied the acid of critical analysis to a hundred and one specific ideas, and his general conclusion, to put it briefly, was that no human being had a right, in any way or form, to judge or direct the actions of any other being. Herein we have, in a few words, that gospel of individualism which all our sages preach today." (One sage, at any rate.) Few of his early writings are as inadvertently revealing as this autobiography in disguise, a fillet of Nietzsche in which the young critic gazed into the abyss, saw his own image, and found it good:

He believed that there was need in the world for a class freed from the handicap of law and morality, a class acutely adaptable and immoral; a class bent on achieving, not the equality of all men, but the production, at the top, of the superman. . . . Such a man, were he set down in the world today, would bear an outward resemblance, perhaps, to the most pious and virtuous of his fellow-citizens, but it is apparent that his life would have more of truth in it and less of hypocrisy and cant and pretense than theirs. He would obey the laws of the land frankly and solely because he was afraid of incurring their penalties, and for no other reason, and he would not try to delude his neighbors and himself into believing that he saw anything sacred in them. . . . His mind would be absolutely free of thoughts of sin and hell, and in consequence, he would be vastly happier than the majority of persons about him. All in all, he would be a powerful influence for truth in his community, and as such, would occupy himself with the most noble and sublime task possible to mere human beings: the overthrow of superstition and unreasoning faith, with their long train of fears, horrors, doubts, frauds, injustice and suffering.

In later life Mencken would persuade himself, and seek to persuade others, that his debt to Nietzsche had been "very slight." Perhaps he really thought so: The author of "The Bend in the Tube" scarcely required much more of a working philosophy. Still, every autodidact needs a yardstick against which to measure himself, and it was in Nietzsche's *Übermensch* that Mencken caught his first glimpse of the role he was preparing to play, the journalist with a hammer for whom Judeo-Christian morality was "something to wield his sword upon—to fight, to wound, to hate."

Along with enthusiastic reviews, a heightened degree of self-awareness, and an entry in *Who's Who*, *The Philosophy of Friedrich Nietzsche* helped bring its author the thing he wanted most: a pulpit from which to address the world beyond Baltimore. In the spring of 1908 he was invited to contribute a monthly book review–article to the *Smart Set*, a down-at-the-heels magazine based in New York. He wangled a day off from the *Sun*, took the train up to Manhattan, met with the editor, and made a deal (fifty dollars a month, plus review copies of new novels, which he could keep or sell). He also made the acquaintance of another up-and-comer who was there for a similar reason: He had been offered the job of drama critic. In the parlance of Hollywood, Mencken and George Jean Nathan met cute: "The stranger thrust out his hand to me and exclaimed, 'I'm H. L. Mencken from Baltimore and I'm the biggest damned fool in Christendom and I don't want to hear any boastful reply that you claim the honor.'" Then as always they would seem strangely sorted—Mencken looked like a fireplug, while his small, sleek, glossy-faced new colleague resembled nothing so much as a well-dressed otter—but they somehow recognized each another as kindred spirits and retired to the nearest bar to swap opinions on life and letters, an enterprise that would keep them happily occupied for the next decade and a half. Had Mencken plowed through the collected works of Nietzsche in German for no other reasons than to be offered a column in the *Smart Set* and to meet Nathan as a result, it would have been worth the trouble.

Most magazines die young and are fast forgotten. The *Smart Set* had a longish run, from 1900 to 1930, but it would now be a footnote to the history of American literature had it not been for Mencken and Nathan. They made the *Smart Set* respectable—insofar as it was possible—and it made them famous. As the playwright S. N. Behrman recalled, "In American colleges when I was an undergraduate (1912–1916), *The Smart Set* magazine, edited by H. L. Mencken and George Jean Nathan, had an electrifying effect. It was an influence. Along with *The New Machiavelli* and *The New Republic*, we swallowed *The Smart Set* as part of our regular nourishment. . . . In the literary courses aspirants wanted to write for *The Smart Set*, as, several decades later, they wanted to write for *The New Yorker*." In fact, it was not until 1914 that they took over the magazine, but their joint presence in the cheaply printed pages of the "Magazine of Cleverness" and the "Aristocrat among Magazines," as the *Smart Set* variously styled itself, was already overwhelming by the time Behrman got around to reading it. Whatever the contents of any given issue might happen to be, Mencken and Nathan were always the stars of the show.

The *Smart Set* began life as a literary pendant to *Town Topics*, a magazine that covered the doings of New York society. Col. William D'Alton Mann, a Civil War veteran turned magazine publisher, intended it as a vehicle for the "aesthetic" writings beloved of English *littérateurs* of the nineties, the Mauve Decade of Oscar Wilde, James McNeill Whistler, and Aubrey Beardsley; each issue contained a novelette, several short stories, poetry, epigrams, and at least one piece in French. The contributors included James Branch Cabell, O. Henry, Jack London, and Frank Norris, as well as scores of writers known only to scholars of turn-of-the-century American literature, and sundry youngsters who longed to make names for themselves (Sinclair Lewis and Damon Runyon sold light verse to the magazine). Whether famous or obscure, all were paid as little as Mann could get away with, making the *Smart Set* a cash cow. Its circulation swelled to 100,000 by the end of 1900, and 165,000 by 1905. Three years later, however, the magazine, through no fault of its own, was on the ropes.

Town Topics had published an innuendo-laden story about Alice Roosevelt, Theodore's daughter, and a subsequent suit for criminal libel revealed that its columnists were willing to keep the peccadilloes of high-bred folk out of print in return for unsecured loans to the colonel. The *Smart Set* had nothing to do with this shakedown racket, but it was tarred by association, and its circulation dropped precipitously. In an attempt to shore up the magazine's reputation for seriousness, Mann decided to add regular columns on books and the theater, and that was where Mencken and Nathan came in.

George Jean Nathan, like the *Smart Set*, would be forgotten were it not for his association with Mencken (though well-read film buffs may also recognize him as the model for Addison DeWitt, the venal, waspish critic in *All About Eve*). In his day, though, it was taken for granted whenever his name was yoked with Mencken's that their work was of like significance. He was born in Fort Wayne, Indiana, in 1882, the son of a wealthy, well-traveled *bon vivant* from Alsace-Lorraine who owned a vineyard in France and a coffee plantation in Brazil. He summered in Europe and studied at the University of Bologna, the University of Paris, Heidelberg University, and Cornell University; fencing was his chief extracurricular activity. He found his way to the *New York Herald* by 1905, began writing on theater shortly afterward (his uncle, a theatrical manager, had introduced him to Sarah Bernhardt when he was ten), and quickly attracted attention with his floridly written reviews.

Nathan inaugurated the *Smart Set*'s theater column in 1909. He had no more patience than Mencken with the invertebrate state of American drama, and devoted his energies to remaking it along more stimulating lines. Charles Angoff, Mencken's assistant at the *American Mercury*, claimed in 1952 that "Nathan's influence . . . during the past thirty years has been enormous," though the long-term effects of his work have proved tenuous: "He, more than any other critic, was responsible for the emergence of the O'Neill drama, which first brought American dramaturgy to serious world attention. He also did much to bring the better European drama to this country. Bahr,

Molnar, the Quinteros, Echegaray, Sierra, Pirandello, Lennox Robinson, Porto-Riche, Sacha Guitry, the Capeks, Georg Kaiser, and a score of others owe whatever hearing they have had among us largely to Nathan." In better-remembered fare, his track record was spottier: He thought *The Great Gatsby* an "inferior" book and Noël Coward's *Design for Living* "a pansy paraphrase of *Candida*," had no use for *Annie Get Your Gun* or *South Pacific*, dismissed Antony Tudor's ballet *Dark Elegies* as "about as close to the obscene and ridiculous as is humanly imaginable," and described T. S. Eliot's *The Cocktail Party* as "bosh sprinkled with mystic cologne."* In addition he offered himself up to posterity as a familiar essayist and epigrammatist ("Men go to the theatre to forget; women, to remember"), but his languid efforts in these genres are now no more read than his criticism, perhaps because they lack Mencken's fierceness.

At the high-water mark of their friendship, Nathan sketched himself with pointillistic exactitude, deftly suggesting the dandyish side of his character:

> He is a man of middle height, straight, slim, dark, with eyes like the middle of August, black hair which he brushes back *à la française*, and a rather sullen mouth. . . . He never reads the political news in the papers. . . . He was born, as the expression has it, with a gold spoon in his mouth. He has never had to work for a living. . . . His telephone operator, at his apartment, has a list of five persons to whom he will talk—so many and no more. He refuses to answer the telephone before five o'clock in the afternoon. . . . He never visits a house a second time in

---

*Mencken had trouble with *Gatsby*, too, though he was far closer to the mark than Nathan: "The story, for all its basic triviality, has a fine texture, a careful and brilliant finish. The obvious phrase is simply not in it. The sentences roll along smoothly, sparklingly, variously. There is evidence in every line of hard and intelligent effort. It is a quite new Fitzgerald who emerges from this little book, and the qualities that he shows are dignified and solid."

which he has encountered dogs, cats, children, automatic pianos, grace before or after meals, women authors, actors, *The New Republic*, or prints of the Mona Lisa. . . . He owns three top hats, fourteen walking sticks, and two Russian wolf-hounds.

The charm of his affectations eluded many people—Edmund Wilson curtly dismissed him as "a pernickety and snubbing little man"—but those were the days when Mencken and Nathan reveled in their shared tastes, as well as their shared conviction that the world was run by and for idiots. Mencken's own aestheticism ran deep—it was not for nothing that he had read Wilde as an adolescent—making it far from surprising that he would warm to Nathan's hedonistic passion for "the surface of life: life's music and color, its charm and ease, its humor and its loveliness." It's possible, too, that he may have envied the poise that went with Nathan's unearned income, not to mention his success with women. The fact that Nathan was a thoroughly assimilated Jew (he converted to Roman Catholicism on his deathbed) would doubtless have intrigued him as well. He seems to have known few Jews in 1908, none more than casually, and Nathan was the first of many friends and colleagues whose equivocal relationship to their Jewishness he would find a source of fascination in the years to come.

The strongest bond of all, though, may simply have been that the two men were contemporaries. Mencken's affection for Baltimore was no pose: It had already given him comfort and success beyond the dreams of most men. But it had little to offer in the way of high culture, nor had it supplied him with an intimate friend who knew as much as he did about the things he cared about most. He had spent the past eight years talking to himself in public, mostly about politics but increasingly about literature and the arts as well. Now he had someone to talk to, a fellow critic of his own age who was well read, well educated, widely traveled, and firmly committed to the belief that "nothing in this world matters very much in the long run and that it is accordingly profitless to get too much worked up over anything."

New friends fix on their similarities, old ones their differences. For all the breadth of his culture, Nathan had no serious interest in the general ideas to which Mencken consecrated the second half of his writing life. "The great problems of the world—social, political, economic, and theological—do not concern me in the slightest," he claimed, and that blithe unconcern was the shoal on which their collaboration was to run aground. By the mid-twenties, Mencken had come to see Nathan's detachment as a kind of provincialism ("His world was limited by New York, and even of New York he knew little save Broadway"). Long after they parted company, he would make his own pen portrait of his dapper ex-friend, this one etched with a rusty nail:

> A slight fellow, of less than average height, he is intensely self-conscious about his physique, and is at great pains to avoid being seen with women who are not smaller than he is. If they are known by sight to all the Jews and whores who hang about the theatres and nightclubs, so much the better, for though he has denied, in recent years, that he is a Jew himself, a typically Jewish inferiority complex is in him, and it gives him great satisfaction to have some eminent (or even only notorious) fair one under his arm.

But such brutalities were unimaginable on the May afternoon when they strolled into the bar of the Beaux-Arts Café, ordered a round of cocktails, and toasted their new jobs. "What's your attitude toward the world?" Mencken asked. "I view it as a mess in which the clowns are paid more than they are worth, so I respectfully suggest that, when we get going, we get our full share."

Mencken got going first, making his debut in the November issue with a lively piece called "The Good, the Bad and the Best Sellers" in which he panned Upton Sinclair's overearnest *The Moneychangers* ("Hordes of the *bacillus platitudae* have entered Sinclair's system and are preying upon his vitals"), praised the "delicious lightness" of *The*

Circular Staircase, Mary Roberts Rinehart's first mystery novel, and took succinct note of *Views and Reviews,* a collection of "early essays by Henry James—some in the English language." Though he later claimed to have developed "a certain amount of stage-fright" while writing the first of his 182 *Smart Set* book columns, it wasn't evident in the finished product. "Mr. Mencken is one of the best-known newspaper men in the country, and his judgment and criticisms of the various new books are well worth reading," the magazine obligingly informed its readers. Well known he wasn't, not even in Baltimore, but "The Good, the Bad and the Best Sellers" was definitely worth reading. Having spent his young years gulping down novels of every sort, he was full of strong opinions about the kind of writing he liked, and more than ready to loose them on the world in copious quantities. During the next four years, he read, by his own believable reckoning, an average of a novel a day, most forgettable and many downright awful.

American literature was choking on its own high-mindedness in 1908, and the enfant terrible who eight years earlier had praised *Huckleberry Finn* as "the greatest novel yet produced by an American writer" was no less firm in his rejection of the neutered evasions of that decorous era, the age of William Dean Howells, Gene Stratton Porter, and Harold Bell Wright. Already he was handing down his first commandment of art: Thou shalt not preach. Upton Sinclair, he announced in his *Smart Set* debut, "has hopelessly confused the functions of the novelist with those of the crusader. . . . he is going the road of Walt Whitman, of Edwin Markham, of the later Zola, all of whom began as artists and ended as mad mullahs." Two issues later, he put forward his own definition of the "real" novel, as opposed to the "near-novels, pseudo-novels and quasi-novels" of the purveyors of insipid trash whose produce was already clogging his mailbox:

Now and then, however, there appears upon the bookstalls a new book which meets upon fair ground those great books of other years which set the bounds of fiction's symphonic form—

a new book which tells, with insight, imagination and conviction, the story of some one man's struggle with his fate—which shows us, like a vast fever chart, the ebb and flow of his ideas and ideals, and the multitude of forces shaping them—which gives us, in brief, a veritable, moving chronicle of a human being driven, tortured and fashioned by the blood within him and the world without—a chronicle with a beginning, a middle and an end, and some reasonable theory of existence over all.

He went on to cite *Huckleberry Finn* and *Lord Jim*, the two books that would ever after be his touchstones of excellence, as novels of the "first rank." Mencken was far from alone in his admiration for Mark Twain's masterpiece, whose directness he hoped might serve as an antidote to the weak tea being dished up by the Anglophiles with triple-barreled names whose well-scrubbed books he disdained. But in 1908, as he would later note with pride, Joseph Conrad was "no more than a name in the United States." It made no matter that Mencken's Conrad was as personal a creation as his Nietzsche, and bore as little resemblance to the original.* Mencken would always be the most personal of critics, cheerfully superimposing his own opinions on those of the writers he liked best, as he readily admitted in the definitive statement of his critical method that he published a few years later. The good critic, he claimed, "is always being swallowed up by the creative artist . . . He is trying to arrest and challenge a sufficient body of readers, to impress them with the charm and novelty of his ideas, to provoke them into an agreeable (or shocked) awareness of him, and he is trying to achieve thereby for his own inner ego the grateful feeling of a function performed, a tension relieved, a *katharsis* attained which Wagner achieved when he wrote 'Die Walküre,' and a hen achieves every time she lays an egg."

In the end what mattered was not that Mencken got Conrad

---

*For him *Heart of Darkness* was full of "harsh roars of cosmic laughter, vast splutterings of transcendental mirth, echoing and reëchoing down the black corridors of empty space."

wrong but that he used the *Smart Set* to inform American readers, confidently and regularly, that the author of *Lord Jim* was a major novelist—the first of his signal achievements as a literary critic, though few contemporary scholars seem to know about it. Still, Conrad was English, after a fashion, and Mencken, as American as Twain himself, longed for a homegrown realist of like stature whom he could use as a stick with which to clobber "the frauds and dolts who still reigned in American letters." Three more years would go by before he found the man he sought, right under his nose.

THEODORE DREISER IS REMEMBERED, BUT NOT READ. ONLY FOUR OF HIS novels remain in print, and of them only the first, *Sister Carrie*, continues to be taught with any frequency in American colleges and universities. A search of the World Wide Web in the fall of 2000 turned up just 6,990 hits for Dreiser, a tenth as many as Ernest Hemingway (and a fourth as many as Mencken). He has his devoted admirers, Saul Bellow prominent among them, but when he is remembered today, it is usually for what Mencken wittily termed his "incurable antipathy to the *mot juste*." No other major novelist writing in English has been so ignorant of the rudiments of style. As his letters reveal, he couldn't even spell "grammar," much less master it. Nearly everyone who writes about him cites the same leaden sentence from *An American Tragedy*—"The 'death house' in this particular prison was one of those crass erections and maintenances of human insensibility and stupidity for which no one primarily was really responsible"—yet it is possible to open any of his books almost at random and find similar specimens. He wrote by the yard, shoveling in half-digested facts by the pound, and his prose was the coarsest of cotton twill, functional but devoid of grace.

If Dreiser was a genius, as so many critics of the twenties and thirties devoutly believed, then he was an idiot savant, capable of creating unforgettable characters on paper but devoid of any other gift save an uncanny knack for persuading women to have sex with him. Even the staunchest of his supporters would be reduced to despair by the impenetrable stupidity with which he dabbled in the great political issues of his time. Mencken spoke of his "insatiable appetite for the not true," while a more recent critic, who praised him as "America's greatest novelist," went on to describe him as a "bore, crank, celery-juice drinker, and member of the select company of morons who believed that Franklin Delano Roosevelt was part Jewish," an indictment that would be funnier were it even slightly exaggerated. Ultimately his passion for politics killed his art stone dead. He spent the last twenty years of his life snuffling out causes like a truffle-hunting pig, and never wrote another readable word.

In 1908, though, Dreiser had yet to finish his second novel, much less discover politics. He was busy earning a living as the managing editor of Butterick Publications, putting out three women's magazines with what by all accounts was unusual skill. Despite the sincerity of his oft-stated belief that human life was meaningless, he was an incorrigible theotrope in search of a more accommodating deity with whom to replace the stern Catholic God of his childhood (he would prove especially partial to Stalin), and though he liked his paneled office and five-figure salary, he could never have dished up sentimental bilge to housewives so successfully had he not believed in it as well. In any case his options were limited: *Sister Carrie* had earned him a grand total of $68.40 in royalties, thereby sending him into a tailspin of depression from which he had only just begun to recover. Instead of finishing his second novel, he put it on the shelf and beat a hasty retreat into the lucrative world of women's journalism, bundling dress patterns and oleaginous prose into slick packages. "We like sentiment, we like humor, we like realism, but it must be tinged with sufficient idealism to make it all of a truly uplifting character," he told a potential contributor in 1909. For all anyone knew, his career as a novelist was over.

It was at this moment that Mencken strolled into Dreiser's office for the first time. The two men were already doing business by mail. Dreiser appreciated Mencken's professionalism and liked his prose style (he said that it "bristled with gay phraseology and a largely suppressed though still peeping mirth," a line that could only have been written by the author of *An American Tragedy*), so he hired him sight unseen to do a series of innocuous articles on child care, produced in collaboration with a Baltimore physician, which eventually made up a book called *What You Ought to Know About Your Baby*, issued under the doctor's name. In March 1908 Dreiser invited Mencken to visit him in New York, and the two men thereupon embarked on what the novelist called an "instantaneous friendship," one destined to endure, however fitfully, until Dreiser's death thirty-seven years later.

What could Mencken have seen in such an oaf? Dreiser's elephantine charm, a quality rarely on display in his novels, comes through with unexpected clarity in his published reminiscence of his first meeting with the man who, more than any other, would help make him famous. So, too, does Mencken's own noisy good humor. Not everyone warmed to it, but for those who did, he was all but impossible to resist:

> [T]here appeared in my office a taut, ruddy, blue-eyed, snub-nosed youth of twenty eight or nine whose brisk gait and ingratiating smile proved to me at once enormously intriguing and amusing. . . . With the sang-froid of a Caesar or a Napoleon he made himself comfortable in a large and impressive chair which was designed primarily to reduce the over-confidence of the average beginner. And from that particular and unintended vantage point he beamed on me with the confidence of a smirking fox about to devour a chicken. So I was the editor of the Butterick publications. He had been told about me. However, in spite of *Sister Carrie*, I doubt if he had ever heard of me before this. After studying him in that almost arch-episcopal setting which the chair provided, I began to laugh. "Well, well," I said, "if it isn't Anheuser's own brightest boy out to see the town."

Predictably enough Mencken remembered things differently, finding his inquisitor to be "a somewhat solemn fellow externally . . .

it was not until I knew him better that I discovered his loutish humor." Nor was he unfamiliar with *Sister Carrie*. Quite the contrary: There can be little doubt that his interest in Dreiser arose from the fact that the older man had written a novel of which Mencken approved almost without reserve. To be sure, they had other things in common. Two German-American journalist-autodidacts who thought life meaningless could scarcely have failed to notice the similarities in their backgrounds and philosophies. But above all, it was *Sister Carrie* that pulled them together. Mencken knew how important a book it was, and suspected what its writer might be able to do for American literature. He longed for an American writer to praise, a man unafraid to face the ugliness of modern life, and Dreiser was that man, a connoisseur of shabbiness and fear. Everything about *Sister Carrie*—its comparative frankness about sexual matters, its refusal to observe the moral niceties of punishing Carrie Meeber for her sins, its fearsomely honest portrait of Hurstwood, a middle-aged man in decline—was as far removed as possible from the genteel tradition Mencken excoriated in the pages of the *Smart Set*. "The American, try as he will, can never imagine any work of the imagination as wholly devoid of moral content," he wrote nine years later in *A Book of Prefaces*. "It must either tend toward the promotion of virtue, or be suspect and abominable." *Sister Carrie* was a deeply moral book, but it offered the soft-mouthed reader no reassuring vision of virtue triumphant, only a harsh portrayal of lost souls caught in the web of life.

It's impossible to understand Mencken's contribution to American letters without understanding what he saw in Theodore Dreiser. More than anything else, what he found lacking in contemporary American literature was an awareness of the common life. You could read a hundred new novels and come away supposing that the people he had met as a police reporter—the saloonkeepers and prostitutes and crooked politicians and murderers—did not exist at all. But Dreiser knew better. If he was not the first American novelist to portray the underside of American gentility, he was among the first to do so with something like complete candor: He wrote in plain language about a world of poverty and desire whose very existence other writers denied. Mencken responded powerfully to this candor; it was the thing he admired most about *Sister Carrie*. Yet he himself came from a completely different world, one where comfort was taken for granted. Dreiser would point this out to him years later, long after their friendship had foundered on the rocks of that difference:

> You see, Mencken, unlike yourself, I am biased. I was born poor. For a time, in November and December, once, I went without shoes. I saw my beloved mother suffer from want—even worry and wring her hands in misery. And for that reason, perhaps— let it be what you will—I, regardless of whom or what, am for a social system that can and will do better than that for its members—those who try, however humbly,—and more, *wish to learn how* to help themselves, but are none-the-less defeated by the trickeries of a set of vainglorious dunces, who actually believe that money—however come by—the privilege of buying this and that—distinguishes them above all others.

To such fulsome self-dramatization Mencken could only respond with embarrassed joking, and so he amiably replied that when they first met, Dreiser was "a rich magazine editor with six or eight secretaries. I recall perfectly that you had on a suit of Hart, Schaffner and Marx (not Karl, but Julius) clothes costing at least $27.50, and used a gold lead-pencil. Thus I simply can't think of you as a proletarian. To me you must always be a kind of literary J. Pierpont Morgan." He tactfully neglected to add that he could have no respect for a hypocrite who preached revolution while harassing his publishers for money his books did not earn. But that was in 1943, by which time Dreiser had ceased to be a serious novelist in anything but name. In 1908 his reputation rested upon *Sister Carrie* alone, and gauchely written though it was, America had yet to produce another novel quite like it.

His interest in Dreiser was not unselfish. (Few of his friendships were, though few of his friends realized it.) In *My Life as Author and Editor*, he spoke of the "considerable self-interest" that had led him to seek out an American novelist to champion. From the start of his long run at the *Smart Set*, he understood that a critic cannot make a name for himself solely by panning bad books. He must also find new writers to praise, "if only for the sake of the dramatic contrast." To make them famous would make him famous, too, and he knew exactly what kind of novelist would turn the trick: "What I needed was an author who was completely American in his themes and his point of view, who dealt with people and situations of wide and durable interest, who had something to say about his characters that was not too obvious, who was nevertheless simple enough to be understood by the vulgar, and who knew how to concoct and tell an engrossing story."

Dreiser filled the bill in every way but one: He was no longer writing fiction. Cowed by the failure of *Sister Carrie*, he had retreated to his fancy office at Butterick Publications to make some money and lick his wounds. Not until 1911 would he publish his second book. Nor was Mencken quite ready to do battle for the American novelist of his dreams. He had not even started writing for the *Smart Set* when he met Dreiser. Indeed, the novelist would claim long after the fact that he had recommended Mencken to Fred Splint, then the editor of the *Smart Set*, though Splint told Mencken otherwise. What is certain is that Dreiser offered Mencken a full-time job at Butterick, which the younger man declined with thanks, and that their odd friendship continued to flourish. Heated arguments were always part of it: Mencken inscribed a copy of his *Nietzsche* to Dreiser "[i]n memory of furious disputations on sorcery & the art of letters." It isn't difficult to imagine what sort of "sorcery" the disputatious unbeliever had in mind, or how exasperating he must have found his new friend's short-lived pseudo-intellectual enthusiasms. But whatever passing temptation he may have felt to wash his hands of Dreiser's "puerile philosophizing," he chose instead to bide his time. Anyone capable of writing *Sister Carrie* was surely capable, sooner or later, of writing another novel as worthy of being praised to the skies by a young critic on the make.

EVEN THE MOST ASTONISHING CAREERS HAVE PERIODS OF RETRENCHment and consolidation, and Mencken was overdue for one by the end of 1908. For all his great gifts, he was no prodigy but a hard worker who became himself not suddenly but gradually, and he would spend the next couple of years catching up with the spurt of growth evident in *The Philosophy of Friedrich Nietzsche* and his *Smart Set* columns. In 1909, he collaborated with Holger A. Koppel, a Danish diplomat based in Baltimore, on new editions of Ibsen's *A Doll's House* and *Little Eyolf* (Koppel made literal translations from the original Norwegian, which Mencken turned into polished English). The following year, he edited an anthology called *The Gist of Nietzsche* and conducted an epistolary debate on individualism with a socialist acquaintance that eventually appeared in book form as *Men Versus the Man*. None of these books won him any particular notice, though *Men Versus the Man* shows how his political thinking had solidified—hardened, really. The law of the survival of the fittest, he declares, is "immutable," thus making socialism an absurdity; human progress is the product of the will to power, and all social arrangements failing to take this fact into account are doomed to failure; inequality is natural, even desirable, both in and of itself and as an alternative to mob rule; the world exists to be run by "the first-caste man." Looking back on *Men Versus the Man* fifteen years later, he found it good enough, if not altogether successful: "The book seems somewhat archaic today but it at least shows one thing clearly, that my politics were firmly formulated so early as 1909."

Having formulated his politics, Mencken put them aside, not because he thought they were unimportant but because there seemed no pressing need for them in the America of 1909. For years the White House had been occupied by men who took it for granted that tariffs should be kept high and the federal government small. Even the reform-minded Theodore Roosevelt, for all his longing to strut on a larger stage, had only so much power to wield, and so much money to spend. In a country without a federal income tax, committed to laissez-faire economics as a matter of quasi-religious principle, true power rested in the hands of businessmen and bankers. Most Americans, Mencken among them, were happy enough with the results, and those who were not had failed to turn their discontent into countervailing power. The populists and the socialists launched one third party after another, and all went down in flames.

By no conceivable stretch of the imagination could Mencken be said to have idealized the big businessmen of his day. "On the contrary," he would write in 1941, "it seems to me that such swine were and are my enemies even more plainly than the Communists, not only because they devoted themselves to robbing me, but also and more importantly because their intolerable hoggishness raised the boobery in revolt, and the ensuing revolt threatened to ruin me even more certainly." Yet he preferred them to politicians. His belief in small government was in no small part a function of his utter disbelief in the good faith of elected officials. To him they were comical boot-lickers who would say and do anything to hold onto their jobs. He had first seen them up close as a small child, touring the saloons of Washington, D.C., with his father, who sold cigars in bulk to "restauranteurs" (as the saloonkeepers of the nation's capital called themselves). One of August's best customers, though he found judges and senators tolerable, had a violent aversion to congressmen: "He addressed them familiarly as George, Jack and Bill, and once I heard him invite one of them to get the hell out of the place, and stay out. My father explained that this was because Congressmen were too numerous in Washington to be of any note; moreover, not a few of them were given to caterwauling and wrestling in barrooms, a habit that he always deprecated." His son would be no more disposed to take the inhabitants of Washington seriously, least of all the ones who longed to be president. Mencken wrote about conventions as if they were circuses, never dreaming that a time might come when the clowns would acquire enough power to affect his own life; when they did, he stopped laughing. But that was a long way off, and as his thirtieth birthday drew near, he was too busy to worry about the grumblings of second-caste men too weak to accept the inevitable. Unlike them he was a rising star, and the *Sun* was ready to let him shine.

Early in 1910 Charles Grasty, for whom Mencken had worked briefly four years earlier at the *Evening News*, put together a syndicate of backers who acquired a controlling interest in the *Sun* and installed him as owner and publisher. Grasty was determined to bring the stodgy *Sun* up to the speed of the evening papers, which, in the pretelevision days of American journalism, were where all the freshest news was to be found. He started an evening edition of the *Sun* in April, and made Mencken the mainstay of its editorial page. He wrote a short article signed "H. L. M." and at least two unsigned editorials each day, edited the letters section, and reviewed plays on the side. His new job was so demanding that it left him no time for any outside writing except his *Smart Set* reviews (he published no other magazine articles in 1910 or 1911, and his correspondence dried up as well). Such backbreaking labor would have turned a lesser man into a writing machine. Instead he used the daily grind as a stone against which to sharpen his mind. Day by day he grappled with idea after idea and issue after issue, amassing the intellectual capital on which he would draw for the rest of his working life.

By then Mencken had settled on most of his favorite subjects, and managed to write about several of them in his first week on the page:

Let us confess it to one another: Baltimore is a good old town. . . . It is, indeed, a cardinal socialistic doctrine, though the orators of the party deny it, that the desire of a weak and inefficient man to seize the things beyond his reach is, in some recondite and incomprehensible fashion, more honorable to the man himself and more pleasing to the Master of the Universe than the desire of a strong and efficient man to keep the things he has acquired. . . . There is no space here to rehearse the astonishing merits of 'Huckleberry Finn' in detail. That task the reader will begin to perform for himself, once he has given the book a third or fourth reading.

Those first articles were impressive, the later ones better still, while the last ones were very nearly as good as anything he ever wrote. Yet next to none of Mencken's two-hundred-odd "H. L. M." columns have survived, save in his own crumbling scrapbooks. Instead of collecting them he recycled their remembered substance into the definitive essays of his maturity, in the process refining his style still further. The Mencken we know is the finished product of that lengthy process of self-creation, the trumpet-blaring troublemaker who exploded on the American scene at the dawn of the Jazz Age, fully formed and larger than life. But it was on the editorial page of the *Evening Sun*, out of sight of the rest of America, that he painstakingly perfected himself.

Despite his anonymity, Mencken's gifts as a columnist did not go unnoticed. Harry Black, a member of the *Sun* board, suggested to Grasty that his talents might be put to better use in a signed column that was more personal in tone. Grasty liked the idea, and "H. L. M." bade farewell to the editorial page on May 6, 1911, with a tribute to the life-enhancing powers of alcohol, "the father and mother of joy." Two days later the page carried a column called "The World in Review," signed "H. L. Mencken"; the title was changed to "The Free Lance" the next day. Mencken's byline had never before been printed in the *Sunpapers*. It would be seen there, six days a week, for the next four and a half years.

Mencken would later describe "The Free Lance" as "a private editorial column devoted wholly to my personal opinions and prejudices." At first each "Free Lance" dealt with two or three separate topics, and though local politics was a staple item ("Sixty cents reward to anyone who will offer one good excuse for the City Council's existence"), the column ranged as widely as its author's fancy. On any given day the readers of the *Evening Sun* might find epigrams ("The truth that survives is the lie that it is pleasantest to believe"), light verse, capsule book reviews, examples of what Mencken was already calling "the American language" ("I went fer to git it, but he taken it away"), even complaints about how people misspell and mispronounce "Nietzsche." It was the sound of a smart young man thinking out loud, and if the results were ephemeral, they still made a great splash.

Though his readers could not have known it, he was doing something far more original than is generally recognized. The newspaper column as an institution was still in its infancy in 1911, and the first "op-ed" page, that of Herbert Bayard Swope's *New York Evening World*, would not be launched for another decade. Many columnists were publishing daily miscellanies superficially similar to "The Free Lance," most of which were made up in part of contributions from their readers, but Mencken appears to have been one of the first modern American newspapermen to write a signed editorial-page column in which he commented regularly on major issues of the day. Part of its effectiveness came from the dullish company it kept. "The Free Lance" ran all the way down the right-hand column of the editorial page, which was otherwise devoted to unsigned editorials, bad poetry, light leaders, and letters to the editor. Because it was the only signed column and the only strongly individual voice on the page (the *Evening Sun* was not yet running editorial cartoons), it stood out all the more. Baltimoreans read Mencken not only because "The Free Lance" was so much more amusing than the rest of the paper, but because it was deliberately controversial: "I never mentioned any contemporary public character, whether political, theological, moral, financial or professorial, without denouncing him as a fraud. I believe today that I was right every time." Mencken also made a point of publishing "all the worst attacks" on "The Free Lance" in the letters-to-the-editor column, which he continued to edit. This shrewdness did not go unrewarded. Until 1911 his name had been familiar only to his fellow professionals and the handful of locals who read him in the *Smart Set*, but "The Free Lance" made him the most talked-about journalist in town.

With notoriety came freedom: He was relieved of his other editorial tasks and given an office of his own. More important, he was free at last to write about whatever he chose, under his own name and in his own ever-recognizable voice. To be sure, the results were fleeting in interest and fragmentary in form. (He never attempted to reprint any of the columns that went into his four fat scrapbooks of "Free Lance" clippings, though he quarried them for the raw material of his next few books.) "The Free Lance" was strictly a local phenomenon, and thus is unlikely to have had any direct effect on the work of such pioneering op-ed columnists as Heywood Broun and Walter Lippmann. All the same it was an indispensable part of the making of Mencken. "Before it had gone on a year," he would claim, "I knew precisely what I was about and where I was heading. In it I worked out much of the material that was later to enter into my books, and to color the editorial policy of the *American Mercury*." Though the "H. L. M." columns deserve more credit than that, he was essentially correct. For all the hard work of the dozen preceding years, it was "The Free Lance" that at long last brought his journalistic apprenticeship to a rousing close.

Local renown stops at the city limits, but Mencken's scrapbooks reveal that he was also making a name for himself outside Baltimore, if not as a newspaper columnist. It was his *Smart Set* pieces that were turning heads. His name, for instance, appeared in the *New Orleans Times-Democrat* early in 1911: "Two men with sharp and naughty pens, who write for *The Smart Set*, are George Jean Nathan and H. L. Mencken." And it would soon be seen in newspapers and magazines all across the United States, yoked with that of the former managing editor of Butterick Publications. While Mencken was pounding out unsigned editorials for the *Sun*, Dreiser was finishing his long-delayed second novel, a *Sister Carrie*–like chronicle of the rise and fall of Jennie Gerhardt, a poor young girl who bears the bastard child of a senator. He had no choice but to get it done as fast as possible: Butterick had fired him for chasing after the teenage daughter of a colleague, and he was running out of money.

The Dreiser-Mencken correspondence tells the rest of the story:

"The truth will out. I have just finished one book—Jennie Gerhardt—and am half through with another. I expect to try out this book game for about four or five books after which unless I am enjoying a good income from them I will quit." (Dreiser to Mencken, February 24, 1911)

"You know I am one of those who hold 'Sister Carrie' in actual reverence . . . I look forward eagerly to 'Jennie Gerhardt' and to her successors." (Mencken to Dreiser, March 3)

"The story comes upon me with great force; it touches my own experience of life in a hundred places; it preaches (or perhaps I had better say exhibits) a philosophy of life that seems to me to be sound; altogether I get a powerful effect of reality, stark and unashamed." (Mencken to Dreiser, April 23)

"If the Harpers [Dreiser's publishers] want my review in advance they can get a proof of it by telephoning to Norman Boyer at the S.S. It is in type by now. God bless all honest men." (Mencken to Dreiser, September 20)

"I confess that my cheeks colored some for I'm not ready yet to believe it." (Dreiser to Mencken, October 17)

The review that made Dreiser blush, "A Novel of the First Rank," appeared in the November *Smart Set*. It was a fifty-gun salute, the kind of review about which starving authors dream, usually in vain:

> If you miss reading *Jennie Gerhardt*, by Theodore Dreiser, you will miss the best American novel, all things considered, that has reached the book counters in a dozen years. On second thought, change "a dozen" into "twenty-five." On third thought, strike out everything after "counters." On fourth thought, strike out everything after "novel." Why back and fill? Why evade and qualify? Hot from it, I am firmly convinced that *Jennie Gerhardt* is the best American novel I have ever read, with the lonesome but Himalayan exception of *Huckleberry Finn*, and so I may as well say it aloud and at once and have done with it. . . . It lacks

the grace of this one, the humor of that one, the perfect form of some other one; but taking it as it stands, grim, gaunt, mirthless, shapeless, it remains, and by long odds, the most impressive work of art that we have yet to show in prose fiction—a tale, not unrelated, in its stark simplicity, to *Germinal* and *Anna Karenina* and *Lord Jim*—a tale assertively American in its scene and its human material, and yet so European in its method, its point of view, its almost reverential seriousness, that one can scarcely imagine an American writing it.

Mencken took due note of the book's shortcomings, writing frankly of "its gross crudities, its incessant returns to C major." But he insisted that irrespective of its flaws, Dreiser's "profound pessimism" and his determination to produce a starkly realistic novel with "no more moral than a string quartet or the first book of Euclid" placed *Jennie Gerhardt* head and shoulders above the work of any other living American novelist, Henry James specifically included.

The most striking thing about "A Novel of the First Rank" is the extremity of its enthusiasm. Even allowing for Mencken's lifelong tendency to overegg the pudding, nine decades' worth of hindsight leads one to wonder whether he really thought Dreiser worthy of direct comparison with Zola, Tolstoy, and Conrad. There is no reason to suppose otherwise. Nowhere in his other writings, published or unpublished, does he say anything to the contrary about *Jennie Gerhardt*, a book whose bleak determinism would have been wholly to his taste ("We're moved about like chessmen by circumstances over which we have no control"). Yet for all the seeming sincerity of his praise, he must also have heard the still small voice of careerism whispering in his inner ear. Could Mencken have been swayed by his awareness of the effect that discovering a major American writer would have on his own reputation? Or did he truly believe *Jennie Gerhardt* to be a bona fide masterpiece that transcended its flaws? All we know for sure is that three years to the month after he made his debut in the *Smart Set*, Mencken had finally found a new American novelist for whom he was willing to bang the gong.

The review was noticed within days—literally—of its publication. Mencken's scrapbook for 1911 contains a clipping from the *New York Post* dated October 25: "It is seldom given to a critic to discover and proclaim a novel of the first rank, but no less is the happy fortune of H. L. Mencken." As he had expected, Dreiser's publishers quoted liberally from "A Novel of the First Rank" in their advertisements (the first time that one of Mencken's *Smart Set* reviews had been used for such purposes), and *Jennie Gerhardt*, to everyone's amazement, began to sell, strongly and steadily. Rare is the artist who can resist hyperbolic praise, especially when it appears in magazines. The Dreiser-Mencken friendship grew progressively more intimate, and in due course the novelist invited the critic to dine *en famille* in New York. Mencken knew something of his friend's extramarital affairs, so it came as no surprise to him when Dreiser vanished after dinner and failed to reappear for a full hour. No sooner was he out the door than Jug Dreiser poured her troubles into her guest's lap, begging him to help rein in her compulsively adulterous husband.

When Mencken's sealed papers were opened thirty-five years after his death, they were found to contain an unvarnished account of the evening, and it tells at least as much about him as it does about the unhappy couple:

> Who, indeed, was I, to pronounce judgment upon a man who was my elder and my better—a great artist, and, in the history of the Republic, almost incomparable? If the needs of his extremely difficult and onerous trade required him to consume and spit out a schoolma'm from the cow country, then it was certainly to be lamented—but that was as far as I could go in logic, or in such ethical theory as I subscribed to. . . . Thus I dismissed her from my mind with hardly more than formal regret, and took my stand with the devil's party, which is to say, with

Dreiser. After all, I was *his* friend and supporter, not *hers*. She
had done nothing to lift me, or even to interest me; indeed, the
most significant thing about her, intellectually speaking, was
that she was a customer of Mary Baker G. Eddy, who seemed to
me to be on all fours with the Fox sisters and Brigham Young.

That was Mencken to the life. He loved to claim membership in
the devil's party, while conducting his own life with the utmost propriety and circumspection. One smiles to learn that he went out of
his way to tell his first biographer that "I never drink alone," just as
one smiles to find in his *New Dictionary of Quotations* these lines from
a beloved author of his youth: "I hope you have not been leading a
double life, pretending to be wicked, and being really good all the
time. That would be hypocrisy."

MENCKEN'S REPUTATION NOW RESTED ON THE TWIN PILLARS OF THE
Evening Sun and the *Smart Set*, both of which changed hands in the
course of a single year. In February 1911, the unscrupulous Colonel
Mann had unloaded the magazine on John Adams Thayer, an
ex–advertising manager who had made his fortune by buying, puff-
ing, and reselling the then-popular *Everybody's Magazine*. By 1912
the *Smart Set's* circulation had declined to somewhere between
eighty thousand and ninety thousand, and Thayer sought to reverse
its fortunes by persuading his best-known regular contributor to
become the magazine's editor.

Unsure of his new boss and preoccupied with "The Free Lance,"
Mencken turned the offer down flat, instead recommending Willard
Huntington Wright, a twenty-five-year-old admirer of Nietzsche
who had served as literary editor of *Town Topics* before moving to the
Los Angeles Times in the same capacity. Mencken liked Wright's book
reviews and was impressed by the fact that he was also in charge of
the *Times*'s book advertising, suggesting that he had a head for business. Thayer took Mencken's advice, and regretted it. Wright's idea
was to make the *Smart Set* the American counterpart of such

modern-minded European magazines as *Simplicissimus* and Ford Madox Ford's *English Review*, to which end he hired Ezra Pound to scout for overseas contributors, ran pieces by Conrad, Beerbohm, Frank Harris, D. H. Lawrence, August Strindberg, and W. B. Yeats, and egged Mencken and Nathan on to ever greater heights of impertinence. He also paid his writers more money. Mencken's fee was doubled to $100 a month, boosting his total income for 1913 to $4,656.48, the equivalent of about $80,000 today and a sum he later described as "no mean revenue for a bachelor in those days."

Mencken and Nathan got on famously with their new editor, collaborating on a slight but amusing book about European nightlife called *Europe after 8:15* (it was published shortly before the outbreak of World War I, and vanished without trace shortly thereafter) and an abortive proposal for a new magazine to be called the *Blue Weekly*. Thayer expressed interest in the latter project, but backed off as soon as he got a look at the dummy, in which Wright declared himself "violently against virtually everything that the right-thinking Americans of the time regarded as sacred, from Christianity to democracy." As for his stewardship of the *Smart Set*, it was a noble experiment but a financial disaster. Thayer discovered at year's end that the circulation had fallen to fifty thousand resulting in a deficit of $20,000. He immediately gave Wright the boot, tried and failed yet again to persuade Mencken to take over the magazine, and spent the next few months sweating as the circulation dropped still further.

Mencken's own celebrity continued to flourish, though he was not yet a household name. It was around this time that the *Los Angeles Examiner* became the second California newspaper to garble his byline, this time misidentifying him as "R. L. Menecken." More to his liking was the author's note that accompanied "Newspaper Morals," his first article for the *Atlantic Monthly*, identifying him as "a member of the staff of the *Baltimore Sun*, and well known in his city on account of the personal column which he conducts daily . . . the author of a life of Nietzsche and of other volumes of a philosophical or social cast." Later that year a Doubleday advertisement relating "what distinguished writers have said of Joseph Conrad" cited him along with H. G. Wells, John Galsworthy, and Ford Madox Ford. But the *Smart Set* was growing shakier by the month, and though he later claimed to have felt sure that other magazines would take an interest in his work, he was too cautious by nature to let go of the bird in his hand. The success of "The Free Lance" had made it possible for him to give up the tiresome chore of editing other people's copy. Now he took it up again—but on his own terms.

With the outbreak of war in Europe and the resulting collapse of the American stock market in the summer of 1914, the *Smart Set* changed owners yet again, this time passing into the hands of a wholesale paper dealer named Eugene F. Crowe, who took over the magazine in order to collect the money it owed him. Crowe put Eltinge F. Warner, the publisher of *Field and Stream*, in charge of his new acquisition, and Warner, needing to find a new editor as fast as possible, turned to George Jean Nathan, whom he had met a few months earlier on a transatlantic cruise. Like Mencken, Nathan had previously been approached by Thayer about editing the magazine, and had said no. This time, he offered a counterproposal: Why couldn't he and Mencken edit the *Smart Set* together?

Mencken came to Manhattan the next day to discuss the matter with Nathan. The sticking point was his refusal to give up "The Free Lance" or move to New York, but the two friends figured out a way around it. Nathan would run the *Smart Set* office, while Mencken would serve as first reader of manuscripts in Baltimore, meeting with Nathan twice a month in New York. In return they would be given a one-third interest in the new Smart Set Company, Inc., and once the magazine settled its debts, each man would receive a one-hundred-dollars-a-month salary in addition to his regular columnist's fee. On August 12 Nathan telegraphed Mencken in Baltimore: MAGAZINE OURS. ASSUME CONTROL TODAY. IMPORTANT YOU COME UP SOON. PAPERS TO BE SIGNED. NEED YOUR DECISION. EUREKA. It was too late for them to make any changes in the October issue, but they did manage to emblazon the cover with a slogan indicative of things to come: "One Civilized Reader Is Worth a Thousand Boneheads."

For all their bravado Mencken and Nathan knew they would have to move carefully in order to rebuild the magazine's circulation without running afoul of the ever-watchful censors. Two weeks after the deal was closed, Mencken wrote to Ellery Sedgwick, who had become the editor of the ultrarespectable *Atlantic Monthly:* "Our policy, I needn't say, is to be lively without being nasty. On the one hand, no smut, and on the other, nothing uplifting. A magazine for civilized adults in their lighter moods. A sort of frivolous sister to the Atlantic. Ideal authors: George Moore, Max Beerbohm, Otto Julius Bierbaum, Maurice Baring and Lord Dunsany, with Dreiser and Joseph Conrad now and then." Corraling such writers would cost money, and the *Smart Set* had none to spare. Warner had laid off all the existing staff except for a single secretary, and moved the office from its five-thousand-dollars-a-year quarters on Fifth Avenue to a thirty-five-dollars-a-month cubbyhole. Time was just as short: Mencken and Nathan had only four weeks to put together the November issue. The same day that they signed the contract, they started rummaging through the magazine's backlog of unpublished articles, hoping to unearth at least a few salvageable items. Instead, they found nothing but "rubbish—the heaped-up bad buys of years, some of them going back to Wright's days as editor and even before." Desperate for copy, they ended up writing more than half of the issue themselves. Mencken contributed his regular book review, "Critics of More or Less Badness," and Nathan his regular drama review, "The Trail of the Lonesome Spine," and they managed to scrounge up a short story by Edna St. Vincent Millay and a poem by Sara Teasdale. For the rest they relied on Mencken's own bulging stockpile of unsold and rejected manuscripts, disguising them with a colorful array of suspicious-looking pseudonyms, including Janet Jefferson, Marie de Verdi, Pierre D'Aubigny, Owen Hatteras, George Weems Peregoy, Sherrard Mullikin, and Raoul della Torre. Mencken salted the pages with "Free Lance"–style fillers ("The Constitution of the United States: the last refuge of scoundrels") and burned off one of his last remaining unpublished novelettes, "The Barbarous Bradley" ("So this was the end! This was the Ethel Mason whose peculiar reddish hair had attracted him, whose pretty paleness had charmed him, whose chance it had been to gobble him, enslave him, marry him!"). Of the 156 pages in the November issue, he personally filled up just short of 40.

It was, all things considered, a fairly embarrassing debut, and the editors knew it. "Our first issue was unquestionably a sorry one," Mencken acknowledged in *My Life as Author and Editor*. But neither man embarrassed easily. The important thing was that they now had a magazine of their own to play with, one that might someday bring them a modest measure of fame, and perhaps even a little folding money. "If the thing goes through," Mencken told Dreiser, "there will be a future in it for both of us." For once in his life, he was guilty of understatement, though one can hardly blame him for failing to foresee that the half-dead, cheaply printed magazine he and Nathan had taken over on the spur of the moment would make them legends in their own time.

4

"THE PURITANS HAVE ALL THE CARDS"

At the Smart Set, *1915–1918*

"I note what you say about your aspiration to edit a magazine," Mencken wrote to William Saroyan in 1936. "I am sending you by this mail a six-chambered revolver. Load it and fire every one into your head. You will thank me after you get to Hell and learn from other editors how dreadful their job was on earth. I wouldn't go back to magazine editing for all the money wasted by the Brain Trust." While his younger self would presumably have disagreed, the job he had taken on was dreadful enough by any objective standard. To arrest the downward spiral of a failing magazine is difficult under the best of circumstances; to try to do it while simultaneously writing a daily newspaper column is a good way to end up in a hospital. Nor was it necessarily to his advantage that he had a collaborator. Every great magazine is ultimately about its editor: his tastes, his quirks, his obsessions. He may set up one or two separate fiefdoms within the larger kingdom and allow them to develop contrasting editorial identities, but unless the magazine as a whole reflects his own sensibility, it will fail to capture the imagination of the public. Yet Mencken and Nathan edited the *Smart Set* jointly, with each man holding unlimited veto power over the other's decisions.

How could two such outsize personalities ever have devised such an arrangement, much less put up with it for ten years? In a pam-

phlet written to lure subscribers, Mencken described it in terms so
sweetly reasonable as to beggar belief:

> I read all the manuscripts that are sent to us, and send Nathan
> those that I think are fit to print. If he agrees, they go into type
> at once; if he dissents, they are rejected forthwith. This veto is
> absolute, and works both ways. It saves us a great many useless
> and possibly acrimonious discussions. It takes two yeses to get a
> poem or essay or story into the magazine, but one no is suffi-
> cient to keep it out. In practice, we do not disagree sharply
> more than once in a hundred times, and even then, as I say, the
> debate is over as soon as it begins. I doubt that this scheme has
> ever lost us a manuscript genuinely worth printing. . . . We
> employ no readers, and take no advice. Every piece of manu-
> script that comes into the office passes through my hands, or
> those of Nathan, and usually through the hands of both of us. I
> live in Baltimore, but come to New York every other week.

It was almost that simple, but not quite. Part of what made the
double-veto system workable was that in 1914 and for a good many
years thereafter, Mencken and Nathan were of like mind about most
things, and where they differed the *Smart Set* did not go. As a matter
of policy the magazine published nothing about party politics (in
which Nathan had no interest) or the world war (about which
Mencken's strong views would in any case soon become unprintable,
as we shall see). It helped, too, that they established from the outset a
mutually satisfactory division of labor. Mencken, working from his
upstairs retreat in Hollins Street, beat the bushes for manuscripts,
while Nathan ran the New York office, did the copyediting, and laid
out each month's issue. Add to all this the fact that they liked each
other enormously, and the success of their balancing act starts to
make more sense.

At first glance Nathan would seem to have been carrying more than his share of the load, but in practice Mencken had the harder job, at least in the beginning. Writing for the *Smart Set* was not yet considered a privilege, and he was only able to offer his contributors two cents a word. It helped that what the *Smart Set* paid, it paid promptly. Mencken made it a point of honor to read and return most incoming manuscripts the same day he got them in the mail, and *Smart Set* authors were paid off as soon as their work was accepted for publication, with checks sent out every Friday. Except for the *Saturday Evening Post*, no other national magazine in the United States paid so quickly (and none does so now). Once the word got out that free-lancers in need of a fast dollar could count on Mencken to turn their pieces around in a day or two, the copy started to flow more freely. But a monthly magazine has to be filled up every four weeks, rain or shine, so he sent out letters to every writer he knew, asking if they had anything in hand that might be suitable for the *Smart Set*. Dreiser offered a batch of one-act plays—though he insisted on haggling over the price, a sign of trouble to come—and others responded in kind. The December issue, which featured Dreiser's *The Blue Sphere*, was noticeably better than its predecessor, and by January the circulation was inching upward for the first time in months.

The magazine was doing well enough by April that Mencken was emboldened to send letters to several authors of his acquaintance, inviting them to endorse the new *Smart Set*. "If you have ever given the resurrected Smart Set a glance," he wrote to Dreiser, "and can do it without injury to your conscience, and have no scruples otherwise, I wish you would dash off a few lines saying that it has shown progress during the past six months and is now a magazine that the civilized reader may peruse without damage to his stomach." The answer he received must have come as a shock: "Under you and Nathan the thing seems to have tamed down to a light, non-disturbing period of persiflage and badinage, which now and then is amusing but which not even the preachers of Keokuk will resent seriously. It is as innocent as the Ladies' Home Journal. . . . Personally I think you are sound as a critic of books and that the magazine as you are doing it has some interest, but not enough to call forth from me the praise you want."

Though Mencken replied mildly ("I am sorry that The Smart Set doesn't please you"), he was stung by Dreiser's reply, and probably surprised by his ingratitude. What must have hurt most, though, was that Dreiser was right: The *Smart Set* promised more than it delivered. Poor editors can only hit up their friends so many times before wearing out their welcome, and Mencken had been forced to hunt for affordable unknowns capable of turning out pieces "civilized in point of view, written with some sense of style, and devoid of all moral purpose." In this manner the magazine acquired a lasting reputation for "fostering and indeed wet-nursing the national letters." Closer scrutiny tells another story, though one early find was indeed spectacular. The *Smart Set* published two stories by James Joyce, "A Little Cloud" and "The Boarding-House," in the May issue, the first time any of Joyce's work had appeared in an American publication. (It would also be the last time he appeared in the *Smart Set*, since Mencken had no use for his later work, dismissing *Ulysses* as "deliberately mystifying and mainly puerile. . . . I have never been able to get over a suspicion that Joyce concocted it as a kind of vengeful hoax.") Several other writers of note found their way into its pages, including James Branch Cabell, Willa Cather, Aldous Huxley, Somerset Maugham, and Eugene O'Neill, the last of whom was a Nathan enthusiasm. But most of them would not publish there until after the war, and most were already solidly established by the time they did so; by and large the regulars were reliable hacks. Mencken himself would discover only one indisputably major young American author, F. Scott Fitzgerald, who made his debut in 1919 with "Babes in the Woods." Until then one can flip through issue after issue of the "aristocrat among magazines" without encountering a single familiar name or a single memorable piece—save those written by the editors themselves.

In 1935 Malcolm Cowley summed up the *Smart Set* from the point of view of a left-wing literary highbrow:

> They were not much interested, I suspect, in [Wright's] program of publishing the most brilliant work of the best European writers. Nathan wanted to say what he thought about the new plays; Mencken wanted to write literary criticisms and pursue his long crusade against rural Baptists. As long as they were given their own pages to fill as they pleased, as long as they could print a few stories they really admired, they were willing to edit the rest of the magazine for people who like fiction with a little tang to it and relish a bit of subtlety now and then—in other words, for drugstore cowboys.

Cowley's tone may have been contentious, but he was still close to the mark. As early as 1915 Mencken was reminding Nathan that "we want to devote more and more space to our own compositions." He never put it so crassly when others were listening, but that was the real reason they took charge of the *Smart Set:* to write what they pleased and see it in print.

In Mencken's case this mainly meant literary criticism, though it is not for his *Smart Set* book columns that he is now mainly remembered, nor should he be. While he insisted that "an author should be judged by what he tries honestly to do, not by what anyone else, whether critic or not, thinks he *ought* to do," his own taste was constrictingly narrow. American literature before Mark Twain was a closed book to him, European modernism a joke. (In one of his columns he made a point of bragging about never having read *The Brothers Karamazov*.) Most of the time he wrote amusingly about bad books of no importance, and on the occasions when he had respectable ones to review, he was, like most working critics, wrong as often as not. For him Edith Wharton's *The Age of Innocence* was "not bad in the sense that Dreiser's writing is often bad, but it lacks

all character, all distinction: any literate person might have done it."
He thought poetry "beautiful balderdash," preferring naturalistic
novels that unflinchingly portrayed "the harsh facts of life," a phrase
into which he packed the whole of his homemade Darwinism.
Indeed, it is not too strong to say that he saw literary criticism less as
an end in itself than as a means of registering "sharp and more or less
truculent dissent from the *mores* of my country," and once he was in
a position to write directly about those mores for a national audience, he devoted less time to reviewing novels. Even at the height of
his influence as a critic, he "seized every chance to get out of the
dream world into the real one," for he was primarily interested not in
imaginative literature but general ideas. "You will escape from literary criticism, too, as I am trying to do," he told the poet Louis
Untermeyer in 1919. "The wider field of ideas in general is too allur-
ing. . . . We live, not in a literary age, but in a fiercely political age."

These myriad defects led René Wellek to render harsh judgment
on Mencken in his magisterial *History of Modern Criticism*. Acknowledging that "even today he has kept a large group of admirers, col-
lectors, editors, and most important, devoted readers," Wellek went
on to say that Mencken "cannot survive as a literary critic. He
belongs strictly to the past." Yet even Wellek acknowledged the role
he played in his own time: "Mencken fulfilled an important function . . . in liberating the American literary scene from the incubus of
complacency and conformity, narrow-minded prudery, blinkering
patriotism, and stifling conventions about beauty." Those achievements loomed larger in 1914 than they do today, when the freedom
of expression for which Mencken fought has come to be seen as a
mixed blessing. Nearly a century later it is hard to imagine that there
was ever a time when the genteel tradition held absolute sway over
American letters, much less that writers and editors who dared to
defy it ran the risk of having their work censored, or even suppressed
altogether.

It was against the primness and vapidity of that tradition that
Mencken set himself, demanding that American writers shake off

their Anglophile inhibitions and immerse themselves in "the infinitely picturesque and brilliant life" of their own time and place, "begirt by politics, railways and commercial enterprise (and no less by revivals, cuspidors and braggadocio)." Edmund Wilson, who grew up reading him in the *Smart Set*, never forgot the tonic effect his calls to arms had on a generation of young readers who had been taught that William Dean Howells was a great novelist: "It took an ex-newspaperman, who wrote his name with simple initials and did not shrink from the inferior paper of a raffish magazine, to denounce the false reputations that the public had been induced to believe in during the nineties and the early nineteen-hundreds. . . . Mencken had the temerity to put his foot through the genteel tradition, and it suddenly turned out that the spell no longer held." Dreiser, Cabell, Cather, George Ade, Ambrose Bierce, James Huneker, Sinclair Lewis, Jack London, Frank Norris, Upton Sinclair: These were the writers to whom Wilson and his contemporaries were introduced by Mencken, and if the roster, Dreiser and Cather excepted, no longer looks so impressive to us now, it is because the Fitzgeralds and Hemingways who came later felt freer still to engage directly with the vulgar, vibrant realities of everyday American life, in part because Mencken had paved their way. He was an incomparable publicist for the writers he liked, as well as a fearless debunker of those he loathed, and the cleansing contempt with which he disposed of the hapless confectioner of the latest middlebrow bestseller was no less exhilarating than the delight he took in *Jennie Gerhardt:* "Its aim is to fill the breast with soothing and optimistic emotions—to make the fat woman forget that she is fat—to purge the tired business man of his bile, to convince the flapper that Douglas Fairbanks may yet learn to love her, to prove that this dreary old world, as botched and bad as it is, might yet be a darn sight worse." With book reviews like that, who cared if the monthly novelette was a bit purple? One read the *Smart Set* to learn from Mencken and Nathan what was fresh and new, and laugh with them at all that was tired and stale. Everything else was filler.

Mencken's breezy style was itself new to American criticism

(though some of his readers would have recognized its not-so-distant origins in the similarly colloquial prefaces of George Bernard Shaw), and it had as much to do with his impact as did his opinions. He would soon prune away some of his youthful extravagance, and his writing would become more effective as a result. But it was already apparent that not since Twain had there been an American writer so unafraid to speak the American language. For now he sounded on paper much like what he was in life, a young man overflowing with ideas. He was also—though few knew it—a man in love.

"A MAN'S WOMEN FOLK, WHATEVER THEIR OUTWARD SHOW OF RESPECT for his merit and authority, always regard him secretly as an ass, and with something akin to pity." So Mencken claimed, but what he thought of his own women folk, or of women in general, is harder to know. He liked a great many women and was liked by them in turn, but preferred to socialize with men. Though he admired certain women writers and published their work in his magazines, he had little use for female intellectuals, and none at all for feminists. He wrote a book called *In Defense of Women*, but it is a heavily ironic "defense" which lends itself to multiple interpretations, none of them straightforward. Like most Victorians, he kept his romantic affairs strictly private. He married at fifty and was widowed at fifty-four, and nowhere in his published writings, his diary, or his unpublished memoirs does he acknowledge any other romantic or sexual connection, though he did make this slightly smug remark to his first biographer: "It is my experience that women are seldom attracted to me at first sight: I get on better with men. But in the long run most women who interest me return the interest." No doubt this is true, but so far as can be established with certainty, he had serious involvements with only two women other than his wife. (As one of his doctors told another of his biographers, he did not have "a particularly strong sex drive.") When Marion Bloom, his first love, married another man, he burned her letters. It is only because she did not reciprocate that we know anything of her relationship with Mencken beyond the barest of details, and only one or two of his letters to her can properly be described as intimate. The rest are arch, flirtatious, even suggestive, but never tender. It is tempting to conclude from all this that women simply did not mean much to Mencken, and it may even be true. Still, Marion meant a great deal to him, at least for a while.

She came from a tiny farm town some forty miles from Baltimore, one of six children of a crippled farmer's daughter and a schoolmaster turned dairyman. Adam Bloom shot himself when Marion was seven years old—depressed beyond enduring, according to family legend, by an itinerant lady evangelist who preached hellfire and damnation—and his death plunged the family into barely respectable poverty. Estelle, the third oldest of the four Bloom sisters, fled the nest in 1904, escaping to Baltimore. Marion, five years younger, followed suit as soon as she could, moving to Washington to start her own search for a larger life; by then Estelle had married and been deserted in humiliatingly quick succession, so she joined Marion there in 1912. Just like the plucky heroines of a Dreiser novel, the two sisters shared an apartment, found jobs, bobbed their hair, smoked cigarettes, and went to bed with men: "They were young adventurers, without background or means, as are so many of the thousands who reconnoiter the great cities, and without many, if indeed any, severe or wholly unyielding moral scruples. . . . Being young and attractive, their place was a rendezvous for certain men of position, all intent upon trips to the theater, card-playing, dancing, dining."

The prose is indeed Dreiser's, but it is not from a novel. He included Estelle and Marion, thinly disguised as "the Redmond girls," in *A Gallery of Women*, the caddish memoir of his love life that he published in 1929. He was to meet the Bloom sisters through Mencken, who in turn had met them in February 1914. Estelle was trying to track down her missing husband so that she could divorce him, and she and Marion had come to the offices of the *Sun* in search of information (and possibly also jobs). How they happened to cross Mencken's path that day is a mystery, but it could have had something to do with the fact that they were both good-looking—the twenty-three-year-old Marion was small and dark, her older sister plumper and prettier—as well as clever.

Beyond this we know nothing, for Mencken left no record of his relationship with Marion other than the letters he sent her. He said little or nothing about her to most of his friends, and would never speak familiarly of her again, whether in conversation or correspondence, once they finally stopped seeing each other for good. But he began writing to her in Washington almost immediately, and took her to dinner for the first time five months later. As his letters reveal, he also became friendly with Estelle:

"Certainly I am coming over for a visit: watch for me in my Sunday clothes. . . . Your invasion was very pleasant: I was in low spirits & needed charming company." (Mencken to Marion Bloom, February 18, 1914)

"My dear Miss Bloom: Beware! Maybe I'll invade your fair city next week. . . . Will you compromise by taking dinner with me down town? Do!" (July 1)

"Dear Miss: The idea is, of course, to let me come to Washington again. I enjoyed the visit immensely." (August 2)

"Once I get loose from my chains I'll dig up some books, autograph them elegantly with poetical inscriptions, and send them to you by Rudolph, my valet. Meanwhile, I kiss your hand with sentiments of the highest esteem." (August 23)

"It is no crime to be married, though intensely uncomfortable. I often regret that I didn't try it when I was younger. Today I grow senile and it is eugenically impossible." (Mencken to Estelle Bloom Kubitz, September 24)

"Dear M: If you are not at home [in Washington] next Monday I'll break down and cry like a baby!" (Mencken to Marion Bloom, October 21)

Marion scribbled a note at the bottom of the last of these letters: "Oct. 21, 1914—Memorable week." Seven years later she would tell Estelle that it had been "the beginning." For Mencken it was a different sort of beginning, one that had been a surprisingly long time in coming. He was thirty-four years old and still living at home with his family, and though one gathers from reading between the lines of *Newspaper Days* that he knew his way around the whorehouses of Baltimore, he seems never to have been in love before. Romance came to him, as it so often does, at the least opportune of moments, when he was spending his mornings writing "The Free Lance" and his afternoons, evenings, and weekends trying to make something good out of the *Smart Set*.

It also came in an unlikely form, for Marion was not the sort of woman with whom Anna Mencken would have expected her oldest son to settle down. Instead, he had fallen for someone not altogether unlike Sister Carrie. "Jennie Gerhardt, like Carrie Meeber, is a rose grown from turnip seed," he wrote in the *Smart Set* less than a year before meeting Marion Bloom. "Over each, at the start, hangs poverty, ignorance, the dumb helplessness of the Shudra—and yet in each there is that indescribable something, that element of essential gentleness, that innate inward beauty which levels all caste barriers." Clearly he was fascinated by the self-made young women who were finding their way into the novels he read and the offices he frequented. Now he was sleeping with one of them. Would their affair deepen into something permanent? Or was Marion Bloom, for all her charms, not quite respectable enough to bring home to Hollins Street?

It may say something about his opinion of Marion as a marriage prospect that he promptly started encouraging her to contribute to the *Smart Set*, even though she had no experience as a writer. "You are 80 times as clever as most of the literary wenches who are pictured in the public prints," he assured her. Perhaps he thought a published author would make a more acceptable mate. That was how she would explain it to Estelle, and to herself, fourteen years later: "I suppose he wanted a writer—nothing was too great or too much for a pot-bellied, slope-shouldered clown. He had me climbing [toward] unreachable stars, lashing and cursing my dumb self because I lacked. Well, that's over. These days I might ask him what he could have accomplished working with my inadequacies." Instead she dutifully turned out epigrams, which he polished as best he could and put into the magazine, and he sent her a steady stream of letters. (He also kept on writing to Estelle, who became a close friend in her own right.) His typewritten gallantries were charming but discreet, as if he were keeping an eye on some future biographer: "Pending proofs that you have read [*The Philosophy of Friedrich Nietzsche*] I shall withhold the remark (springing straight from the heart) that I view you with sentiments of the highest affection."

Mencken had little time to spare for gallantry, however. The *Smart Set* was receiving so many unsolicited manuscripts by the middle of 1915 that he and Nathan decided to siphon off some of the less satisfactory ones into a magazine for "the morons," skimming off the profits to keep the *Smart Set* afloat. *Parisienne Monthly Magazine*, a fifteen-cent pulp devoted to trashy short stories and novelettes set in France, became so successful that they launched two similar publications, *Saucy Stories* (in 1916) and *Black Mask* (in 1920). All were edited anonymously and with the utmost contempt—Mencken referred to them as "the louse magazines"—though *Black Mask*, which specialized in "homicidal fiction," would later become famous for having served as an incubator for the hard-boiled detective stories of Raymond Chandler and Dashiell Hammett, the second of whom also made it into the *Smart Set*. By then, though, the editors had sold their interest in all three magazines to Warner and Crowe, splitting $20,000 for *Parisienne* and *Saucy Stories* and $40,000 for *Black Mask*, a total of $660,000 in today's dollars. It wasn't enough to make either man rich, but it was more money than Mencken had ever seen in his life, and "sufficed to relieve Nathan and me from want permanently."

Parisienne took off so fast that Mencken, seeing it would soon put the *Smart Set* on a more stable footing, decided it was now feasible for him to give up "The Free Lance." His labors had already drained him to the point of exhaustion, so much so that in August he checked into Johns Hopkins Hospital for a week, complaining of "a persistent sensation in the tongue." This strange complaint was the first officially recorded manifestation of his lifelong hypochondria—according to his medical records, the doctors blamed it on "psychoneurosis"—but it was also a sign that he was asking too much of himself, and on October 23, 1915, he bowed to the inevitable and wrote his last daily column for the *Sunpapers*.

Abandoning "The Free Lance" was his decision, made "on my own motion entirely." Aside from being worn out, he thought it was time for a change: "I had an uneasy fear, without doubt well grounded, that if I kept on with the Free Lance I would become a mere columnist, bound down to a routine job and with no energy left for better things; worse, that I would degenerate into a town celebrity, a local worthy—something that I have always tried most diligently to avoid." But he also suspected that the decision might soon be taken out of his hands. Anti-German feeling in Baltimore and the rest of the country had been on the rise since May 7, when a U-boat torpedoed the *Lusitania* off the coast of Ireland, killing 128 Americans. Mencken had never made any secret of where his own sympathies lay. He was for Kaiser Wilhelm II and against the British, whom he believed had manipulated Germany into going to war. "If I ever get out of my present morass," he wrote to Dreiser, "I shall begin the serious study of German, to the end that I may spend my declining years in a civilized country." Such sentiments, he knew, would become unpublishable, even illegal, once the United States entered the fray, and since he expected Woodrow Wilson to "horn into the war soon or late," he preferred to walk away from "The Free Lance" while it was still an asset to the *Evening Sun* instead of waiting until it became a liability.

Mencken's love of Germany should not be mistaken for nostalgic fidelity to the land of the Menckenii. Throughout his life, he would speak with scorn of Baltimore's German immigrants, whom he dismissed as "ignoramuses of the petty trading class." He had learned nothing of German *Kultur* from August Mencken or F. Knapp's Institute, and he did not visit Germany for the first time until 1908, a year after he published his Nietzsche book. But that visit was enough to pique his interest, and by 1914 he was writing about Germany with an ardor for which the kindest word is "naive." In "The Mailed Fist and

Its Prophet," an essay he published that year in the *Atlantic Monthly*, he heaped praise on the "new Germany" dreamed of by Nietzsche and founded by Bismarck, a country that, though nominally democratic, was in practice "a delimited, aristocratic democracy in the Athenian sense—a democracy of intelligence, of strength, of superior fitness . . . a new aristocracy of the laboratory, the study, and the shop." The men at the top were a "superbly efficient ruling caste" of unsentimental, scientifically minded meritocrats who disdained politics and had no use for "the *kaffeeklatsch* view of life" or the "mysticism and simple piety" of their forefathers. They forged a superstate of supermen who were better at everything than anyone else—better scientists, better scholars, better poets and playwrights and composers—and having done so, they found in the writings of Nietzsche a philosophy that explained their achievements to themselves: "Go through *Thus Spake Zarathustra* from end to end, and you will find that nine-tenths of its ideas are essentially German ideas, that they coincide almost exactly with what we have come to know of the new German spirit. . . . It is a riotous affirmation of race-efficiency, a magnificent defiance of destiny, a sublime celebration of ambition."

Mencken's discovery of what he called his "race consciousness" was, he implied many years later, something like a conversion: "It suddenly dawned on me, somewhat to my surprise, that the whole body of doctrine that I had been preaching was fundamentally anti–Anglo-Saxon, and that if I had any spiritual home at all it must be in the land of my ancestors." Yet his newfound Anglophobia could not have come as much of a surprise, for what he saw (or thought he saw) in Germany went straight to the hard heart of his long-settled view of human nature. The years he had spent editing the *Herald* had taught him that all men were created unequal, and his reading of Nietzsche left him certain that the strong ones—among whom he numbered himself—would naturally prevail over their inferiors unless blocked from doing so by external forces. Foremost among those forces, he felt, were Christianity and democracy. Christianity had imposed Nietzsche's "slave morality" on the human race, while democracy, by vesting sovereignty in the tyrannical will of the mob of weak-minded slaves, prevented the minority of unbelieving supermen from remaking the world along meritocratic lines. Put them together and the result was puritanism, the life-denying regime that Mencken took to be nothing more than envy empowered, famously defining it as "the haunting fear that someone, somewhere, may be happy."

Mencken's critique of the spiritual narrowness of puritanism would become central to his fame. It was so magnetically expressed that it even attracted the sympathetic attention of religious believers who disagreed with him about nearly everything else. ("I have so warm an admiration for Mr. Mencken as the critic of Puritan pride and stupidity," G. K. Chesterton wrote in 1930, "that I regret that he should . . . try to make himself out a back number out of mere irreligious irritation.") But as he repudiated English puritanism, so he more comprehensively rejected the American way of life, which he took to be deeply rooted in the puritan scheme of things. Alienation, he declared, was his birthright: "I have always lived in the wrong country." Having come to the conclusion that Germany was everything America was not, he proceeded to say so repeatedly in "The Free Lance," assuring the citizens of Baltimore that "German notions of what is good and bad, what is right and wrong, what is effective and ineffective, will prevail in the world." That was the unlikely conclusion he had drawn from his extensive but superficial reading of Nietzsche: He had come to believe in the Victorian idea of progress, only with a German accent.

How did he reconcile his admiration for an authoritarian state such as Wilhelmine Germany with his oft-stated belief in "liberty up to the extreme limits of the feasible and tolerable"? For him there was nothing to reconcile: "Liberty, of course, is not for slaves; I do not advocate inflicting it on men against their conscience." But this was intellectual sleight of hand, and did nothing to resolve the contradiction. Was he being disingenuous when he later claimed to be "strongly in favor of the sort of free democracy visualized by Jefferson," or had he simply not thought the thing through? It is impossible to say, though we know he was not a "liberal," by which he and the rest of his generation meant what is now known in American political parlance as a "classical liberal," a pure John Stuart Mill–type libertarian. We know it because he said so on numerous occasions (a fact of which many of his present-day admirers are unaware). In 1923 he confessed himself to be "unable to make the grade as a Liberal. Liberalism always involves freeing human beings against their will—often, indeed, to their obvious damage, as in the cases of the majority of Negroes and women." On another occasion he defined a liberal as "one who is willing to believe anything twice." But he was also not a political scientist, nor does he appear to have read at all widely in the classics of political philosophy, and intellectual consistency was never his strong suit.* It could be that he thought Jeffersonian democracy was the best the United States could hope for, or that it was the necessary precondition for the emergence of a German-style "democracy of strength." In any case the "Germany" of "The Mailed Fist and Its Prophet" was a country he knew mostly from books. He had visited it twice but never lived there, or anywhere else besides Baltimore. It was the land of his dreams, and existed nowhere else. The real Germany, as events were to reveal, was a very different place, a country whose philosophy was an expansive idealism that under pressure would collapse into the kind of totalitarianism Irving Babbitt had in mind when he observed that "the last stage of sentimentalism is homicidal mania."

Mencken was no totalitarian, though he could sound like one: "The Germans offer the world an ethic that, whatever its harshness, is at least based upon truth. . . . If, to set it up, whole principalities and races of men have to die, the world will still be getting it at no more than the immemorial market rate for such commodities." He was thinking of blood and iron, not gas chambers or gulags, just as he was incapable of envisioning the long-term consequences of the post-modern embrace of Nietzsche's attempt to move beyond traditional concepts of good and evil. Yet both were unintended consequences of

---

*He seems, for instance, never to have read Tocqueville, who was no less critical of democracy but wrote about it far more insightfully.

his own ideas taken to extremes, and he was nothing if not an extremist. "It always amuses me to be denounced as a conservative, as happens almost every time I am mentioned at all," he wrote in a private memorandum dated 1941. "I am actually an extreme radical, and if I had (or desired) the job of making over the world most of its existing institutions would be destroyed." So he fancied himself, sitting in the upstairs office of his comfortable home on Hollins Street, pounding out article after article in which he chipped away at the cornerstones of the culture that had made possible the bourgeois comfort in which he delighted. It was not the only whirlwind he would live to reap.

THE DEMISE OF "THE FREE LANCE" DID NOT STOP HIM FROM SPEAKING his mind in the *Evening Sun*, or anywhere else. It took a world war to do that. Two days after his last "Free Lance" ran, he embarked on a new series of editorial-page articles, starting with two sets of "notes for a proposed treatise upon the origin and nature of Puritanism," in which he explored in greater detail the relationship between Christianity and democracy, taking every opportunity along the way to praise Germany and damn England: "At the bottom of Puritanism one always finds envy of the fellow who is having a better time in the world. At the bottom of democracy one finds the same thing. This is the cause of a fact commonly observed: that the Puritan is usually a democrat, and *vice versa*. . . . England is the mother-country of Puritanism, and will be its first victim." He spent the next eight months wandering among his favorite themes—the novels of Conrad, the music of Schubert, the idiocies of politicians. Then, in July, he returned abruptly to the subject of puritanism, publishing two more installments of notes on its origins and nature, followed by three consecutive columns about Theodore Dreiser. Neither topic was chosen casually. Dreiser had fallen afoul of the censors, and was in urgent need of help.

Mencken's relations with Dreiser had grown more distant in the five years since he reviewed *Jennie Gerhardt*. Dreiser published two more novels, *The Financier* (1912) and *The Titan* (1914), both of which Mencken praised in the *Smart Set*. As a handsome token of his appreciation, Dreiser presented him with the manuscript of *Sister Carrie*. But Mencken was concerned about his friend's growing garrulousness ("D. got drunk upon his own story and ran amuck," he wrote to a friend about *The Financier*), and revolted by his promiscuity and arrogance. Dreiser then made the mistake of writing *The "Genius,"* a self-serving, obviously autobiographical novel about an artist whose wife failed to understand his need to copulate at will with other women. Published late in 1915, this 350,000-word song of himself received the reviews it deserved—even Mencken was unable to conceal his distaste, though he tried—and sold poorly. In due course, it came to the attention of John S. Sumner, the lawyer who ran the New York Society for the Suppression of Vice, the organization in charge of enforcing the state obscenity laws. Sumner went through the 736 pages of *The "Genius"* and found seventy-five "lewd" and seventeen "profane" passages (none of which would raise the most prudish of eyebrows today). He took the list to Dreiser's publishers, who agreed on the spot to pull all remaining copies of the novel out of bookstores across the country.

Dreiser instantly got in touch with Mencken, who had been dodging the censors ever since he and Nathan launched *Parisienne*. Despite his reservations about *The "Genius"* and its author, Mencken was determined not to let the puritans have their way with the man he still considered to be America's greatest living novelist. Three days after Sumner dropped his bombshell, Mencken sent Dreiser a letter asking for "an exact statement of the passages" in question and offering cool-headed advice in return:

> The thing to do with the moralists in the case of such a valuable property as "The Genius" is to offer some sort of compromise and so force them into the position of negotiating with you. In the end, if the thing is properly managed, it will be unnecessary to take out more than a few sentences. . . . The country is in a state of moral mania and the only thing for a prudent man to do

is to stall off the moralists however he can and trust to the future for his release. If this attitude may seem to be pessimistic, please don't forget that it is born of extraordinarily wide experience. My whole life, once I get free from my present engagements, will be devoted to combatting Puritanism. But in the meantime, I see very clearly that the Puritans have nearly all of the cards. They drew up the laws now on the statute books and they cunningly contrived them to serve their own purposes.

Dreiser, being Dreiser, refused to compromise: "A fight is the only thing . . . I hope and pray they send me to jail." So Mencken, presumably sighing deeply to himself at his friend's folly, began the task of sweet-talking other authors into signing a "Dreiser Protest" he had drafted and persuaded the normally timorous Authors' League of America to sponsor: "Some of us may differ from Mr. Dreiser in his aims and methods, and some of us may be out of sympathy with his point of view, but we believe that an attack by irresponsible and arbitrary persons upon the writings of an author of such manifest sincerity and such high accomplishments must inevitably do great damage to the freedom of letters in the United States, and bring down upon the American people the ridicule and contempt of other nations."

Just as cannily worded were the hundreds of letters he sent in support of the Dreiser Protest, in some cases swallowing his pride to do so (as well as spending some three hundred dollars out of his own pocket—nearly five thousand dollars in today's money—on "postage, stationery and secretarial aid"). To William Allen White, the celebrated editor of Kansas's *Emporia Gazette*, went a note whose tact might have surprised the readers of the *Evening Sun*:

May I suggest that your signature would help to give force to a revolt against a censorship which grows more exacting and arbitrary daily, and which threatens, if American authors do not stand firmly against it, to lay an intolerable burden upon American letters. . . . I am no admirer of "The 'Genius'" and have printed very unfavorable reviews of it. But it is surely one thing for a book to be attacked by criticism and quite another thing for it to be assaulted by persons who boast openly that they care nothing for artistic values whatever. . . . I know how much Dreiser has given up to do his work. I know that he is a sincere artist and an honest man.

White signed, as did many others, but the list of American writers who chose not to support Dreiser was also long, and included more than a few of the novelists whom Mencken had long been castigating in the *Smart Set* (one of them, predictably, was William Dean Howells). Not that Sumner cared who did or didn't sign. His mind was made up. "It is not for any limited group of individuals," he told the Authors' League, "to attempt to force upon the people in general their own particular ideas of what is decent or indecent. . . . We believe that American Letters can survive and hold its place in the literary world without the necessity for a descent by authors into the vicious side of life for material for their productions." As Mencken had warned, the puritans had all the cards, and then some. *The "Genius"* would remain out of circulation for the next six years.

Having done his best for Dreiser, Mencken turned his attention back to the war. He talked the *Sun* into sending him to Germany as a correspondent and left for Europe in December, determined to see for himself what was going on in the land of his dreams. He kept a diary of the trip, which he preserved for the rest of his life, together with a huge scrapbook full of tearsheets of his stories, unpublished manuscripts, and galley proofs, and various other souvenirs, including copies of the house ads the *Sun* ran for his dispatches:

MENCKEN IS NOT NEUTRAL. He is pro German. He has told you so in his Free Lance column time and again. But this will not prevent him giving you the real report of actual conditions in wartime Germany. Mencken is first of all a great reporter. He was a reporter before he was an editor or a critic. He was a reporter before he began to deliver those sledge-hammer blows to pretense and sham in the now famous Free Lance Column of The Evening Sun. . . . His great admiration for everything German will help him get the "inside story."

The Germans gave him the red-carpet treatment, even letting him spend five days at the Russian front, where at one point he drew enemy fire ("I remember a moment in a German switch trench when it seemed to me to be all up with Henry"). But his visit was cut short when the kaiser declared unrestricted submarine warfare against all shipping going into or out of Great Britain, causing the Wilson administration to break off diplomatic relations and forcing Mencken to head for home while he still could. He went by way of Switzerland, France, Spain, and Cuba, filing stories from Havana about the ongoing Cuban revolution before returning to Baltimore on March 14. Three weeks later the United States declared war on the Central Powers, and Mencken's byline vanished from the *Sunpapers*. His withdrawal was voluntary: He knew the *Sun* would be neither willing nor able to print what he thought about the war, and he preferred not to put his friends in the position of having to ask him to quit. Tipped off that 1524 Hollins Street might be searched, he placed "all my more embarrassing papers" in a strongbox, buried it in the backyard, laid a stretch of brick pavement over the freshly dug hole, and settled in to face the wrath of his countrymen.

Mencken had good reason to expect trouble that spring. "Once lead this people into war and they'll forget there was ever such a thing as tolerance," Woodrow Wilson himself had warned, and he was right. In the United States as in Great Britain, those were the days when everything German was guilty until proved innocent, when sauerkraut was renamed "liberty cabbage," the music of Bach,

Beethoven, and Brahms was summarily struck from orchestral pro-
grams, and people with Saxon-sounding surnames changed them or
lived to regret it. It was the worst imaginable time to be known as an
admirer of Germany, much less to have been praising its glories day
in and out on the editorial page of the *Evening Sun*, and it was only
because Mencken kept his head down for the next twenty months
that he suffered nothing more unpleasant than periodic harassment
and the loss of faithless friends. The Justice Department put him
under surveillance, and War Department censors opened his mail,
but they learned nothing save that he was getting more than his
share of letters from Hun-hating Baltimoreans who remembered
what he had said in "The Free Lance" about the coming triumph of
German *Kultur*. "During the last war," J. Edgar Hoover wrote in a
1941 memo included in Mencken's fifty-seven-page FBI file, "there
was no indication according to the records of this bureau that [he]
was involved in espionage activities."

When it was all over, Mencken took his usual jovial view of what
had happened, at least for public consumption. He told a friend:

> The war actually treated me magnificently. Its first shock gave
> me a business chance which brought me leisure, enabled me to
> escape from daily journalism, got me enough money to make
> me secure, and so helped me to write five books. Moreover, the
> ensuing festivities filled me with new ideas, greatly changed my
> aims, and flooded me with such an amount of material that I'll
> never be able to use a tenth of it. Still more, I had a lot of capi-
> tal entertainment—some rough, gaudy debates, some curious
> adventures with spy-hunters, and even a taste of life in the field.
> So I'd be an ass to complain.

So he said, and to a great extent it was true. The unending pressure of
work had exhausted him both physically and spiritually, leaving him,
"for the first and last time in my life," unsure about the course of his
career. War forced him off the treadmill of newspaper journalism,

giving him the leisure to put together one of his most significant books, *A Book of Prefaces*, and one of his best, *In Defense of Women*. But while World War I may have been good to Mencken, it did not leave him unscarred. "There has never been an occasion when the agents of the United States showed any awareness of the existence of the individual citizen, H. L. Mencken, that they did not try to do me in," he later wrote in an unpublished autobiographical note. "In so far as my personal relations with it offer any evidence the government I live under is devoted exclusively to extortion and oppression."

The knowledge that his own government had been looking over his shoulder forever changed the way he wrote about the world. Dreiser may have been ready to go to jail rather than see *The "Genius"* censored, but that was not Mencken's way: "I can't understand the martyr. Far from going to the stake for a Great Truth, I wouldn't even miss a meal for it." Never again would he speak so frankly of his vision of an ideal society as he did in "The Mailed Fist and Its Prophet" or his *Evening Sun* columns of 1915. On top of everything else, he was appalled by the completeness with which the "aristocratic democracy" of prewar Germany had given way to the chaos of the Weimar Republic ("Unless the Germans shoot all of their present democratic statesmen, Germany will sink to the level of the United States"). More to the point, he valued his creature comforts too much to dodge the rocks of the mob. From now on he would stick to criticizing the manners and morals of his native land, letting his readers draw their own conclusions about what sort of regime might suit him better. In a way this half-calculated ambiguity was to his decided advantage, for it made him a better artist: Mencken was never at his best when pounding the pulpit. It also made possible the popularity that came to him after the war, when intellectuals saw in him what they wanted to see, and no more. Still, it must have galled him to know that the puritans had gotten the best of him—that he was not quite brave enough to say what he thought.

The new Mencken made his debut in June in the pages of the *New York Evening Mail*, by then the only newspaper prepared to publish him (the managing editor was a friend), and then only with the stipulation that he not write about the war. Instead, he produced more than a hundred articles in the manner of his old "H. L. M." columns, only more self-assured and sparkling. Among them were the *Ur*-versions of several pieces he would later revise, expand, and include in his self-anthologies, including three of his most enduring essays, "The Sahara of the Bozart," a wickedly amusing survey of the cultural sterility of the postbellum South; "A Neglected Anniversary," a mock-serious history of the bathtub that was taken for true by a large percentage of its less sophisticated readers; and "Critics and Their Ways," in which Mencken put forth the notion that a good critic is a catalyst who "provoke[s] the reaction between the work of art and the spectator. . . . Out of the process comes understanding, appreciation, intelligent enjoyment—and that is precisely what the artist tried to produce."

Next came *A Book of Prefaces*, which he remembered ever after as "the most important book, in its effects upon my professional career, of any in my long line." It contained appreciations of Conrad, Dreiser, and James Huneker, a now-forgotten critic of music and letters whom he knew, liked, and esteemed as the first American critic to break free from the genteel tradition, followed by a lengthy essay called "Puritanism as a Literary Force" in which he set forth his case against the puritans in the most polished and controlled prose he had published to date:

American literature is set off sharply from all other literatures. In none other will you find so wholesale and ecstatic a sacrifice of aesthetic ideas, of all the fine gusto of passion and beauty, to notions of what is meet, proper and nice. . . . Books are still judged among us, not by their form and organization as works of art, their accuracy and vividness as representations of life, their validity and perspicacity as interpretations of it, but by their conformity to the national prejudices . . . Any questioning of the moral ideas that prevail—the principal business, it must be plain, of the novelist, the serious dramatist, the professed inquirer into human motives and acts—is received with the utmost hostility. To attempt such an enterprise is to disturb the peace—and the disturber of the peace, in the national view, quickly passes over into the downright criminal.

As in his *Evening Mail* articles, Mencken took care to disguise the German triumphalism lurking just beneath the surface of his disdain for American culture, and though his sole purpose in doing so was to stay out of jail, the book profited immensely as a result. In focusing the spotlight of his anger solely on domestic moralism and its discontents, he was writing about what he knew from personal experience: the touristy fantasies of his earlier writings about Germany were nowhere evident in this robust indictment of the all-American philistines who had banned *The "Genius"* and taught their fellow citizens to say "limb" instead of "leg."

A Book of Prefaces was, Mencken boasted, "the most headlong and uncompromising attack upon the American *Kultur* ever made up to that time." But it is rather too headlong, too patently a book with a message. At times it borders on downright earnestness: The chapter on Dreiser even contains a chart showing which of the great man's works were to be found in the public libraries of major American cities. (*Sister Carrie*, it seems, could be checked out by readers in New York, Chicago, Philadelphia, Los Angeles, San Francisco, and Kansas City, but not Boston, Washington, or Baltimore.) Nor did it display Mencken the critic to his best advantage. The Conrad chapter, like his Nietzsche book, is enthusiastic but shallow. "Theodore Dreiser" is more convincing, though it was stitched together from *Smart Set* reviews, and the seams are visible. Still, it's interesting to see Mencken trying to square his admiration for *Jennie Gerhardt* and *The Titan* with his dislike of *The Financier* and *The "Genius,"* just as it is revealing that the essay's overall effect is as much negative as positive. He would see little of

Dreiser for some time after *A Book of Prefaces* came out, and it isn't hard to guess why. "We remained on good terms," he told a journalist a few years later, "so long as I was palpably his inferior—a mere beater of drums for him." Now he was marching to a different tune. "There are passages in it so clumsy, so inept, so irritating that they seem almost unbelievable; nothing worse is to be found in the newspapers," he wrote of The *"Genius."* "Nor is there any compensatory deftness in structure, or solidity of design, to make up for this carelessness in detail."

A few reviewers were astute enough to spot the book's weaknesses. Randolph Bourne, writing in the *New Republic*, found *A Book of Prefaces* "less interesting and provocative than the irresponsible comment [Mencken] gives us in his magazine. How is it that so robust a hater of uplift and puritanism becomes so fanatical a crusader himself?" Prescient in a different way was Stuart P. Sherman, who blew the whistle on Mencken's Germanophilia in the pages of the *Nation*: "His continuous laudation of a Teutonic-Oriental pessimism and nihilism in philosophy, of anti-democratic politics, of the subjection and contempt of women, of the *Herrenmoral*,* and of anything but Anglo-Saxon civilization, is not precisely and strictly *aesthetic* criticism; an unsympathetic person might call it infatuated propagandism." Sherman was doing nothing more than telling the truth, but to publish such truths in the middle of a war was no laughing matter, least of all for a man whose mail was being opened by federal agents, and Mencken never forgave him for "hitting below the belt." Few other reviewers made any mention of Mencken's pro-German past, though, and *A Book of Prefaces* was both taken seriously and read fairly widely. He knew it to be a pivot in his career, a sign that he had recaptured the sense of professional purpose that had left him in the years just before the war: "It gave me a kind of authority that I could never have got from my *Smart Set* reviews alone, and

---

* Nietzsche-speak for "morals of the master race."

brought me a large number of new readers." It also brought him a
new publisher, one who in time became a close and faithful friend.

AT TWENTY-FIVE, ALFRED KNOPF WAS THE MENCKEN OF AMERICAN
bookmen. Tall, slender, and energetic, he cut a dashing, even flashy
figure, and he had fresh ideas about how to make and sell books. He
wanted them to look tasteful and elegant, and he advertised them with
flair. Knopf had first sought Mencken out in 1913, and the two men
had been in touch ever since, but Mencken had been reluctant to
entrust his books to a new publishing house, though there was more to
it than that. "I had little if any prejudice against Jews myself," he wrote
in *My Life as Author and Editor*, "and in fact spent a great deal of my
leisure in their company, but they were rare in the publishing business
and rather resented by the *Goyim*, and there was little indication that
they would ever be successful. . . . [I]t also seemed to me, in the early
days of our acquaintance, that [Knopf] showed a certain amount of the
obnoxious tactlessness of his race." But as Mencken got to know
Knopf better, he changed his mind about the young publisher. Not
only was he printing good books, but he turned out to be the kind of
Jew Mencken liked, a cultured music lover who, like so many other
successful Jews of German descent, was himself something of a social
anti-Semite. ("He realizes himself that there are now too many Jews in
his office," Mencken recorded approvingly in a 1946 diary entry.)
Another point in Knopf's favor was that he turned up his nose at
Dreiser's private life, telling Mencken, "I don't want any author on my
list that I'd hesitate to invite to my house." Best of all he was willing,
even eager, to publish Mencken's books at a time when most of the
bigger houses were unwilling to go near so controversial a figure.

So Mencken gave him *A Book of Prefaces*, and Knopf made the
most of it, running ads meant to suggest that the wild man of the
Smart Set was now older and wiser, though still far from stodgy:
"The book is anything but academic criticism. It is not intended for
the classroom, but for the reader who genuinely loves books. More
sober in manner than the author's Smart Set reviews, which for years

have been followed regularly by a host of interested readers, it is yet refreshingly free from all the usual critical pose and platitudinizing. . . . At last a critic who is neither a college professor nor a sophomore." Thanks in considerable part to Knopf's marketing, *A Book of Prefaces* sold a total of 16,596 copies before finally going out of print in 1933—more than respectable for a book of criticism.

The success of *A Book of Prefaces*, though it helped establish Mencken, could not change the fact that his native land was still at war with the country he loved best. On July 8, 1918, the same day that the *Evening Mail* ran a Mencken column on hay fever (from which he was a lifelong and ever-complaining sufferer), Edward Rumely, the paper's owner and publisher, was arrested and charged with having purchased the *Mail* with funds put up by the German government.* Mencken had nothing to do with Rumely's alleged misconduct, nor had his columns for the *Mail* been more than usually controversial, but the new management, taking no chances, bought out his contract and gave him the boot without further ado. Given his track record, the surprising thing is that he stayed afloat as long as he did, but there was no getting around the fact that for the first time in nineteen years, the best newspaperman in the United States had no newspaper for which to write. Not a single editor anywhere in the country was willing to publish his work.

Two months later the war robbed Mencken of something at least as important as his berth at the *Evening Mail*: Marion Bloom went to Europe to serve as a nurse's aide.

Their affair had grown complicated in the preceding three years, though his genial letters, as gallant (and careful) as ever, show little sign of change. He continued to see Marion in Washington and, later, New York, where she and Estelle moved in 1916. She continued to write epigrams and short pieces for the *Smart Set* and at one

---

*The charges appear to have been untrue, but he spent a year in jail anyway.

point even asked for a job, though the budget, he claimed, was too tight to allow for expanding the staff. Mencken continued to be friendly with Estelle, so much so that he made the mistake of introducing her to Dreiser, who promptly annexed her as his secretary-mistress, a relationship that lasted for three years and brought her much grief. It is from Dreiser's diary entries that we know that Mencken and Marion spent nights together at the New York apartment she shared with Estelle, and that he preferred to have Estelle on hand to provide a patina of respectability, a scruple that amused the libertine novelist no end: "Mencken is there with her sister and likes to have her stay there nights when he is there, 'for looks.' The cautious conventionalist!"

Yet Marion was unwilling to settle for what Mencken so cautiously offered. She kept a carbon copy of one of the letters she sent him in 1916, explaining her side of a fight they had had, and her anguish contrasts sharply with his amiability: "You infuriate me by hiding a tenderness behind your cruelty. Why then, be either? . . . I've tried so hard to be what you want me to be and the result is you hurt me." Was he as detached in their meetings as he was in his letters? Did she expect a proposal of marriage that he failed to make? Whatever the offense, he apologized immediately, if not ardently ("Let me know how many days I am to serve in jail, and I'll go to the gate and give myself up to the warden"), and they continued to see each other, but something had gone wrong between them—and stayed wrong.

Not long afterward Mencken published one of Marion's "storyettes" in the *Smart Set*. Unlike most of her contributions, "Reflection" was signed:

> He sat in her studio arguing heatedly. Contrary to alleged feminine habit she remained silent, smiling lovingly at the rich tones of his voice as he argued around and against the point in question, in bewildering confusion. . . .
>
> Her tender smile encouraged him to continue and she watched as if fascinated his boyish trick of twisting and pulling

his short thick fingers until the joints cracked. Her eyes followed the matronly figure as he paced the room in his awkwardly fitting suit, yet so wholly typical of him.

"What a beautiful, beautiful voice he has," she thought passionately, "and how I love him! But Heavens, how much I will find to hate about him when I have ceased to love him!"

Might he have meant the publication of the story as a gesture of reconciliation? Certainly no one who knew the parties involved could have failed to guess the identity of the dumpy, long-winded knuckle cracker in the ill-fitting suit. As for Marion, one can only imagine what it must have meant to her to be able to tease the famous man she loved in the pages of his very own magazine.

Marion kept on trying to make a place for herself in Mencken's world. She set up shop as a literary agent in 1917, and with his help obtained a few prestigious clients, among them H. G. Wells, about whose novels he had written frequently in the *Smart Set* (at first favorably, then less so). The agency did reasonably well, but as its fortunes improved, Mencken began to spend less time with her. It was a habit she had learned to recognize. He wanted her, but there was a strict limit to the degree of intimacy he was prepared to accept. "Can he so easily close a door on those years?" she wrote despairingly to her sister. "It is true, he never loved me but I tried so hard to be satisfied with what he did give. And I can't blame him in any way because he made me so wildly happy. If both legs were cut off I couldn't miss him more."

Unable to break through the wall of his ambivalence, Marion gave up the fight. Early in 1918 she closed her New York office, enlisted in the Army Medical Corps, and reported to Washington for training. Jolted by her sudden decision, Mencken had an equally sudden change of heart, and they began seeing one another regularly again. At the same time he went to work on *In Defense of Women*, a short book cobbled together from essays he had previously published in the *Smart Set* and the *Evening Mail*. In it he "defended" the race of women by showering them with slyly backhanded praise, though his irony was often difficult to distinguish from sincerity:

> Perhaps one of the chief charms of women lies precisely in the fact that they are dishonourable, *i.e.*, that they are relatively uncivilized. . . . A man thinks that he is more intelligent than his wife because he can add up a column of figures more accurately, and because he understands the imbecile jargon of the stock market, and because he is able to distinguish between the ideas of rival politicians, and because he is privy to the minutiae of some sordid and degrading business or profession, say soap-selling or the law. But these empty talents, of course, are not really signs of a profound intelligence; they are, in fact, merely superficial accomplishments.

In Defense of Women is the first of Mencken's major books that can still be read from cover to cover with unfeigned pleasure, not only because it is so funny but because its tart observations about women sound a prophetic note: "Now that women have the political power to obtain their just rights, they will begin to lose their old power to obtain special privileges by sentimental appeals. Men, facing them squarely, will consider them anew, not as romantic political and social invalids, to be coddled and caressed, but as free competitors in a harsh world. When that reconsideration gets underway there will be a general overhauling of the relations between the sexes, and some of the fair ones, I suspect, will begin to wonder why they didn't let well enough alone."

What Marion thought of the book went unrecorded. She and Mencken exchanged only a few letters in 1918, for they were seeing each other constantly, knowing that their time was short. In September her unit was mobilized and transferred to New York in preparation for the trip to Europe. She and Mencken spent their last night together at the Algonquin Hotel. Then she set sail for France.

Three months later she received a letter written on Algonquin stationery, mailed the day she left New York. When she finally read it, she was filled with "wild joy." Mencken sent her many more letters, but few were as direct as this one, and none as affecting. The woman he loved had gone away to serve in a war he hated. For once he let slip the mask of geniality, and showed a glimpse of the man within:

> I shall not forget, my dear—this last visit, nor any of the others. You will believe how much I have loved you when the bad dream is over and we are all secure and happy again. You have been very good to me.
>
> Don't mope on the voyage, or in France. More new experiences will do you good. You are going abroad younger than I did—the first time. You will have some loud laughs.
>
> I kiss your hand.
>
> Yrs
H

A Mencken Album

H. L. Mencken preserved hundreds of photographs in his scrapbooks. Here are eleven of the most evocative.

"A young reporter to whom all the major catastrophes of mankind were still more or less novel": Mencken at his desk in the city room of the Baltimore *Herald*, 1901.

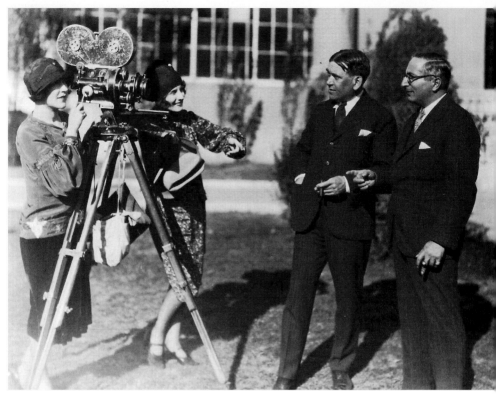

At play in Hollywood with Aileen Pringle, Norma Shearer, and Louis B. Mayer, 1926.

"A cub studying art": Posing for Nikol Schattenstein, New York, 1927.
The suspenders were a gift from Rudolph Valentino.

Hubris meets Nemesis: Writing a dispatch to the Baltimore *Sun* at the Democratic National Convention, 1932. Note the two-fingered typing technique.

To Mr and Mrs Otto Schellhase
Saturday Night Club — April 1937.

The Saturday Night Club in an uncharacteristically somber moment, 1937.

For Alfred Knopf
This was taken
by Mwsths Van
Vechten at a
temperature of
94° Fahrenheit.

Portrait of the artist with a cigar, New York, 1932. The photograph was taken by
Carl Van Vechten and the inscription, to Alfred Knopf, is
in Mencken's distinctively sharp-cornered hand.

"I formerly was not as wise as I am now":
Sara Haardt Mencken and husband on vacation in Jamaica, 1932.

"I'll have her in mind until thought and memory adjourn":
On board the SS *Bremen*, June 15, 1935, two weeks after Sara's death.

Man at work: Mencken in his Hollins Street office in the forties.

Among the believers: Mencken at the Progressive Party Convention, 1940,
four months before his stroke.

"Not too much life. It often lasts too long":
Coming home from Johns Hopkins Hospital, 1951.

5

"DON'T WAIT UNTIL THE FIREMEN ARRIVE"

Becoming a Legend, 1918–1923

Wars are as much about waiting as fighting, and H. L. Mencken found himself with time on his hands as World War I wound down. It was an unfamiliar sensation—and a disagreeable one. The practice of daily journalism accustoms a writer to seeing his thoughts in print shortly after he thinks them; in time it becomes a kind of addiction. "There is something delightful about getting an idea on paper while it is still hot and charming, and seeing it in print before it begins to pale and stale," Mencken said at the end of his writing life. Now he was cut off from his drug of choice, silenced by wartime censorship and fearful editors. He still had a magazine to get out, but the self-imposed limits under which the *Smart Set* was edited (no war, no politics) were draining the savor from the job. And Marion was gone, perhaps never to return, though even if she did, it would be to a country that had been changed beyond imagining by Woodrow Wilson.

Because Mencken wrote comparatively little about Wilson, we tend not to link the two men in our minds in the way that we think reflexively of Mencken and Coolidge, or Mencken and FDR. Yet Mencken regarded him as a scoundrel virtually without peer. Unfazed by the fact that he had been elevated to the presidency by a minority of voters (Theodore Roosevelt having fractured the

Republican majority by running on a third-party ticket), Wilson wielded power with the certitude of a college president turned reform politician who believed devoutly in God and himself, not necessarily in that order. Assisted by the ratification in 1913 of the Sixteenth Amendment, authorizing the first peacetime federal income tax, he moved decisively to remake the United States along progressive lines, establishing the Federal Reserve System and the Federal Trade Commission, passing the Federal Farm Loan Act, and introducing the eight-hour workday. Still, it was the entry of the United States into World War I that allowed him to go where no Democrat had gone before. By 1919 the top income-tax rate had skyrocketed from 7 to 77 percent; the Espionage and Sedition Acts had made it illegal for Americans to utter "disloyal, profane, scurrilous or abusive" words about the men who ruled them; and "the manufacture, sale or transportation of intoxicating liquors" had been made a criminal offense.

Mencken responded to Prohibition by selling his car and using the proceeds to purchase a large stock of "the best wines and liquors I could find," stored in a homemade basement vault whose door bore a custom-painted sign emblazoned with a skull and crossbones: "This Vault is protected by a device releasing Chlorine Gas under 200 pounds pressure. Enter it at your own Risk." He found other aspects of the Wilson regime harder to swallow. In his political journalism he affected detachment from the day-to-day follies of America's leaders. "We live in a land of abounding quackeries," he wrote, "and if we do not learn how to laugh we succumb to the melancholy disease which afflicts the race of viewers-with-alarm." But his sense of humor had never been so severely tested by a sitting president, and it is possible that his fury might have gotten the better of him had he been free to write about Wilson in 1918, as it did when he wrote about FDR in 1935. Once he was finally free to publish his uncensored opinion, he made not the slightest pretense of being amused: "Between Wilson and his brigades of informers, spies, volunteer detectives, perjurers and complaisant judges, and the Prohibitionists and their messianic delusion, the liberty of the citizen has pretty well vanished in America. In two or three years, if the thing goes on, every third American will be a spy upon his fellow citizens."

For now, though, the newspapers were closed to him, while the *Smart Set* was subject to scrutiny of the strictest kind. "The new espionage act," he explained to a correspondent, "gives the Postmaster General almost absolute power to censor American magazines. He may deny the mails to any one that he doesn't like, for any reason or no reason. No crime is defined; no hearing is allowed; no notice is necessary; there is no appeal." Nor was censorship the only thing that had him worried in the fall of 1918. As bachelors he and Nathan were eligible for the draft, and one or both of them might well be called up before the war ended, thus placing their magazine in jeopardy. Unable to blast away at the man he held responsible for his plight, Mencken distracted himself by building a brick fence in the backyard of Hollins Street (bricklaying was one of his few hobbies) and writing a book about a subject well removed from the immediate concerns of the censors.

From his earliest days as a newspaper columnist, he had taken a keen interest in what he called "the American language," knowing that his own style was a distinctive commingling of the traditional rigor of British English with the flexibility of the American vulgate ("I have a hand for a compromise dialect which embodies the common materials of both languages"). Starting in 1913, he published in the *Smart Set* a series of six articles about the American national character, devoting one, "The American: His Language," to an analysis of American English. He concluded that "American is now so rich in new words, new phrases and old words transferred to new objects that it is utterly unintelligible to an educated Englishman" and was fast evolving into "an entirely new language." This openness to innovation struck him as yet another aspect of the artless utilitarianism he saw as central to the American temper: "Always [the American's] one desire is to make speech lucid, lively, dramatic, staccato, arresting, clear—and to that end he is willing to sacrifice every purely aesthetic consideration. He judges language as he judges poetry, not at all by its grace of form but wholly by its clarity and poignancy of content."

He had meant to spin the six articles into a book, but the war stopped him from publishing so broad-gauged an attack on his fellow countrymen. Instead he used "The American: His Ideas of Beauty" and "The American: His New Puritanism" as the basis for the last chapter of *A Book of Prefaces*, while "The American: His Language" was expanded into *The American Language*, a book-length study of "certain salient differences between the English of England and the English of America as practically spoken and written—differences in vocabulary, in syntax, in the shade and habits of idiom, and even, coming to the common speech, in grammar." These differences he took to be reflections of "a general impatience of rule and restraint, a democratic enmity to all authority, an extravagant and often grotesque humor, [and] an extraordinary capacity for metaphor" that were collectively characteristic of "the spirit of America" (not to mention his own spirit, which was more typically American than he preferred to acknowledge).

Neither Mencken nor Knopf expected *The American Language* to sell, so the print run was limited to fifteen hundred copies. But the first edition sold out within a year, and Mencken was inundated with letters from readers whose corrections and amplifications he incorporated into a second edition published two years later. This "revised and enlarged" edition also contained such welcome extras as a sharp-eared translation of the Declaration of Independence into "American" that became one of his best-remembered burlesques: "When things get so balled up that the people of a country got to cut loose from some other country, and go it on their own hook, without asking no permission from nobody, excepting maybe God Almighty, then they ought to let everybody know why they done it, so that everybody can see they are not trying to put nothing over on nobody." The letters kept coming—he would receive at least one letter a day about *The American Language* for the rest of his working life—and so did the revised editions, each of which sold better than its predecessor. The fourth edition of 1936 was chosen by the Book-of-the-Month Club as a Main Selection, remaining a staple of the club's dividend offerings well into the seventies.

Anyone seeking to understand the popular appeal of *The American Language* can find a plausible explanation in Mencken's preface to the first edition: "This study . . . shows a certain utility. But its chief excuse is its human interest, for it prods deeply into national idiosyncrasies and ways of mind, and that sort of prodding is always entertaining." Yet it was no mere list of amusing Americanisms. In all its various editions, the book was an exhaustively documented examination of the myriad ways in which American English had grown away from the mother tongue, whether by altering old words or coining new ones. These differences had long been noted on both sides of the Atlantic, but never before had anyone gone to the trouble of cataloging and analyzing them in such copious and illuminating detail:

> According to a recent historian of the American merchant marine, the first schooner ever seen was launched at Gloucester, Mass., in 1713. The word, it appears, was originally spelled *scooner*. To *scoon* was a verb borrowed by the New Englanders from some Scotch dialect, and meant to skim or skip across the water like a flat stone. As the first schooner left the ways and glided out into Gloucester harbor, an enraptured spectator shouted: "Oh, see how she scoons!" "A scooner let her be!" replied Captain Andrew Robinson, her builder—and all boats of her peculiar and novel fore-and-aft rig took the name thereafter. The Dutch mariners borrowed the term and changed the spelling, and this change was soon accepted in America. The Scotch root came from the Norse *skunna*, to hasten, and there are analogues in Icelandic, Anglo-Saxon and Old High German.

Mencken was exploring uncharted territory—no similar work had been published for forty-seven years—and though he was an amateur scholar, there was nothing amateurish about his scholarship.

The American Language is full of sentences that could have been penned by the dullest of pedagogues ("Next after the use of the broad *a*, the elision of *r* before consonants and in the terminal position is the thing that Americans are most conscious of in English speech"). The autodidact's passion for exactitude led him to acquire "virtually every book and pamphlet on the subject ever printed, in whatever language"; by the forties he was contributing three or four monographs a year to *American Speech*, the journal of the American Dialect Society. What started as a distraction from the war ended as a lifelong obsession. The very last magazine piece he published was a *New Yorker* article about the etymology of "O.K.," by which time the study of American English was well on its way to becoming a full-fledged academic cottage industry, fathered by a man who had never bothered to hide his loathing for the academy. In 1963 Knopf published a one-volume abridgement of *The American Language* by Raven I. McDavid, Jr., a distinguished scholar of linguistics who noted in his introduction that "linguistic science in North America, and particularly the study of the varieties of English spoken in the Western Hemisphere, has developed remarkably since the first publication of 'The American Language' in 1919. . . . A list of those whose concern with linguistics was stimulated by correspondence about 'The American Language,' or whose research was furthered directly by Mencken himself, would include most of those who have made the study of American English a respectable discipline in its own right."

A half century after Mencken brought out the second supplement to the fourth edition of *The American Language*, the study of linguistics has moved off in other directions, and the magnum opus that helped establish him as a writer and thinker of substance is now one of his least read works (though Knopf has kept it in print). Still, anyone interested in the man himself will find much of interest tucked in among the seemingly endless lists of euphemisms, loan words, place-names, and terms of abuse that make up the bulk of the 325,000-

word fourth edition and its two equally fat supplemental volumes. Few pages go by without an engagingly quirky touch or two. Even the simplest definitions are often topped with a twist of opinion, *à la* Dr. Johnson: "Every Puritan is not necessarily a *wowser:* to be one he must devote himself zealously to reforming the morals of his neighbors, and, in particular, to throwing obstacles in the way of their enjoyment of what they choose to regard as pleasures."

Mencken's Anglophobia, inflamed beyond endurance by the war, receives a thorough airing, especially in the later editions:

> In England all branches of human endeavor are bent to the service of the state, and there is an alliance between society and politics, science and literature that is unmatched anywhere else on earth. Though this alliance, on occasion, may find it profitable to be polite to the Yankee, and even to conciliate him, there remains an active aversion under the surface, born of the incurable rivalry between the two countries, and accentuated perhaps by their common tradition and their similar speech. . . . What begins as an uproar over a word sometimes ends as a holy war to keep the knavish Yankee from ruining the English *Kultur* and annexing the tattered relics of the Empire.

So does his skeptical view of American democracy, which bobs unexpectedly to the surface in a discussion of "The Hallmarks of American":

> The American is not, of course, lacking in a capacity for discipline; he submits to leadership readily, and even to tyranny. But, curiously, it is not the leadership that is old and decorous that commonly fetches him, but the leadership that is new and extravagant. He will resist dictation out of the past, but he will follow a new messiah with almost Russian willingness, and into the wildest vagaries of economics, religion, morals, and speech. A new fallacy in politics spreads faster in the United States than

anywhere else on earth, and so does a new revelation of God, or a new shibboleth, or metaphor, or piece of slang.

And he goes out of his way to pay tribute to the man without whom *The American Language* would never have been written:

> The business of introducing the American language to good literary society was reserved for [Samuel] Clemens—and Clemens had a long wait before any of the accepted authorities of his generation recognized that he was not a mere zany . . . but a first-rate artist. . . . He was the first American author of world rank to write a genuinely colloquial and native American.

More than anything else it is this quirkiness that makes *The American Language* worth reading today (especially in the skillfully abridged one-volume version). Rarely has a scholar, amateur or otherwise, succeeded in writing a book so revealing of his own habits of mind—and so blessedly free of pedantry—without compromising his essential seriousness. Small wonder the first edition received reviews favorable enough that "they gave me some uneasiness." It reassured Mencken when he began to get "tart letters from Anglomaniacs who believe that it is felonious for us to have our own language." He never liked to please too many people at a time.

WITH THE PUBLICATION OF *THE AMERICAN LANGUAGE*, MENCKEN became a full-time Knopf author, but he had flirted briefly with another publisher prior to tying the knot. *In Defense of Women* and *Damn! A Book of Calumny*, a collection of miscellaneous essays from the *Smart Set* and the *Evening Mail* assembled at around the same time, were brought out by Philip Goodman, a New York advertising man who longed to go into business for himself as a book publisher. He had an ingenious idea—selling books in drugstores, which were then in the process of becoming the convenience stores of their day—and talked Mencken into signing on as one of his first authors. The venture was ahead of its time, but the two men would remain close friends for fifteen years, until World War II pushed them apart. Like Knopf, Goodman was a Jew of German descent, and he and Mencken sent each other hundreds of letters in which they exchanged elephantine jocosities about a Lake Wobegon–type troupe of imaginary German-Jewish characters: "Ten years ago Adolf was driving a wagon for Knefely, the cheese man. How he studied double entry bookkeeping under Old Man Kurtz, cashier of the Burghardt Brewery, and became a master of Bauverein finance—this is a story that would dim your eyes." But he was a thoroughly inept publisher, and *In Defense of Women*, which came out at the end of of 1918 (with Goodman's first name spelled "Ppilip" on the title page), sold only a thousand copies.

A new book by the author of *A Book of Prefaces* should have been more readily marketable, for the War to End War had finally ended in November, and Mencken, against all odds, had come through it unscathed, unprosecuted, financially secure, and fairly reputable. Censorship notwithstanding, his previous books had been noticed, sometimes unfavorably but usually respectfully, and he had become a pundit at large who was always good for a quote. One sign of his growing fame was the "review in rhyme" of *In Defense of Women* that Howard Dietz, himself soon to become famous as a Broadway lyricist, wrote for the *New York Evening Sun*: "H. L. Mencken will surprise you/With the smartness from his pen:/Read this volume, we advise you—/Read it once and then again." Before long few Americans would need to be told twice.

He was still a newspaperman without portfolio, but the *Smart Set* gave him a megaphone through which he could hail the "150,000 civilized Americans" who were his imagined audience, and the coming of peace loosened some of the constraints under which he and Nathan had been laboring. In April they launched a new department called "Répétition Générale," a jointly written miscellany of notebook-style entries in which they held forth at will on subjects not suited to their regular columns, and though "there were some things in it that grossly violated the pruderies, and especially the patriotic pruderies, of the first post-Armistice year," it soon became one of the magazine's most popular features. Important new faces were starting to appear: F. Scott Fitzgerald made his *Smart Set* debut in September, Willa Cather in October, Sherwood Anderson in November.* At long last, the magazine was starting to fulfill, however fitfully, the promise of its perky rejection slip: "The Smart Set is addressed to sophisticated persons. Try to avoid the trite, the eloquent and the maudlin."

Nathan later said that these were "the gala days" in Mencken's life, and his memories of the *Smart Set* in the early days of Prohibition were indeed festive. Though Mencken affected to despise the city he called "Sodom and Gomorrah," he made the most of his visits there; at one point, according to Nathan, he even had a marble slab shipped from Baltimore and set up a saloon-style free lunch for office callers. The renewed pleasure they were taking in their collaboration comes through in a leaflet called "Suggestions to Our Visitors," one of the many spoofs that Mencken printed up and handed out to friends:

1. The editorial chambers are open daily, except Saturdays, Sundays and Bank Holidays, from 10.30 a.m. to 11.15 a.m.

2. Carriage calls at 11.15 a.m. precisely.

3. The Editors sincerely trust that guests will abstain from offering fees or gratuities to their servants. . . .

6. Visitors are kindly requested to refrain from expectorating out of the windows.

7. The Editors regret that it will be impossible for them, under any circumstances, to engage in conversations by telephone. . . .

---

*Cather had lately become one of Mencken's enthusiasms. He praised *My Ántonia* in February as "sound, delicate, penetrating, brilliant, charming. . . . Beneath the swathings of balderdash, the surface of numskullery and illusion, the tawdry stuff of Middle Western *Kultur*, she discovers human beings embattled against fate and the gods, and into her picture of their dull struggle she gets a spirit that is genuinely heroic, and a pathos that is genuinely moving."

9. Solicitors for illicit wine merchants are received only on Thursdays, from 12 o'clock noon until 4.30 P.M. . . . .

16. Choose your emergency exit when you come in; don't wait until the firemen arrive. . . .

25. The Editors cannot undertake to acknowledge the receipt of flowers, cigars, autographed books, picture postcards, signed photographs, loving cups or other gratuities. All such objects are sent at once to the free wards of public hospitals.

26. Positively no cheques cashed.

The rest of the country, however, was not quite ready for an unwatered dose of their cynicism. Disillusion may have been just around the corner, but idealism was still the order of the day. "Everything for which America fought has been accomplished," President Wilson told his fellow countrymen on Armistice Day. "It will now be our fortunate duty to assist by example, by sober, friendly counsel, and by material aid in the establishment of just democracy throughout the world." It is easy to imagine what Mencken would have written about the armistice message had he been free to do so, but the Espionage and Sedition Acts continued to cramp his style, and patriotic Americans had in any case grown accustomed to making short work of Hunnish mockers. Within days of the publication of *The American Language*, the Eighteenth Amendment was ratified, making the Wartime Prohibition Act permanent; a few months later Wilson's Attorney General, A. Mitchell Palmer, would embark on a full-scale campaign against "foreign-born subversives and agitators" that led to the expulsion of nearly six thousand aliens. Meanwhile the Ku Klux Klan was mushrooming into a nationwide organization with a membership well in excess of four million. No one wanted to hear that such phenomena might not be so easily reconciled with the lofty ideals of "just democracy," much less that many of the things for which Americans had fought and died on foreign soil had yet to be even partly accomplished. It was too soon for the truth.

All this was to be brought home to Mencken when Marion Bloom returned from Europe in February. Hard though her life had been up to then, the war showed her things for which nothing could have prepared her. She tramped through the mud of France and saw "boys dying like flies . . . it was no uncommon thing to recognize a patient by his wounds rather than by his name." Mencken was shocked by her condition ("The horrors and privations of Brest nearly finished her," he told a friend), and both of them realized that their relationship had reached a turning point. Marion preserved among her papers a draft of a letter she sent to him a month after their reunion, and it leaves little doubt of her agitated state of mind at the beginning of 1919:

> I begin to see I do not want to marry you. It would be a ghastly ambition. I see you a quaking, shaking mass of rebellion and flesh and there is no romance in the vision. . . . You want to throw all the responsibility of a matrimonial decision on me so that I can be blamed if it goes not go right, i.e., the way you outline. . . . You firmly believe that matrimony would interfere with your career, yet you can't stop playing with its charming provocations. You want to be married and you want to be single. . . . You confuse me terribly with these eternal ponderings—the carefree gesture that is always withdrawn in fear that you *may* be taking a chance.

His reply was untypically frank: "You have made a piece of literature, and you have penetrated to some unbearable truths. Of this, more anon. The next time we meet I want to hear you for an hour—a solo a cappella. That will be either our last real talk—or our first." But Marion had changed in ways that he was not prepared to accept, and the letters he sent over the next few months show that he knew it:

"She lashes herself into a fury of self-pity and then empties the whole thing on me. I can't stand it any longer. I am constantly ill, have a great amount of hard work to do, and am months behind with it. For six months this endless weeping and wailing has kept me on a red-hot stove. I simply can't go on with it any longer." (Mencken to Estelle Bloom Kubitz, May 28)

"I can't understand your attitude. You need money badly, you are living far beyond your income, and yet you seem willing to gad about endlessly. I despair of making you give any serious thought to your affairs. . . . Get your ideas into some sort of coherence, as to what you are going to do, and let me hear them." (Mencken to Marion Bloom, c. July 19)

"What is this Christian Science stuff that Marion is unloading on me by the gallon? Has she really fallen for that buncombe? I am amazed. There is nothing in the world that riles me more than Christian Science. I know all about it from snout to tail. . . . Its fundamental ideas are idiotic. I can't imagine Marion taking such stuff seriously. Or is she simply kidding me? I surely hope so." (Mencken to Estelle, September 10)

"I am surely not going to try to disconvert her. It is plain that Christian Science has made her more contented and probably helped her health. But it happens to be my pet abomination." (Mencken to Estelle, September 16)

As ludicrous as it may sound in the telling, Marion's "conversion" to Christian Science was no joke, at least not to Mencken. It was inconceivable that she should have embraced a doctrine at which he had been sneering since his earliest newspaper days. Aside from everything else, he was a hypochondriac who fancied himself a rationalist, a combination of traits uniquely ill suited to cohabitation with a woman who now believed that all disease was an illusion and thus should not be entrusted to the care of doctors. (According to his brother August, Mencken's attitude toward doctors was "roughly parallel to a Catholic's attitude toward priests. That is, he looked on a doctor as a superior person whose word or opinion was not to be questioned. . . . If a doctor told him to jump off a roof, he would have jumped.") Whatever he still felt for Marion—and they continued to see each other off and on for some time to come—it was thus inconceivable that he would choose to spend the rest of his life with her. Though he loved her, he could not accept her as she was, and while another man might have come to terms with their differences, he was too stubborn to try.

Long after they parted he would continue to write about Mary Baker Eddy, founder of the Church of Christ, Scientist, with particular venomousness. In *A Mencken Chrestomathy*, the anthology of his own writings that he prepared shortly before suffering the stroke that ended his career, he led off a chapter called "Quackery" with a 1927 newspaper column that summed up his views on Christian Science with brutal directness: "Is it desirable to preserve the lives of children whose parents read and take seriously such dreadful bilge as is in 'Science and Health'? . . . Such strains are manifestly dysgenic. Their persistence unchecked would quickly bring the whole human race down to an average IQ of 10 or 15. Being intelligent would become a criminal offense everywhere, as it already is in Mississippi and Tennessee." His romance with Marion was dead by the time he wrote those words, but in truth it had come to an end in 1919, leaving only a protracted attempt to deny the inevitable.

Mencken and Marion continued to see each other throughout the early twenties, even talking on occasion of marriage, though they came no closer to bridging the gap of backgrounds and sensibilities that separated them. "We have got along the past few years by the simple device of avoiding certain subjects," he told her in 1922. "Every time they have crept in there has been a row. But under changed conditions it would be utterly impossible to keep them out. Do you blame me if I hesitate to take on a Civil War?" Their intimacy waxed and waned, but it was never anything she could count on, and she suspected him of dating other women on the side. His letters to her remained jovial and detached, her letters to Estelle anguished and confused: "I'd bet a lot that my lack of virginity is the cause of my trouble with H. . . . He wants a virgin to wife, with all the dash of our broad thinking, backed up with the exterior things of wealth, position, etc."

In 1922 Knopf brought out an expanded second edition of *In Defense of Women* incorporating new material Mencken had published in "Répétition Générale." Even more so than its predecessor, this version can be read as an attempt by an eminent Victorian to make sense of the modern woman with whom he had gotten himself entangled, and the penultimate chapter, "The Eternal Romance," could have been a half-affectionate, half-sardonic reply to Marion's earlier portrait of the critic as a buffoon in a badly cut suit:

> It is the close of a busy and vexatious day—say half past five or six o'clock of a winter afternoon. I have had a cocktail or two, and am stretched out on a divan in front of a fire, smoking. At the edge of the divan, close enough for me to reach her with my hand, sits a woman not too young, but still good-looking and well-dressed—above all, a woman with a soft, low-pitched, agreeable voice. As I snooze she talks—of anything, everything, all the things that women talk of: books, music, the play, men, other women. No politics. No business. No religion. No metaphysics. . . . I listen to the exquisite murmur of her voice. Gradually I fall asleep—but only for an instant. At once, observing it, she raises her voice ever so little, and I am awake. Then to sleep again. . . . I ask you seriously: could anything be more unutterably beautiful?

He told Marion that she was the woman at the edge of the divan, and she claimed to be pleased. Two decades later he would deny it, writing in *My Life as Author and Editor* that "the model for my portrait . . . was a half-Jewish lady who had been a successful actress and was later to have a brief time of glory as a movie star." Marion's name appears nowhere in that eighteen-hundred-page manuscript.

Even in a memoir left to the Enoch Pratt Free Library on condition that it not be opened until thirty-five years after his death, he was unwilling to speak of one of the most momentous relationships of his life.*

Marion began seeing another man, a dashing World War I veteran named Lou Maritzer who was determined to marry her, and she told Mencken she was thinking of saying yes. He made a half-hearted effort to talk her out of it, but she was tired of being taken for granted, and in the summer of 1923 she became Mrs. Lou Maritzer. Estelle phoned Mencken to give him the news, and he wrote to Marion at once: "What am I to say? That I wish you the happiness that no one deserves so much as you do. That I have got out some old letters of yours, and read them and burned them, and now sit down and smoke my pipe, and wonder what in hell it is all about." To Estelle he was franker:

> Little will that gal ever suspect how near she came to cooking, washing and ironing for the next President of the United States. But certainly you must know what happened: the Christian Science fever simply scared me to death. Worse, my general rampaging in late years scared *her* to death. The thing would have been hopeless. I believe that she will be happy with Maritzer. He seems to be a charming fellow, and he is very romantic. She lives in a world of dreams—many of them beautiful. If she had settled down to writing she would have made a great success. I am 43, and all things begin to seem far off. But I shall not forget.

---

*Only in a single, chilling letter of 1948 does she reappear momentarily. Mencken heard from a correspondent who had seen a catalog from the auction house of Parke-Bernet that listed various Mencken-related items put on sale by Marion, among them a typescript of *A Book of Prefaces*. Asked for information about her, Mencken replied, "Marion Bloom was a woman I knew about twenty-five years ago. She was an occasional contributor to the *Smart Set*. For years past she has been living in Washington. I don't recall seeing her since 1930 or thereabout. . . . I'll be on my way to Florida before a copy of the P-B catalogue could reach me, so please don't bother to send it."

The marriage would disintegrate in a year and a half, and Mencken and Marion resumed their correspondence. His verbal gallantries filled her with hope, but it was vain. Never again was she the woman in his life. That place would be filled by a genteel young lady from Alabama who was everything Marion Bloom was not.

AS ALWAYS, WORK WAS HIS EVER-PRESENT COMFORT, AND HOWEVER DIS-tressed he may have been by the souring of his romance with Marion, he had more agreeable things to think about in 1919, starting with his next book. He had been inching up the greasy pole of fame for a decade, and now he was putting together a collection that would finish what *A Book of Prefaces* and *The American Language* had started. It contained both "butcheries of some of the elder demigods" and "onslaughts upon some of the reigning favorites of 1919," and would bear on its cover the perfect title, *Prejudices*.

According to Mencken the idea came from a Cleveland bookseller who clipped his *Smart Set* reviews and put them in folders "for easy reference by people coming to the shop." The man suggested to Alfred Knopf that they might make a good book. Knopf and Mencken were both skeptical at first, just as they had been with *The American Language*, but they changed their minds, and by early 1919 the first volume of *Prejudices* (there would be five more) was in the works. Most of it came from the *Smart Set*, but Mencken took the opportunity to polish and revise the original pieces—a practice he had begun with *A Book of Prefaces* and would continue thereafter—and the end product profited greatly from his meticulous editing. The contents, he told a correspondent, were chosen with the purpose of irritating as many halfwits as possible: "Such books are mere stinkpots, heaved occasionally to keep the animals perturbed. The real artillery fire will begin a bit later."

What Mencken meant by "artillery fire" was something more ambitious than a compilation of rewritten book reviews, but he also knew that *Prejudices* was worthy of serious consideration in its own right, if only because it cut so sharply against the grain of received opinion. Arnold Bennett, Ralph Waldo Emerson, Robert Frost, William Dean Howells, Jack London, and H. G. Wells were among those present, and though he had friendly things to say about some of them, others were treated with something less than the respect to which they were accustomed:

> The Emerson cult, in America, has been an affectation from the start. Not many of the chautauqua orators, vassarized old maids and other such bogus *intelligentsia* who devote themselves to it have any intelligible understanding of the Transcendentalism at the heart of it. . . . Frost? A standard New England poet, with a few changes in phraseology, and the substitution of sour resignationism for sweet resignationism. Whittier without the whiskers. . . . The prophesying business is like writing fugues; it is fatal to every one save the man of absolute genius. The lesser fellow—and Wells, for all his cleverness, is surely one of the lesser fellows—is bound to come to grief at it, and one of the first signs of his coming to grief is the drying up of his sense of humor.

Even George Bernard Shaw was subjected to a thoroughoing reconsideration, causing the lavish praise of 1905 to give way to a more astringent verdict: "Always the ethical obsession, the hall-mark of the Scotch Puritan, is visible in him. His politics is mere moral indignation. . . . [H]e always sees his opponent, not only as wrong, but also as a scoundrel."

For *Smart Set* subscribers *Prejudices: First Series* would contain no surprises. Its effect, as Mencken intended, was cumulative: "Most of the contents were not new . . . but their impact, coming out in small installments, had been much less than their impact in one blast." Just as important, reviewers who had been keeping up with him month by month now had a prime opportunity to consider him at leisure and at length, and they did so in greater numbers—and with greater enthusiasm—than ever before. What they found was a book as contrarian as *A Book of Prefaces* but not nearly as preachy, written in a style full of slapstick vigor. That may well have been the most surprising thing about Mencken's *Smart Set* reviews: Never before had American literary criticism been so much *fun*.

Not all the notices were favorable, but even the negative ones were phrased in such a way as to make the book sound interesting. The *Boston Evening Transcript*, staidest of all American papers, dismissed it as a "smartly phrased series of cheap and unimportant comments upon contemporary literary topics"—and then changed course a week later, calling it "a book full of sound, hard sense." Typical of the younger critics was Burton Rascoe, who wrote in the *Chicago Tribune:* "There is no man living in these states who knows the rising, the future, generation of literary men half so well as Mr. Mencken. There is no one who has one thumb so constantly on the pulse of the present day literary tendencies, and the other to his nose at the artificial and ephemeral." Mencken's scrapbooks began to swell with similar praise. Everything written about him prior to 1918 had taken up barely more than a single 300-page volume, but the clips for 1918 alone filled more than 150 pages, those for 1919 well over 200. The publicity, good and bad, made a difference. The first *Prejudices* sold 1,678 copies by the end of the year, and 14,034 more by the time it finally went out of print in 1942. Then as now, essay collections sold poorly, so Knopf had every reason to be happy with the results. "I have discovered something," he told Mencken one day in 1920. "It is that H. L. Mencken has become a good property." As much as anything else, it was *Prejudices: First Series* that made him one.

It also made him something more. In 1942 Alfred Kazin recalled "the first trombone blasts from Mencken's *Prejudices* in 1919 that sounded the worldliness and pride of the new emancipation." Even after he had put book reviewing behind him, he would be remembered as the foremost critical advocate of the Lost Generation, whose members returned from World War I to find "all Gods dead, all wars fought, all faiths in man shaken." So said F. Scott Fitzgerald in *This Side of Paradise*, and like so many other young American writers, he was sure that Mencken spoke for him. Sherwood Anderson inscribed a copy of *The Triumph of the Egg* to "Henry L. Mencken, champion of men who need champions." Indeed, Mencken genuinely admired many of the up-and-comers whose work he and Nathan were publishing, Fitzgerald in particular. (He praised *This Side of Paradise* as "the best American novel that I have seen of late . . . original in structure, extremely sophisticated in manner, and adorned with a brilliancy that is as rare in American writing as honesty is in American statecraft.") But he also thought them naive—he had never had any faiths to shake—and as they began to declare themselves politically, he found their idealism half-baked. Nor did he want anything to do with their self-destructive carousing: When it came to dissipation Mencken talked a better game than he played. In any case his interest in criticism was already slacking off, and though he would continue to pour his prodigal energies into the *Smart Set* for another four years, *Prejudices: First Series* was as much an end as a beginning. He wanted to do something else, and it was the *Sun*, from which he had been exiled for three years, that gave him the break for which he longed.

Early in 1920 the paper acquired a new publisher, and Paul Patterson, who had been close to Mencken for several years, was determined to bring him back into the fold. He offered his old friend a spot on the editorial page of the *Evening Sun*, as well as proposing to make him that paper's semiofficial gray eminence. In addition to writing a signed column, Mencken would be paid an annual retainer of $3,000 to sit on the editorial board—to serve, in effect, as the *Sun*'s Wizard of Oz, providing backstage advice without any managerial responsibility. "I have no definite duties," he explained to Isaac Goldberg in 1925. "The theory is that it is a useful thing to have at call a man familiar with the workings of the paper and yet removed from its policies." No doubt it would also be a useful thing for the paper to have the best-known writer in Baltimore in its employ once again, but Mencken accepted Patterson's invitation with honest pleasure, not least because he cared about the *Sun*. "My own connection with the paper is thin, and rather inconvenient to me," he wrote to Ernest Boyd, an Irish diplomat and part-time writer who struck up a friendship with Mencken when he was posted to Baltimore as British vice-consul. "I'll probably pull out, once some sort of a new organization is completed." That was bosh. Patterson was determined to improve the *Sunpapers'* quality, and while Mencken questioned how much could be done—"A first-rate journalist doesn't want to live in Baltimore," he told Boyd, presumably with a straight face—he meant to do what he could.

Mencken returned to the *Sun* on February 9 with a column called "A Carnival of Buncombe" (his preferred variant of *bunk*) in which he disposed of the current crop of presidential candidates as if they were a half-dozen debutante novelists, ripe for the slaughter: "All of the great patriots now engaged in edging and squirming their way toward the Presidency of the Republic run true to form. That is to say, they are all extremely wary, and all more or less palpable frauds." For the first few months he wrote at odd but increasingly frequent intervals, taking time off in June to cover the presidential nominating conventions, a quadrennial chore in which he reveled, filing dispatches that combined hard news with sketches of Prohibition-era politicians at play. ("The genial dampness continues," he wrote from the Democratic convention in San Francisco. "No honest delegate need go unkissed by stimulants, provided he has a good nose and is not full of fears that every side street is peopled by highwaymen.") After that he began writing regularly for the Monday paper, and from then on, except when he was on vacation or covering an out-of-town story, he would appear on its editorial page each week for the next eighteen years.

His columns run about sixteen hundred words each, more than twice as long as a modern-day op-ed piece. Yet they are not really familiar essays in the traditional sense, since they are typically (though by no means always) inspired by some specific news event. Not unlike the two-thousand-word "Our Notebook" pieces that G. K. Chesterton contributed each week to the *Illustrated London*

News from 1905 to 1936, Mencken's *Evening Sun* articles occupied a middle ground between the leisurely periodical essays of the
eighteenth and nineteenth centuries and the contemporary newspaper column, in which a journalist comments concisely on a topic
of immediate interest. But while many of Chesterton's *Illustrated
London News* essays were collected into books, the Monday Articles,
as they came to be called, are less well-known than any other portion of Mencken's mature body of work. He rewrote several of
them for later publication in magazines and books, but the originals
were not collected during his lifetime, or syndicated in other newspapers. Though they sometimes attracted national attention, the
Monday Articles were produced for the exclusive edification of the
citizens of Baltimore, and to those fortunate enough to read them
week by week, they would be a regular source of provocation and
delight.

No doubt the dull design of the *Evening Sun*'s editorial page
helped make the Monday Articles stand out (it was a seven-column
sea of small type, unadorned by art of any kind), but even now it is no
chore to flip through Mencken's scrapbooks and see what was on his
mind on any given Monday between 1920 and 1938. He wrote about
Schubert and hot dogs, the Ku Klux Klan and nightclubs, rarely letting a week go by without saying something interesting—or irritating. One week it would be about capital punishment:

> Let us have our hangings hereafter, not in the jail yard, but in
> Druid Hill Park, with any law-abiding citizen free to see the
> show that, as a taxpayer, he has to pay for. The lake, drained,
> would make an admirable amphitheatre. A quarter of a million
> spectators could be accommodated outside the rail; have the
> show on Sunday afternoon, and at least a quarter of a million
> would be on hand. The movie rights would be worth some-
> thing; perhaps the city and the widow would split 50-50. I sug-
> gest an embellishment. Bring in the jury that condemned the
> chief performer and chain its members to the scaffold, that they

may miss no detail. And make the judge mount the steps with the condemned and his pastor.

The next time around it might be censorship:

The Jews of Detroit, outraged by the anti-Semitic nonsense printed by Henry Ford in his Dearborn *Independent*, have had an ordinance passed barring that curious journal from the news-stands. . . . I am certainly not anti-Semitic and never read Ford's paper, but I carry away from the Detroit episode a suspicion that he must have mingled some truth with his libels, else the yells would have been less raucous. No sane man objects to palpable lies about him; what he objects to is damaging facts. Perhaps if I read Ford I'd dismiss his case as without merit, and so, maybe, would every other fair man—but the Jews, with singular fatuity, now seem to be doing their best to make it impossible for me and other fair men to read and gag at him.

For the most part, though, the Monday Articles, especially in election years, were about politics. Of the thirty-three published in 1920, about half dealt specifically with politics and politicians, both national and local, including ten about the presidential campaign. Four were about literature and the arts, one about foreign affairs, and the rest covered such topics as Prohibition, radicalism, and economics. But as often as not Mencken found himself returning to what another *Sun* columnist called "the great game of politics," and it is probably for this reason that so few of the Monday Articles have been collected. Most deal with a cast of long-forgotten characters, and the ones that range farther afield tend to be repetitive when read in bulk. He liked to flog the same horses, and while he usually contrived to put a fresh spin on his stock routines, it was rare that he had anything completely new to say about the sempiternal stupidity of politicians, or the evils of puritanism.

On the other hand the Monday Articles were never meant to be read in bulk, and it was no accident that Mencken chose not to collect them. He used them as a notebook in which he tinkered with lines of argument and figures of speech that would later take their definitive shape in the pieces he wrote for the *Smart Set* and the *American Mercury* and revised still further for inclusion in his self-anthologies, the six *Prejudices* and two *Chrestomathies*. A lifetime of deadlines had schooled him in the pragmatism of the professional journalist: He did his best in the time available, knowing there would always be another chance to do better. Here, for example, is the conclusion of the column that he wrote in 1933 to mark the death of Calvin Coolidge:

> We suffer most when the White House bursts with ideas. With a World Saver preceding him (I count out Harding as a mere hallucination) and a Wonder Boy following him, he begins to seem, in retrospect, an extremely comfortable and even praiseworthy citizen. His failings are forgotten; the country remembers only the grateful fact that he let it alone. Well, there are worse epitaphs for a statesman. If the day ever comes when Jefferson's warnings are heeded at last, and we reduce government to its simplest terms, it may very well happen that Cal's bones now resting inconspicuously in the Vermont granite will come to be revered as those of a man who really did the nation some service.

And here is the same passage reworked for the *Chrestomathy* in 1948:

> We suffer most, not when the White House is a peaceful dormitory, but when it is a jitney Mars Hill, with a tin-pot Paul bawling from the roof. Counting out Harding as a cipher only, Dr. Coolidge was preceded by one World Saver and followed by two more. What enlightened American, having to choose between any of them and another Coolidge, would hesitate for

an instant? There were no thrills while he reigned, but neither were there any headaches. He had no ideas, and he was not a nuisance.

Both versions are pure Mencken, but the first was written by the great newspaperman, the second by the great essayist.

"WHAT THE TOTAL OF MY PUBLISHED WRITINGS COMES TO I DON'T know precisely," Mencken said in 1948, "but certainly it must run well beyond 5,000,000 words." While he liked to throw around numbers with lots of zeroes, this one seems more plausible than most (the Monday Articles alone amount to about a million words). The newspaper columns, the *Smart Set* reviews, the books and essays and letters—when did they all get written? Where did he find the time to do so much so well? The answer, said August, was that he arranged the whole of his life with his work in mind: "Everything else was supplementary to it, but work was the important thing, and everything else had to revolve around it and fit into it. . . . He led a life different from other people, but that was the bottom of it."

The daily details of that life were much the same in 1920 as they had been in 1910, for Mencken had become set in his ways long before he turned forty. He arose each morning at eight, took a cold bath, breakfasted with Anna, Gertrude, and August (Charlie Mencken had married and moved away to Pennsylvania), then went back upstairs to his third-floor office-bedroom at nine and sifted through the incoming mail. A near-compulsive correspondent, he wrote dozens of letters each morning, all of which he stamped and mailed himself (his daily stroll to the corner mailbox was among his few forms of exercise). He would soon start dictating his correspondence to a stenographer, but his other writings were banged out by hand on a Corona portable typewriter, a process that never failed to amuse those who saw his two-fingered style up close: "Mencken carried it to the extreme of parody, hitting the keys only with his tiny forefingers and spacing with his right elbow, a routine that made him look like a bear cub imitating a drum majorette." He corrected proofs, received visitors, ate lunch, paid an afternoon visit to the *Sun* at two o'clock sharp—one young staffer claimed that you could set your watch by his arrival—and came home afterward to read and nap on the office couch. (Whenever possible he read lying down.) After dinner he returned to his room to write, chewing juicily on an unlit five-cent cigar as he covered page after page of the cheap copy paper he had been poaching from the *Sun* since 1906. "My manuscript done on the typewriter is almost perfect," he boasted without exaggeration. "I doubt that I average one correction to a page. The heavy work I did for the *Herald* in my early days taught me to think before writing; it is my experience that most newspaper reporters work in the other direction." Precisely at ten o'clock he would put down his tools. Sometimes he went out for a late-evening drink in a speakeasy with friends, but just as often he put on his glasses and handmade silk pajamas, retired to the sleeping porch behind his office, and spent an hour or two reading himself to sleep by gaslight.

Books were his closest companions. He never left Hollins Street without one: "When I travel I read the whole way. If I feel badly and can't work, I read. Even on street-cars I carry a paper or book." At home he was surrounded by six thousand of them, and new ones came in each day's mail. He received a thousand review copies a year, along with fifty-odd magazines each month, including the *Congressional Record*, which he examined closely in order to see what the rascals in Washington were up to. Inevitably most of his reading was for professional purposes, and when he found the time to read for pure pleasure, it was "usually a scientific book," though he occasionally revisited an old favorite: "The last was Poe's criticism—very good stuff. Before that I read Emerson's Essays. I re-read Pepys and Boswell now and then. But Addison, whom I devoured as a boy, now bores me. I seldom read poetry. Of modern books the ones I re-read oftenest are 'Huckleberry Finn' and 'Lord Jim.' I often read a chapter or two of the Bible. Kipling was my god at 18, but now he seems thin to me. But I still like some of his later poetry."

After books he loved music best—indeed, there was no other art form for which he cared at all—and he varied his routine each week by attending the meetings of the informal group known as the Saturday Night Club. In 1902 he had begun playing chamber music regularly with three friends. One was a fellow journalist, Joe Callahan of the *Herald*, but the others were civilians: Albert Hildebrandt was a violin dealer and maker, Samuel Hamburger a haberdasher turned electrician. They met in the back room of a piano store a block from the *Herald*, devoting "two hours to actual playing and the other two to the twin and inseparable art of beer-drinking." In time they acquired a dozen like-minded companions and began giving impromptu readings of works for symphony orchestra performed by whatever combination of instruments happened to be on hand, usually a string quartet, a sprinkling of winds and brass, and four-hand piano, with Mencken playing *secondo* to the *primo* of Max Brödel, a medical illustrator who became one of his best friends. (Mencken acquired Brödel's drawing of a cancer of the uterus, and is said to have upset his mother greatly by threatening to hang it in their Hollins Street home.) The playing took place at Hildebrandt's shop, the drinking in a private room of an obliging hotel or saloon.*

"The club has always had a few non-performing members," Mencken recalled, "but they have been suffered only on condition that they never make any suggestions about programmes or offer any criticism of performances. . . . [T]hey are welcome if they keep their mouths shut and do not sweat visibly when the flute is half a tone higher than the first violin, but not otherwise." The rule was waived for Willie Woollcott, an ink-and-paste manufacturer who is now remembered, if at all, as the brother of Alexander Woollcott, the drama critic who founded the Algonquin Round Table and was caricatured by Moss Hart and George S. Kaufman in *The Man Who Came to Dinner*. (Mencken stayed at the Algonquin Hotel whenever

---

*During Prohibition, the members took turns hosting the postmusic parties in their homes.

he visited New York, but had nothing to do with the Round Table, whose members he regarded as "literati of the third, fourth and fifth rate.") Willie composed the club anthem, a mock-patriotic ditty entitled "I Am a 100% American," whose opening lines Mencken included in his *New Dictionary of Quotations:* "I am a one hundred per cent. American!/I am, God damn, I am!" In addition it was Willie who in 1922 persuaded his colleagues to play Beethoven's first eight symphonies in a single sitting, pausing only for refreshments. Mencken's own account of this well-lubricated marathon, collected in *Heathen Days,* strongly suggests that the refreshments consumed between the playing of the Second and Third Symphonies derailed the proceedings, though he claimed to have lasted through the scherzo of the Sixth Symphony before falling asleep.

Nobody ever claimed that the Saturday Night Club was a band of virtuosos—its performances were by all accounts loud, enthusiastic, and slapdash—but several members were reputable professional musicians who for reasons of their own found pleasure in playing with amateurs. Gustav Strube organized the Baltimore Symphony and was its first conductor; W. Edwin Moffett was the orchestra's principal bassist; Louis Cheslock, Israel Dorman, and Theodore Hemberger all taught at the Peabody Conservatory of Music; Adolph Torovsky was the bandmaster of the U.S. Naval Academy. None of these men would have wasted time on a roomful of drunken incompetents, and it is clear from Cheslock's diary that the club staggered through an imposing amount of first-class music each week. On a typical evening in 1934 the members performed Weber's *Freischütz* Overture, Tchaikovsky's *Nutcracker* Suite, and Schumann's *Rhenish* Symphony, a program comparable in length and specific gravity to what one might hear at an orchestral concert, before adjourning to Schellhase's, a popular German restaurant, for clams, roast beef on rye, beer, and conversation. Cheslock also left behind several descriptions of Mencken's piano playing, and taken together they leave no doubt of his intense involvement in the act of music making: "Henry's humming as he plays sounds like the groaning of a man in agony who tries to stifle the sounds to keep face. . . . From the first notes it was obvious that [he] knew his music. With each new shift in tempo and tonality he was the soul of the ensemble. Sudden *pianissimo*—and he hushed the bass and horn. He signaled each entry . . . at the end he let out an enormous whoop." When a performance went off the rails, he would utter "a lusty *shit!*"

"The world presents itself to me, not chiefly as a complex of visual sensations, but as a complex of aural sensations," Mencken wrote in *Minority Report,* and outside of the joy he found in writing itself, it was music that provided him with one of the few outlets he had for the acute aesthetic sensitivity that he concealed in order to make his way among the hard-drinking men who worked at the *Herald* and the *Sunpapers.** When he wrote about music, as he sometimes did in his Monday Articles, it was as a practicing amateur performer who quite literally knew the score; the surviving manuscripts of his own compositions and arrangements are set down in the clean, confident penmanship of a professional. His tastes were even narrower in classical music than in literature. He liked Bach, Haydn, Mozart, Beethoven, the nineteenth-century Austro-German romantics, and not much else. "So far as I can make out," he announced placidly in 1925, "Stravinsky never had a musical idea in his life." But within the bounds of his interests, he always knew what he was talking about. Not even George Bernard Shaw wrote more memorably about Beethoven's *Eroica* Symphony, Mencken's favorite piece:

There is no sneaking into the foul business by way of a mellifluous and disarming introduction; no preparatory hemming and

---

*That sensitivity is also on display in the many diary entries in which he writes of the beauties of the garden he kept in the backyard of 1524 Hollins Street. As usual with Mencken, though, it was double-edged. One of his chief objections to the poor Appalachian whites who invaded his Union Square neighborhood during World War II was that they never showed "any feeling for beauty. . . . Not one of them has ever been known to cultivate a flower."

hawing to cajole the audience and enable the conductor to find his place in the score. Nay! Out of silence comes the angry crash of the tonic triad, and then at once, with no pause, the first statement of the first subject—grim, domineering, harsh, raucous, and yet curiously lovely—with its astounding collision with that electrical C sharp. The carnage has begun early; we are only in the seventh measure. In the thirteenth and fourteenth comes the incomparable roll down the simple scale of E flat—and what follows is all that has ever been said, perhaps all that ever will be said, about music-making in the grand manner.

Most men have more than one face, and Mencken was no exception. The mainstay of the Saturday Night Club, the boon companion who frolicked with Nathan on his twice-monthly visits to New York, was the same Mencken who spent most of his days and nights curled up with a book or hunched over a typewriter, hammering out testy memoranda in which he assured posterity that he was less friendly than he looked ("My gregariousness is satisfied by a relatively small group of old friends. I see them constantly, and enjoy them very much, but I am seldom tempted to widen the circle. . . . People in general interest me, and if they are odd enough amuse me, but I do not like them"). Yet when Albert Hildebrandt died in 1932, he devoted an entire Monday Article to a tender reminiscence of a man whose name few of his readers would have recognized: "The late Albert Hildebrandt, who died last Thursday, had barely turned sixty, but he really belonged to an older Baltimore, and it was far more charming than the Rotarian Gehenna we endure today."

Mencken, too, belonged to that older Baltimore, and he liked to act as though it still existed. Yet he knew it did not, for the *Sunpapers* told him so. Baltimore had changed since the war, and so had the rest of the United States. The same issue of the *Evening Sun* that carried his first Monday Article also told the story of how a Kentucky mob seeking to lynch a "negro murderer" who had been convicted of murdering a ten-year-old girl (the jury having completed its deliberations in fifteen minutes flat) was greeted by machine-gun fire when it stormed the courthouse. The Senate was debating the Treaty of Versailles, and France was preparing to recognize the Soviet Union; the society page announced the first general session of the Inter-racial Conference of Baltimore, at which "problems now facing whites and negroes" were to be discussed. (TOO MUCH EMOTION IN BUSINESS WOMEN/LOVE OF BECOMING CLOTHES ALSO DRAWBACK, DECLARES NATIONAL FEDERATION, read a headline elsewhere in the paper.) The General Electric Company was offering a fifteen-million-dollar issue of twenty-year 6 percent gold debenture bonds, brokered by J.P. Morgan & Co. Advertisements hawked Bayer Aspirin, Bromo Quinine Tablets, electric washing machines, and Kellogg's Krumbled Bran "to keep regular." Sirloin steak cost 22 cents a pound, gasoline 28 cents a gallon. A pair of "ladies' mercerized lisle stockings" went for 39 cents, a pair of pants for $2.95.

A few years later a housewife in Muncie, Indiana, would explain to an inquiring scholar how World War I had affected her daily life:

> I began to work during the war, when everyone else did; we had to meet payments on our house and everything else was getting so high. The mister objected at first, but now he don't mind. . . . We've built our own home, a nice brown and white bungalow, by a building and loan like everyone else does. We have it almost paid off and it's worth about $6,000. No, I don't lose out with my neighbors because I work; some of them have jobs and those of them who don't envy us who do. . . . We have an electric washing machine, electric iron, and vacuum sweeper. I don't even have to ask my husband any more because I buy these things with my own money. . . . We own a $1,200 Studebaker with a nice California top, semi-enclosed.

Most of us think of flappers and flagpole sitters when we think of the Roaring Twenties, but that semienclosed Studebaker summed up the decade as truly as anything Scott Fitzgerald ever wrote. The purchasing power of American wage earners, which had stayed more or less flat from 1890 to 1915, went up 20 percent between 1915 and 1926. Year by year, there was more money to go around and more things to buy with it, not just for gold-hatted Jay Gatsby but also for the middle-class folk of the Middle West he had fled. They were moving off the farm and into town, and when they got there, they wanted to buy convertibles and washing machines, to join the Rotary Club and listen to dance music on the Victrola and go out to dinner and a movie on Saturday nights. They believed in the power of technology—the twenties were the decade of Henry Ford and Charles Lindbergh—and they nodded in agreement when their next president, an amiable, poker-playing newspaper publisher from Marion, Ohio, told them in his inaugural address that what the country needed was "not nostrums but normalcy." As far as they were concerned, Warren G. Harding, whom they sent to the White House the same year that Mencken returned to the *Sun*, had it just about right. The war had knocked all the Wilsonian idealism out of them, and now they wanted a louder, livelier, more abundant life.

Mencken wanted something different, if not so different as his admirers supposed: He wanted the life he already had. It says much about the leading journalist of the Jazz Age that he hated jazz, which he found "crude and childish." He confessed to his readers in the spring of 1921 that he had succumbed to "an old longing, suppressed for years," and attended his very first movie. He never listened to the radio, never drove a car (the one he sold in 1918 to buy liquor was the only one he ever owned), never flew on an airplane, and used the telephone "as seldom as possible" (though his number was listed, and people called him incessantly). Two trips to Manhattan each month were more than enough to slake his thirst for the noisy bustle of contemporary life. Otherwise he was content to spend the whole of his existence reading and writing, playing Beethoven and Brahms on Saturday nights, and working in the backyard of Hollins Street all

Sunday long. "My surroundings have helped to make me complacent—and complacency is perhaps the most marked of all my characteristics," he said. "I can grow angry and fretful on occasion, but it is seldom for long."

It was a sedate life for a literary lion, which was what he was in the process of becoming, to some extent in spite of himself. Early in 1921 *Vanity Fair* paid tribute to him on the grounds that "he wrote one of the first and best books on Nietzsche in English; because he has for years been almost our only bold assailant of the national Puritanism; and, finally, because he has taken his place as one of our most important living critics and has become a sort of godfather to the younger generation of American writers." Four months later Edmund Wilson, who probably wrote the unsigned paragraph in *Vanity Fair*, expressed himself even more strongly in a signed article in the *New Republic*, calling Mencken "a genuine artist and man of first-rate education and intelligence who is thoroughly familiar with, even thoroughly saturated with, the common life. . . . Mencken is the civilized consciousness of America, its learning, its intelligence and its taste, realizing the grossness of its manners and mind and crying out in horror and chagrin." Years later he claimed to have been made uncomfortable by such praise. "I had suddenly become, by the end of 1920, a sort of symbol of all the disillusionment following World War I, and was credited with a leadership in dissent that I did not want, and tried constantly to avoid," he wrote. No doubt a part of him really did shrink from being cast in such a role, if only because he knew how ill suited he was to play it, though it would take the better part of a decade before he was unmasked. In the meantime the plaudits rolled in, and whatever he thought of them, he saw each and every one. The press clippings pasted in his scrapbooks of the period are dated in his own distinctive hand.

While he may have been a symbol, he was also a hardworking journalist, and 1920 had been one of his busiest years. The *Smart Set* continued to appear each month, and *Black Mask*, the latest money-making venture of Mencken-Nathan, Inc., was launched in April. Meanwhile, in addition to starting his Monday Articles, he brought out *Prejudices: Second Series* and published three other books, a robust translation of Nietzsche's *Antichrist* and two amusing collaborations with Nathan, *The American Credo: A Contribution Toward the Interpretation of the National Mind* and *Heliogabalus, a Buffoonery in Three Acts*. The "buffoonery" was a mildly bawdy historical play, *The American Credo* a facetious collection of alleged articles of American faith ("That many soldiers' lives have been saved in battle by bullets lodging in Bibles which they have carried in their breast pockets") introduced by a hundred-page "preface," signed by both men but written by Mencken, excoriating "the readiness of the inferior American to accept ready-made opinions. . . . Thus the *Boobus americanus* is led and watched over by zealous men, all of them highly skilled at training him in the way that he should think and act."

As if all this were insufficient to keep Mencken busy, he immersed himself in the continuing housecleaning of the *Sunpapers*, helping to draft a lengthy memorandum in which he set forth his editorial vision:

> No man of active, original and courageous mind, unless he be rich or of established position, is any too safe in the United States today. The desire for draconian laws has passed beyond the stage of interference with private acts; it now seeks to challenge even secret thoughts. And this enterprise, instead of being checked by a judiciary jealous of the common rights of man, is actually fostered by a judiciary that seems to have forgotten the elemental principles of justice and rules of law. Here is an opportunity for a great newspaper that seeks to lead rather than to follow in politics. . . . It cannot protest when striking steel-workers are deprived of the right of free assemblage and remain quiet when violent legislation impairs the inviolability of contract. It cannot denounce one man for robbing the government and stand by silent when another man is robbed by the government.

Those words had a smack of authority, coming as they did from a man who had been spied on by his own government, and now that the war was over, Mencken meant to put them to the test. Believing that "it is the prime function of a really first-rate newspaper to serve as a sort of permanent opposition in politics," he vowed to nudge the *Sunpapers* into playing that part. That task was more attractive than the continuing effort of breathing life into the *Smart Set*. He hated its pulpy paper and frilly covers, and as he lost interest in novels, he found himself bored by the job of producing a readable book column each month, though there was one last surprise up his sleeve: He would "discover" another major American novelist.

Mencken and Nathan met Sinclair Lewis at a Manhattan cocktail party in October. Lewis was then known only for his banal contributions to the *Saturday Evening Post*, and Mencken had taken no note of his books (though he published one of Lewis's earliest short stories, "I'm a Stranger Here Myself," in 1916). Nor did he make a favorable first impression on either editor, for he was, as usual, falling-down drunk, an offense he proceeded to compound by buttonholing Mencken to inform him that "he had lately finished a novel of vast and singular merits full worthy of my most careful critical attention." Mencken took the train back to Baltimore the next day, having first grabbed a bundle of review copies to read along the way. One of them happened to be a set of galley proofs of Lewis's novel; curious to see what sort of book so awful a man might have written, he took a look. In Mencken's telling of the story, he was so excited by the time the train reached Philadelphia that he called a Western Union boy and sent a wire to Nathan. "I forget the exact text," he recalled, "but it read substantially: 'That idiot has written a masterpiece.'" The book was *Main Street*."

As was his wont, he embellished the tale slightly. He sent Nathan a letter, not a telegram, but the text was more or less as he remembered it: "Grab hold of the bar-rail, steady yourself, and prepare yourself for a terrible shock! I've just read the advance sheets of the book of the *Lump* we met at Schmidt's and, by God, he has done the job! It's a genuinely excellent piece of work. Get it as soon as you can and take a look. I begin to believe that perhaps there isn't a God after all." To Lewis he wrote less effusively, calling *Main Street* "the best thing of its sort that has been done so far. . . . I'll review it in the January Smart Set, the first issue still open." He was as good as his word, and then some. On January 3 he published a Monday Article in which he called Lewis's bitter satire of small-town life "a phenomenon of the first order, and vastly more significant than a dozen books by Edith Wharton." Later that month he upped the ante in the *Smart Set:* "It presents characters that are genuinely human, and not only genuinely human but also authentically American; it carries them through a series of transactions that are all interesting and plausible; it exhibits those transactions thoughtfully and acutely, in the light of the social and cultural forces underlying them; it is well written, and full of a sharp sense of comedy, and rich in observation, and competently designed."

That review placed Sinclair Lewis on the map of American letters. It was the last time Mencken would put the whole of his critical muscle behind an unknown novelist, and it is significant that the novelist on whom he chose to bestow his favors was a satirist who shared his hatred of the sheer ordinariness of "Middle Western *Kultur.*" A year and a half later Lewis would publish *Babbitt,* an even more ferocious portrait of *Boobus americanus,* and Mencken would praise it in language that made the link between the two authors explicit: "I know the Babbitt type, I believe, as well as most; for twenty years I have devoted myself to the exploration of its peculiarities. Lewis depicts it with complete and absolute fidelity." What is most revealing about this review is the nature of Mencken's admiration for Lewis. It would seem that he thought *Babbitt* to be less a novel than a species of journalism—and it's hard to escape the conclusion that he now thought that the highest form of praise.

FEW MEN PASS THE AGE OF FORTY UNTOUCHED BY DOUBT. FOR ALL HIS worldly success, Mencken was still unmarried, still a hypochondriac (and one, moreover, whose father had died suddenly at the age of forty-four), still cobbling books out of columns, and still publishing a magazine of whose limitations he was acutely aware, though he and Nathan continued to scare up good stories from time to time, among them Somerset Maugham's "Miss Thompson" (April 1921) and Fitzgerald's "The Diamond as Big as the Ritz" (June 1922). His disaffection can be tracked in the pages of *My Life as Author and Editor*: "The year 1921 was a busy one for me, but I accomplished very little of any lasting significance. . . . We turned up relatively little stuff in 1921 that really lifted us. . . . Editorially, 1922 was an average year."

The magazine mattered far more to Nathan, who was glad to assume an increased share of responsibility for its editing, thus allowing his restless partner to spend more time pursuing other journalistic interests. Between 1921 and 1923 Mencken covered the inauguration of President Harding and the Dempsey-Carpentier heavyweight fight for the *Sunpapers*, revised *In Defense of Women*, published two revisions of *The American Language* and a third volume of *Prejudices*, and paid a two-month-long visit to England and Germany, all in addition to writing his regular book reviews and Monday Articles. But he was still unsatisfied, for he wanted to remake the *Smart Set* and there was simply no way to do it: the money wasn't there. "I am sick of the Smart Set after nine years of it," he told a correspondent in 1923, "and eager to get rid of its title, history, advertising, bad paper, worse printing, etc." What he hated above all was its parochialism. He wanted to edit a journal of *ideas*, and that the *Smart Set* could never be. From the beginning of his joint tenure with Nathan, it had been the "magazine of cleverness," arch and sophisticated. To escape that hateful image he would have to build a magazine from the ground up—but who would supply the cash?

The answer, it turned out, was Alfred Knopf. Early in 1923 he broached the possibility of launching a magazine with Mencken as the editor; Mencken insisted that Nathan be part of the package, and by May negotiations had started in earnest. At first Nathan suggested the alternative possibility of buying the *Smart Set* outright, no doubt realizing that his partner was eager to move in a different editorial direction and hoping that this would serve as an anchor. But Knopf decided that it made no sense to pour more money into an already infirm magazine, and Mencken was quick to grab the reins of the new venture. "I might want to change the Table of Contents page completely," he wrote to Nathan. "I must be free in all such matters. But I agree freely to the essential thing: to print your name exactly as my own is printed."

Just whose magazine would it be? According to this letter it was taken for granted from the outset that Nathan would be coeditor of the new magazine, at least nominally (Mencken's name would come first on the cover, Nathan's on the masthead). Their plan was to run it the same way as the *Smart Set*, with Mencken remaining in Baltimore, Nathan running a small office in New York, and both men sharing veto power over the contents. Knopf agreed in writing that they would have complete editorial control, and they each received one-sixth of the issued stock and were to be paid five thousand dollars a year in salary (the first year's salary to consist of additional shares of stock). Most people would have said that all this was as it should be; in the eyes of the world they were joined at the hip. Back in 1920 the popular newspaper columnist Damon Runyon had referred in print to "Mr. Mencken, or Mr. Nathan, I never can tell one barrel from the other when they fire simultaneously in their brisk magazine, the Smart Set." Around the same time, Berton Braley, a newspaperman-poet, published "Three—Minus One," a parody of Eugene Field's "Wynken, Blynken, and Nod" that summed up their collaboration:

> And the two they talked of the aims of Art,
>   Which they alone understood;
> And they quite agreed from the very start
>   That nothing was any good

> Except some novels that Dreiser wrote
>> And some plays from Germany.
> When God objected—they rocked the boat
>> And dropped him into the sea,
> "For you have no critical facultee,"
>> Said Mencken
>> And Nathan
>> To God.

Mencken would later claim to have already had his doubts about Nathan: "I aimed all my professional shots, not at New York, but at the country in general and the world beyond, whereas he was pretty well confined to the local theatre-going audience of Broadway, and in fact has nothing to offer any other." Perhaps he really was as skeptical as that, but he wasn't saying so, at least not on paper. Nowhere in his letters of the period does he even hint that he and Nathan were anything but the closest of friends. But one thing was suggestive of the break to come: In most of the letters he wrote to spread the word about the magazine, *we* had become *I*. "I am preparing to assassinate the *Smart Set* . . . and to start a new magazine—something far grander and gaudier—in brief, a serious review, but with undertones of the atheistic and lascivious," he told Max Brödel. He began soliciting articles from a wider range of authors, including Dreiser, Edmund Wilson, James Weldon Johnson, and Carl Van Doren. To Upton Sinclair, who had figured in his first *Smart Set* review, he sent a letter laying out his priorities: "I shall try to cut a rather wide swathe . . . covering politics, economics, the exact sciences, etc., as well as belles lettres and the other fine arts. . . . In politics it will be, in the main, Tory, but *civilized* Tory. You know me well enough to know that there will be no quarter for the degraded cads who now run the country."

By September, the first manuscripts began to trickle in, and the outlines of the inaugural issue were taking shape. It was Nathan's idea to call it the *American Mercury*, and though Mencken disliked the title, he finally came around. "What we need is something that looks highly respectable outwardly," he wrote to Dreiser in September. "The *American Mercury* is almost perfect for that purpose. What will go on inside the tent is another story. You will recall that the late P. T. Barnum got away with burlesque shows by calling them moral lectures." A handsome-looking dummy issue was prepared by Knopf's designers, and Mencken, who had been grumbling for years about the *Smart Set*'s typography, found it "very simple and beautiful—quite the best thing ever seen in the Republic," though he wondered briefly whether the use of Garamond as a typeface might seem "a bit Frenchy."

Mencken and Nathan announced on October 1 that they were departing the *Smart Set* to launch the *American Mercury*. Ten days later they issued a formal statement, explaining that "their desires and interests now lead them beyond belles lettres and so outside the proper field of the *Smart Set*." The first page of the December issue bore this simple "editorial announcement": "The undersigned retire with this issue from the editorship of THE SMART SET." In the same issue the publishers announced that the *Smart Set* would "henceforth devote itself to fiction of much wider appeal than that which it has offered in the last decade. . . . Love, adventure, mystery, romance—the entire range of human dreams and emotions will be recaptured in the fiction form." (So much for Maugham, Fitzgerald, and Cather.) From that moment on it ceased to be of interest to posterity, though it soldiered on for another six years, publishing the homespun verses of Edgar A. Guest and the hellfire sermons of Billy Sunday. The circulation briefly jumped to six figures, an ironic testament to what Mencken and Nathan had wrought between them, but the Great Depression finally put paid to the "Aristocrat among Magazines," and it closed its doors for good in 1930.

Mencken's valedictory appeared in his last *Smart Set* book review, "Fifteen Years":

> Glancing back over the decade and a half, what strikes me most forcibly is the great change and improvement in the situation of

the American imaginatve writer—the novelist, poet, dramatist and writer of short stories. . . . Today, it seems to me, the American imaginative writer, whether he be novelist, poet or dramatist, is quite as free as he deserves to be. He is free to depict the life about him precisely as he sees it, and to interpret it in any manner he pleases. . . . The important thing is that almost complete freedom now prevails for the serious artist—that publishers stand ready to print him, that critics exist who are competent to recognize him and willing to do battle for him, and that there is a large public eager to read him.

The word "free" is the fulcrum of this passage. When the Associated Press asked him to supply a statement for use in his obituary, he expressed in it only one conviction: "I have believed all my life in free thought and free speech—up to and including the utmost limits of the endurable." As he looked back over his fifteen years at the *Smart Set*, in the course of which he never missed a single issue, his long crusade for freedom of literary expression loomed even larger than the part he had played in telling his countrymen about Conrad and Dreiser. Having declared victory over the puritans (a bit prematurely, as he would discover when the *Mercury* ran afoul of the censors), he retired from the field. At forty-four he had lost his taste for the medium about which he had been writing uninterruptedly since 1908. He claimed to have read some four thousand novels for the *Smart Set*, and that, he not unreasonably felt, was enough. He would continue reviewing books, but his field of interest shifted from fiction to nonfiction, and on the rare occasions when he did write about new novels, his criticisms would be mostly wrongheaded and sometimes dull. In any event he had said all he had to say about the place of the novel in contemporary American life. Rarely had he engaged in antiquarianism: He wrote mainly about the novels of his time, and in particular about what they told him of the world he lived in. The second-rate novelists of his day told him only that America itself had a second-rate culture, and he had made his definitive statement on

that subject in "The National Letters," the long opening essay of
Prejudices: Second Series, in which he declared that "[w]hat ails the
beautiful letters of the Republic, I repeat, is what ails the general culture of the Republic—the lack of a body of sophisticated and civilized public opinion, independent of plutocratic control and superior
to the infantile philosophies of the mob." Now he wanted to write
not about the literary manifestations of that ailment, but the ailment
itself.

This shift of focus is mirrored in the changing contents of the
Prejudices collections. Where *Prejudices: Second Series* had led off with
"The National Letters," *Prejudices: Third Series* opens with a similarly
extended essay called "On Being an American" in which Mencken,
in one of his most spectacular displays of verbal fireworks, hands
down a comprehensive indictment of American culture:

> Here the general average of intelligence, of knowledge, of com-
> petence, of integrity, of self-respect, of honor is so low that any
> man who knows his trade, does not fear ghosts, has read fifty
> good books, and practices the common decencies stands out as
> brilliantly as a wart on a bald head, and is thrown willy-nilly
> into a meager and exclusive aristocracy. And here, more than
> anywhere else that I know of or have heard of, the daily
> panorama of human existence, of private and communal folly—
> the unending procession of governmental extortions and chi-
> caneries, of commercial brigandages and throat-slittings, of
> theological buffooneries, of aesthetic ribaldries, of legal swin-
> dles and harlotries, of miscellaneous rogueries, villainies, imbe-
> cilities, grotesqueries, and extravagances—is so inordinately
> gross and preposterous, so perfectly brought up to the highest
> conceivable amperage, so steadily enriched with an almost fab-
> ulous daring and originality, that only the man who was born
> with a petrified diaphragm can fail to laugh himself to sleep
> every night, and to awake every morning with all the eager,

unflagging expectation of a Sunday-school superintendent touring the Paris peep-shows.

All of Mencken is in this passage. He had said such things many times before, but never before in such a way, briskly sweeping the reader along from one outrageous assertion to the next, his high emotion fully transformed into high comedy by the exaggeration, the startling use of incongruous words like "amperage," the long chains of blunt nouns that send the reader down a boiling torrent of gusto, only to slam headlong at the end into a burlesque metaphor. It is revealing, too, that this essay originated in a 1920 Monday Article, wherein his vision of America as vaudeville show was first set forth in larval form. He would not have published it in the *Smart Set*, in whose decorous pages his attacks had to be carried out by stealth, disguised as mere book reviewing. Here one sees him effecting the "escape from criticism" he had foretold in his 1919 letter to Louis Untermeyer. Indeed, "On Being an American" could have served as a prospectus for the *American Mercury*, in which Mencken would turn away from the world of art to jeer at every aspect of the common life of postwar America, secure in the knowledge that his subscribers would laugh with him.

His passing doubts about the future had now vanished. The prospect of writing about the American scene for a nationwide audience now stirred him so much that he barely had time to think about the magazine he was leaving behind, or anything else—not even Marion Bloom, whose marriage in August wrenched from him only one short letter in which he admitted his pain. Otherwise all his thoughts were of the magazine whose impending launch filled him with an excitement that was widely shared. F. Scott Fitzgerald wrote to a friend in December that "the real event of the year will, of course, be the appearance in January of *The American Mercury*."

One day that fall Mencken had lunch at Marconi's with a prospective contributor, a bright young man named James M. Cain, who had worked as a reporter for the *Sun* and was now trying to make a name for himself as a freelancer. The meal lasted for four hours, during which Mencken did most of the talking, holding forth on books, music, authors, and ideas. Cain went home with an assignment for the first issue of the *Mercury*. "I felt exactly like a boy who had had his baseball autographed by Babe Ruth," he remembered. That was how it felt to meet H. L. Mencken in 1923, on the eve of his emergence as an American legend.

"I AM MY OWN PARTY"

At the American Mercury, *1924–1928*

One day in the summer of 1924, Herbert Hoover asked Calvin Coolidge whether he had seen an article in the current issue of the *American Mercury*. "You mean that one in the magazine with the green cover?" said the president of the United States to his secretary of commerce. "I started to read it, but it was against me, so I didn't finish it."

Nowadays more people have heard of the *Mercury* than seen a copy. This must also have been true in Mencken's day, but it is hard to read about the United States in the mid-twenties without receiving the impression that except for Silent Cal, everyone in the country was either reading the *Mercury* or complaining about it. Within weeks of its launch it was being talked about not merely in Washington and New York but in places as far-flung as Huntington, West Virginia, where the local paper told of its fame: "Every flapper carries one of the green-backed journals conspicuously when it is necessary for her to travel in a train or other public conveyance, and if possible she lugs it along the streets with articles designed for more exterior cosmetic use. To be surprised carrying *The American Mercury* or caught reading it at odd moments is now absolutely The Thing." Long after it had faded from view, novelists seeking to evoke that long-lost age of flappers, jazz, and bathtub gin would make strategic mention of the sedate-looking little journal with the emerald-green cover. In *Point of*

No Return, John P. Marquand slipped it into a description of the home of Mrs. Smythe Leigh, director of the Clyde Players, the little-theater group of a small Massachusetts town circa 1929:

> Clyde was a dear, poky place, full of dear people, but one could always open one's windows to the world. One could bring some-thing new to Clyde, and this was what she always tried to do . . . a few reproductions of modern pictures, a bit of Chinese brocade, a few records of Kreisler and Caruso, and the *American Mercury* and the *New Republic* and of course *Harper's* and the *Atlantic*, and the *New Statesman* and *L'Illustration*. All one had to do was open one's windows to the outer world—and the surprising thing was the number of congenial spirits who gathered if you did it.

It was never the intention of Mencken and Nathan to gratify Mrs. Smythe Leigh and the congenial spirits of the Clyde Players, much less Calvin Coolidge and Herbert Hoover. In an undated flier they warned potential subscribers that "it is doubtful if any member of 'the civilized minority' can long afford to ignore" the *Mercury*. A magazine devoted to the proposition that most Americans were uncivilized would have been doomed to failure in 1918, but by 1924, with Coolidge in the White House and *Babbitt* in the bookstores, Mencken's broad-brush critique of American philistinism had found a sympathetic audience. Disillusioned by the war and its aftermath, American intellectuals, both real and *manqué*, decided that theirs was a land of barbarians. The notion was hardly original—Henry James had said so long ago—but instead of removing themselves perma-nently to Europe, there to write about the emigrant life, most of the new American writers got their passports stamped in Paris and Lon-don, then came home to chronicle the defects of life in these United States. In 1922 Harold Stearns published a book-length symposium called *Civilization in the United States: An Inquiry by Thirty Americans* in which he declared by way of introduction that there wasn't any, decrying America's "emotional and aesthetic starvation, of which the mania for petty regulation, the driving, regimentating, and drilling, the secret society and its grotesque regalia, the firm grasp on the unessentials of material organization of our pleasures and gaieties are all eloquent stigmata." It was an indictment worthy of Mencken, and sure enough, he was one of Stearns' thirty Americans, along with Nathan, the literary critic Van Wyck Brooks, and Ring Lardner, a short-story writer with a keen eye for American grotesquerie (and a Mencken-like ear for American argot). Forty years before Lionel Trilling gave it a name, the adversary culture was born.

How big was it? In 1915 Mencken speculated that the *Smart Set* had a potential audience of 150,000 readers; eight years later, he told Dreiser that it would be "wonderful" if 12,000 of them subscribed to the *Mercury*. It was Alfred Knopf who seems first to have grasped the fact that the new counterculture was becoming fashionable, and that a monthly magazine devoted to debunkery in the Mencken manner might be good business. (It would not have escaped his attention that *Main Street* had sold 390,000 copies by the end of 1922.) Even Knopf set his sights low at first: He cautiously ordered an initial press run of 5,000 copies, of which 3,033 went to charter subscribers, and set the price at fifty cents a copy, fifteen cents more than the *Smart Set*. But the *Mercury* started selling briskly as soon as it reached newsstands, and within days he was forced to order a second printing, then a third. "We have word this morning that the subscription department was 670 subscriptions behind—that is, behind in entering them up," Mencken wrote to Philip Goodman on December 28. "Knopf has bought 30 new yellow neckties and has taken a place in Westchester County to breed Assyrian wolfhounds." In the end, 15,000 copies of the January issue were sold.

Most of the people who bought it had probably also read the *Smart Set*, and they would have found much in the *Mercury* that was familiar, if not necessarily reassuring. Nathan was still covering the theater, while Mencken reviewed the latest books. "Répétition Générale" had been renamed "Clinical Notes," but it remained the same mixture of tart notebook entries and limp quasi-aphorisms ("The older I grow, the more I am persuaded that hedonism is the only sound and practical doctrine of faith for the intelligent man"). As for the *Smart Set*'s latest innovation, "Americana," it was transplanted to the *Mercury* unchanged. A collage of unintentionally funny press clips about life among the booboisie, organized by state and prefaced with Mencken's not-quite-deadpan introductions, it soon became one of the new magazine's most popular and widely quoted departments:

ALABAMA

FINAL TRIUMPH OF CALVINISM IN ALABAMA, OCTOBER 6, 1923:

Birmingham's exclusive clubs—and all other kinds—will be as blue hereafter as city and State laws can make them. Commissioner of Safety W.C. Bloe issued an order today that Sunday golf, billiards and *dominoes* be stopped, beginning tomorrow.

Nevertheless the *American Mercury* was not the *Smart Set*, as its editors took care to inform their readers. In an unsigned four-page editorial-prospectus, they spent the first two and a half pages talking about political and philosophical matters, explaining that the *Mercury* would be above all things anti-utopian: "The Editors are committed to nothing save this: to keep to common sense as fast as they can, to belabor sham as agreeably as possible, to give a civilized entertainment. . . . They will not cry up and offer for sale any sovereign balm, whether political, economic or aesthetic, for all the sorrows of the world." Only then did they get around to mentioning that "[i]n the field of the fine arts THE AMERICAN MERCURY will pursue the course that the Editors have followed for fifteen years past in another place . . . to welcome sound and honest work, whatever its form or lack of form, and to carry on steady artillery practise against every variety of artistic pedant and mountebank."

Having thus disposed of the arts, they concluded by announcing that the *American Mercury*, as befit its title, would devote itself primarily, if not exclusively, to life in America:

There are more political theories on tap in the Republic than anywhere else on earth, and more doctrines in aesthetics, and more religions, and more other schemes for regimenting, harrowing and saving human beings. . . . To explore this great complex of inspirations, to isolate the individual prophets from the herd and examine their proposals, to follow the ponderous revolutions of the mass mind—in brief, to attempt a realistic presentation of the whole gaudy, gorgeous American scene—this will be the principal enterprise of THE AMERICAN MERCURY.

The time was ripe for such an enterprise, and Mencken and Nathan had no shortage of targets from which to choose. In addition to two short stories, four short poems, and a critical essay on Stephen Crane, the first issue of the *Mercury* contained pieces about Abraham Lincoln, George Santayana, American historians and aesthetes, architecture in Manhattan, the English language, the naval disarmament treaty of 1922, communism, spiritualism, law enforcement, and the endocrine glands. Few of these pieces would have been thought suitable for the *Smart Set;* even Mencken's own book reviews now took a dramatically different tack. Instead of holding forth about the latest novels, he wrote about John Burgess's *Recent Changes in American Constitutional Theory,* Charles Conant Josey's *Race and National Solidarity,* and Nikolai Rimsky-Korsakov's *My Musical Life.* He had kept his promise: The *Mercury* cut a very wide swath indeed.

Plenty of other publications covered at least as wide a range of topics, among them Mrs. Smythe Leigh's beloved *Atlantic* and *Harper's,* not to mention the latest entry in the American magazine sweepstakes, a weekly "newsmagazine" called *Time* that was just ten months old when the *American Mercury* made its debut. Yet none of them was generating a fraction of the noise stirred up effortlessly by Mencken and Nathan. What was it about the first issue of the *Mercury* that made it The Thing? Three-quarters of a century after the fact, the answer is not immediately apparent, though it definitely wasn't the lead article, a stiffly written essay called "The Lincoln Legend" in which one Isaac R. Pennypacker suggested that there might have been slightly less to Abraham Lincoln than met the eye ("He was fortunate in that his armed opponents were poorly equipped for what they undertook"). But, then, little else in volume 1, number 1 was guaranteed to catch the casual eye. Of the twenty-one writers other than Mencken and Nathan whose bylines are listed in the table of contents, only one, Dreiser, is at all well known today, and his contribution was a folio of mediocre poems almost certainly purchased in order to get him into the first issue as cheaply as possible. James Branch Cabell, James Huneker, and Carl Van Doren were reasonably familiar names in 1924; Ernest Boyd, Isaac Goldberg, and Ruth Suckow were *Smart Set* standbys. For the rest there were four professors and a professor's daughter, three newspapermen, a forgotten poet, a chemist, an architect, an amateur Civil War historian, and two anonymous writers, and while their subject matter was varied, the tone most of them took was not. Look at Harry Elmer Barnes's "The Drool Method in History":

> The general tendency of the human race to stampede when confronted by the truth is nowhere more evident than in its reaction to history. . . . First, every orthodox American history book must start off with Gobineau's dogma of the superiority of the Aryans, the sole builders of civilization, and then show how all able-bodied and 100% Americans are members of the noblest of all the Aryan tribes: the Anglo-Saxon sub-division of the great Nordic Blond people.

Or Ernest Boyd's "Aesthete: Model 1924":

> When he went to Harvard—or was it Princeton or Yale?—in the early years of the Woodrovian epoch, he was just one of so many mute and inglorious Babbitts preparing to qualify as regular fellows.

Or L. M. Hussey's "The Pother About Glands":

Bernard Shaw, seeking to demonstrate that there must be a definite pathological reason why any man should diverge from the morals of a Scotch Presbyterian, attributed an hypertrophied pituitary to Oscar Wilde and asserted him to be a victim of giantism.

Anyone even casually familiar with Mencken's work would surely have suspected that he had a hand in all these pieces, and several others besides. He was a chronic rewriter, though he usually gave in when challenged by authors with backbones. Most, however, were happy—or at least willing—to go along with his blue-penciling, and one unidentified *Mercury* contributor would later confess in the pages of *Time* that "I saw my article appear in print colored with such words as privatdozent, geheimrat, bierbruder, and hasenpfeffer, which mystified my friends because I don't know German." Nor did this uniformity of tone go unnoticed in the press. "We do not expect Mr. Mencken to understand anything in America (except its language)," the *New Republic* editorialized in February, "but we did expect, though we are now prepared to admit we had no business to, that not all of the Mercury's complaint would be uttered in the same tone of voice."

The *Mercury's* single-voiced consistency was part of its appeal. Even though there were two names on the cover, it was Mencken, not Nathan, who pulled most of the readers into the tent. Like Charles Dickens's *Household Words* and G. K. Chesterton's *G.K.'s Weekly*, the *Mercury* was bought by people who wanted to hear what its contributor in chief had to say, and if the other pieces in each issue sounded like mock Mencken, so much the better. It is revealing that the *Mercury* was an instant success on college campuses. "The undergraduates must have their gods," declared the *Daily Iowan*, the student newspaper of the University of Iowa. "Without someone to follow, someone to direct their thoughts or rather someone whose thoughts they may borrow for their own, they would be lost. Right now H. L.

Mencken is on the pedestal before which they bow down.... [H]e has been adopted as the guiding outlaw of a host of students each with his little red flag and pocket full of pebbles."

It's no less revealing that most of those who criticized the *Mercury* aimed their fire straight at Mencken. One Tennessee editorial writer called him "a native born American with an alien and hostile slant apparent in his hatred and contempt for everything American.... Mencken's metier may be judged, we think, from the fact that he is a student of Nietzsche and a great admirer of Germany and her supermen who were brought low by following Nietzsche's philosophy." The *Knoxville Sentinel* came surprisingly close to the truth. Mencken was, indeed, a student of Nietzsche and an admirer of Germany, and though he did not hate "everything American," he held in fathomless contempt much of what the *Sentinel* stood for, including Christianity, democracy, and the gospel of service. (He would have dismissed as irrelevant the fact that he lived with and uncomplainingly supported his mother and two of his three siblings.) Upton Sinclair once asked him, apparently in all sincerity, why the *Mercury* contained no "constructive" criticism. "The uplift has damn nigh ruined the country," he snorted. "What we need is more sin."

In February the *Mercury* scored a coup, the first publication of Eugene O'Neill's play *All God's Chillun Got Wings*. James M. Cain, who would later write such scandalous thrillers as *The Postman Always Rings Twice* and *Double Indemnity*, cracked the *Mercury* that same month with "The Labor Leader" ("He is recruited from people of the sort that nice ladies call common . . . the sort that mop up the plate with bread"), inaugurating a series of "American portraits." Mencken editorialized about Prohibition and the judiciary and reviewed a dozen books, ranging from Marie Stopes's *Contraception: Theory, History and Practice* to Willa Cather's *A Lost Lady*. Nathan made short work of Shaw's *Saint Joan*, dismissing it as "an *affaire flambée*. . . . One looks for brilliant illumination and one finds but pretty, unsatisfying candle light." Boobs from coast to coast bashed themselves in "Americana," and the chorus of Mencken clones sang in unison on such topics as "The Ku-Kluxer" and "Heredity and the Uplift," expressing opinions all but indistinguishable from those of their editor, set forth in prose that aped his now-familiar mannerisms. That was Mencken's formula for success, and it was working like a charm: The circulation climbed to 26,287 in February and 36,634 in March, and Knopf hired ten new advertising and circulation clerks to handle the incoming mail. Everyone seemed pleased with the results, except for Bible Belt editorial writers who found the *Mercury* atheistic, liberal intellectuals who thought it too skeptical of the possibility of meaningful reform—and Nathan, who had come to the conclusion that his coeditor was hurtling down the wrong track.

They had already had their first quarrel, and it was a bad one. Nathan insisted that the *Mercury* publish *All God's Chillun Got Wings*, which Mencken loathed; when Mencken refused to back down, Nathan went over his head to Knopf, who ruled that the play should go into the next issue. In the short run it was a shrewd call, for O'Neill was already a major name in American drama, and the publication of his latest play would be hailed as one of the *Mercury*'s most astute decisions (and one for which Mencken long shared the credit). At the same time, though, it necessarily meant the end of the double-veto policy that had made it possible for two such strong-minded men to collaborate without strife at the *Smart Set*. Now all their editorial decisions would be subject to mutual negotiation, with Knopf, the publisher, acting as a court of last resort. The old arrangement had worked because the two men were mostly of one mind about the direction of the *Smart Set*. Not so the *Mercury*. After Mencken died, Nathan summed up their differences as follows: "The newspaperman in Mencken [had] superseded the literary man . . . a much more serious attitude infected him. His relative sobriety took the alarming form of a consuming editorial interest in politics and a dismissal of his previous interest in *belles-lettres*, which had been so great a factor in the propensity of our former periodical."

This was true, but it does not explain why his long friendship with

Nathan had started to unravel. *My Life as Author and Editor* breaks off
in mid-1923, and though Mencken wrote elsewhere about the *Mercury*'s early years, he offered no detailed account of his estrangement
from Nathan, only a brief explanation of why they finally parted ways:

> When the *American Mercury* at last came out (its first issue was
> dated January, 1924) I expected to move into smoother waters,
> but its extraordinary and unexpected success kept me on the
> jump. Worse, I was quickly confronted by the discovery that
> my colleague as editor, George Jean Nathan, was completely
> incompetent. . . . After the first few issues of the magazine I
> read all the manuscripts, carried on nearly all the negotiations
> with authors, and prepared manuscripts for the printer—this
> last often a laborious task. In addition I wrote editorials for each
> issue, a series of book reviews that sometimes ran beyond 5000
> words, and most of the briefer notices in the Check-List. The
> magazine was a striking success, and I was sustained by that
> fact, but the labor of getting it out was really formidable.

This bald statement raises more questions than it answers. It fails,
for instance, to explain the exact nature of Nathan's "incompetence";
instead, Mencken merely asserts that his co-editor "turned out to be far
too narrow in his range of ideas to be useful to a general magazine."
Presumably he meant that Nathan was uninterested in tracking down
the kinds of pieces Mencken now wanted to publish, or in doing the
rewriting needed to make many of them publishable. If so, it posed a
problem for Mencken, who was still putting in long hours at the *Sunpapers* in addition to his work on the *Mercury*. He continued to write
his Monday Articles; in June he covered the Republican and Democratic conventions in Cleveland and New York. At the same time he
was assembling a fourth volume of *Prejudices* and the first of two book-
length anthologies of "Americana" items, as well as making preliminary
notes for yet another book, this one a treatise on democracy and its discontents. Even for a man who thrived on hard labor, that was a crush-

ing load. Mencken would later remember 1924 as "the busiest [year] of my whole life." He needed all the help he could get, and if Nathan was unable or unwilling to oblige, he would get it somewhere else.

Was that really all there was to their quarrel? One looks instinctively for a more satisfying explanation of why Mencken should have turned on his closest colleague. Could he have been mortally offended when Nathan broke their long-standing agreement and appealed directly to Knopf to publish *All God's Chillun Got Wings*? "A magazine is a despotism or it is nothing," Mencken wrote in his diary in 1939. "One man and one man only must be responsible for all its essential contents." That was not how the *Smart Set* had worked, nor was it the way the *Mercury* was supposed to work. Perhaps he had simply grown weary of sharing power—and credit—with another man. We can't know for sure, but we do know that sometime in the summer of 1924, Mencken came to the conclusion that Nathan was not pulling his weight as coeditor of the *American Mercury*, and decided to force him out.

In October, Nathan received a letter from his old friend putting things in the plainest possible language:

> After a year's hard experience and due prayer, I come to the conclusion that the scheme of The American Mercury, as it stands is full of defects, and that to me, at least, it must eventually grow impossible. . . .
>
> Therefore, I propose the following alternatives:
>
> 1. I will, as of January 1st next, take over complete control of the editorial department, put in a managing editor to run the magazine and operate the magazine as Sedgwick operates the *Atlantic*; or
>
> 2. I will retire from all editorial duties and responsibilities, and go upon the same footing that other contributors are on.
>
> My inclination at the moment is to choose no. 2. I can see nothing ahead, under the present scheme, save excessive and uninteresting drudgery, and a magazine growing progressively feebler.

Four days later he followed up this bombshell with another letter, gentler but no less firm, in which he tried to make Nathan see things his way: "Our interests are too far apart. We see the world in wholly different colors. When we agree, it is mainly on trivialities. . . . I see no chance of coming close together. On the contrary, I believe that we are drifting further and further apart. I cite an obvious proof of it: we no longer play together. Another: when we sit down to discuss the magazine itself we are off it in ten minutes."

Nathan remained unmollified. He complained in an undated reply of the "unreasonable and often insulting attitude you have displayed toward me." But he knew he was holding a losing hand—Knopf was backing Mencken—and so he folded it, though not before putting up what Mencken called a "violent and long-continued resistance" that made his removal "very difficult and unpleasant." Part of what made it so difficult was that in addition to being the *Mercury*'s founding coeditor and second most frequent contributor, he was also one of the magazine's stockholders. But Mencken usually got what he wanted, and he wanted to be rid of Nathan. By year's end all that remained was to negotiate the terms. "What will you take, in cash, for your interest in the magazine as it stands, including all expectations of salary (if you have any), and every sort of right and title?" he wrote to Nathan on Christmas Day. "I suggest including absolutely everything, save only the matter of contributions to the magazine."

Within a few more days the knot was untied. Nathan resigned his coeditorship, leaving Mencken in sole charge of the *Mercury*. He agreed in May to sell his interest in the magazine to Knopf, and two months later it was announced that he would henceforth be the magazine's "contributing editor." At that time Mencken stopped collaborating with him on "Clinical Notes," which was moved to a spot adjacent to the theater column, thus providing Nathan with his own corner of the *Mercury*, which he occupied for another five years. Though the two men were still yoked in the public eye, their private relationship cooled off, then froze up, a development that Mencken blamed on their angry negotiations over Nathan's financial stake in the magazine. "In friendship there must be no thought of money," he told Isaac Goldberg not long after the break. "An hour after Nathan and I came to a situation in which our financial interests were in conflict, the association of 17 years was at an end."

Of that Mencken made sure there could be no possible doubt. As Knopf remembered it Nathan's name "was removed from the directory in the lobby" and his desk relocated to "an outer office where the stenographers worked." The gala days were over. Nathan tried not to show how badly he had been wounded, but everyone privy to the facts seems to have understood it. "It seems perfectly clear to me from the record," Knopf wrote, "that the one thing Mencken wanted was a complete divorce from Nathan, and it was just as obvious to me—though there was no written record to support this—that the one thing Nathan didn't want was that kind of a divorce. After all, the theater and [Mencken] were, I think, the two great experiences of his life."

Knopf's choice of words is revealing, perhaps unintentionally so. What happened between Mencken and Nathan is strongly reminiscent of the collapse and dissolution of a marriage, just as their first meeting in 1908 had resembled love at first sight. Neither man would have appreciated the comparison (Mencken's unpublished memoirs contain ample evidence of his dislike of homosexuals, as do Nathan's published reviews), but their friendship had once been genuinely close, even intimate, and it is not pleasant to imagine how Nathan would have felt had he read the unpublished memorandum of 1939 in which Mencken set down his thoughts on the rupture:

> One of my convenient habits has been that of dropping completely people who begin to bore me. I think I inherited it from my mother. When anyone for any reason offends me I simply purge them from my list of acquaintances and give them no more thought. This has saved me a great deal of nuisance, and it has also made me some bitter enemies. . . . When I began to realize how badly it had hurt [Nathan] to be dropped from The American Mercury I ceased to see him and have, in fact, not

laid eyes on him for at least five years. I receive plenty of evidence that this fact annoys him excessively. Unquestionably, it involves a certain amount of unpleasantness, but it seems to me that [the] unpleasantness of seeing him would be worse. We exchange notes now and then and are officially on good terms, but he knows very well that I am no longer interested in him.

Little about the *Mercury* would change after Nathan's demotion. Its course was already set, and Mencken was well pleased with what he had wrought—so pleased that it never occurred to him to wonder whether Nathan might have been right. Dreiser thought so. "I always felt," he told Nathan in 1933, "that opposed to [Mencken's] solid philosophic, economic, sociologic and historic interests, your lighter touch was important and, for general literary as well as magazine purposes, made for a more charming and, to my way of thinking, almost equally valuable publication, the old *Smart Set*." These words should be taken with a shakerful of salt, for Dreiser, too, had by then been dropped by Mencken, and was collaborating with Nathan on a new magazine, the *American Spectator*, intended to serve as a sprightlier alternative to the now-moribund *Mercury*. But he had been right in 1915 when he complained about the innocuousness of the *Smart Set*, and he was no less right to complain about the *Mercury*'s heavy-handedness.

Whether Nathan could have helped Mencken stay in touch with the new artistic developments that were sweeping Europe and America in the mid-twenties is another matter. Neither man was much interested in artistic experimentation, and 1924 was a year of violent ferment in the arts. It was the year of *The Magic Mountain* and *A Passage to India*, of Noël Coward's *The Vortex* and Buster Keaton's *Sherlock, Jr.*, of Arnold Schoenberg's psychological opera *Erwartung* and Bronislava Nijinska's bisexual ballet *Les Biches*. Bix Beiderbecke and Sidney Bechet made their first recordings in 1924, and George Gershwin's *Rhapsody in Blue* was premiered at New York's Aeolian Hall. Under Mencken the *Mercury* would grapple with some of these things—it was the first general-interest magazine to cover jazz in anything like a serious way—but modernism mainly went unremarked in its pages. When Mencken felt the need to tell his readers about modern painting, he turned to Rollin Kirby, editorial cartoonist of the *New York World*, whose "Cézanne and His Crowd" was as much a manifestation of American philistinism as anything in "Americana": "I get a sense of a clumsy, frustrated man with a deliberate theory concerning planes and geometric forms attempting to fit a picture into some sort of mosaic of pyramids and arbitrary shapes. It is doubtless ingenious, but nothing more."* Nor was the editor of the *Mercury* much more knowledgeable about such matters, as he proved when he reviewed the autobiography of Isadora Duncan, the inventor of modern dance:

Isadora simply loved to prance around in a shift; all the rest was afterthought. . . . It gave the world, and especially the world of artists, pleasant shock to see the shift waving and billowing to the tunes of Chopin and Tschaikovsky; there was another shock later on when it began to flap to the tune of Wagner and Brahms. It was an era of painfully correct ballet-dancing, and to worn-out tin-pan music. Here, at least, was something new—and straightway it came converted into something portentous. But its meaning, at bottom, was exactly that of any other dancing, which is to say it had scarcely any meaning at all.

That one review sums up much of what was wrong with the *Mercury*. Mencken was more interested in boob-baiting than in covering

---

* It is amusing to read this piece side by side with "The Asses' Carnival," a Monday Article published on January 31, 1921, in which Mencken attacked the members of the U.S. House of Representatives for their lack of "general culture": "What they know of the arts and sciences is absolutely nil. Find me 40 men among the 435 in the House who have ever heard of Beethoven, or who know the difference between Cézanne and Bouguereau, or who could give an intelligible account of the Crimean War, and I'll give you a framed photograph of the Hon. Josephus Daniels."

the remarkable range of cultural developments taking place in America, and when he did take note of homegrown American artists, he like as not wrote about them as if they were merely another species of boob. Middle-class Americans, after all, were doing a great deal more in the twenties than going to church or making bathtub gin. For one thing, they were reading the *American Mercury*. Yet Mencken, like the highbrows of a later generation, was too quick to denigrate the significance of the emerging middlebrow culture of which his own magazine was a manifestation, reluctant though he might be to admit as much. He would have sneered at the news that his fellow countrymen ranked Dickens, Longfellow, Shakespeare, and Tennyson among the "ten greatest men" in history. In 1924 Sinclair Lewis wrote a piece for the *Nation* called "Main Street's Been Paved" in which he returned to his Minnesota hometown to find a pair of working-class girls wearing "well-cut skirts, silk stockings, such shoes as can be bought nowhere in Europe [at the price], quiet blouses, bobbed hair, charming straw hats, and easily cynical expressions terrifying to the awkward man. . . . Both their dads are Bohemian; old mossbacks, tough old birds with whiskers that can't sling more english than a musk-rat. And yet, in one generation, here's their kids—real queens." That was a side of the United States in the twenties the *Mercury* preferred not to cover.

Having seen comparatively little of his native land—most of his domestic travels had been to and from presidential conventions—Mencken may not have been the most qualified of observers of the American scene, but he had certainly spent more than enough time in New York City to have had some awareness of what was going on there in 1924. The problem was his own lack of curiosity. It is impossible to imagine him dropping by Carnegie Hall for the premiere of Aaron Copland's Organ Symphony, strolling into Alfred Stieglitz's exhibition of the latest watercolors by John Marin, or even paying a visit to the Casino Theatre to see the Marx Brothers in *I'll Say She Is*. While it is too easy simply to say that he was as much of a philistine as the philistines whose ignorance he loved to denounce, it is not altogether untrue. A rejection letter he sent to Ernest Boyd reflects this: "Somehow, I have a feeling that this is a bit too refined and philosophical for the Merkur. It is excellent stuff, but it is not rowdy enough. Nobody is insulted."

It didn't help that the *Mercury*, like the *Smart Set* before it, paid its contributors only two cents a word, a low rate even by the more modest standards of the day. (After Nathan left, Mencken never spent more than $2,500 a month on payments to authors.) To be sure, there were compensations of another sort. One admiring contributor described Mencken as "a new kind of editor . . . one who gives immediate attention to your manuscript, pays spot cash, encloses return stamped envelope with the proofs and gives you second serial rights without asking." Another observed that Mencken and Nathan were "remarkable for the definiteness and promptness of their decisions." Add to this the cachet of appearing in the most talked-about magazine of the moment, and it becomes easier to see why so many people were willing to write for a publication that paid so little. As an editor Mencken may have been a butcher, but he treated his carcasses with kindness. Even his letters of rejection were charming rather than demoralizing: "I append my usual hopes that other ideas, of a swell character, are engaging you."

But it takes more than charm to fill the pages of a monthly magazine. Mencken did not pay well enough to consistently attract established talent, and most of the second-tier journalists he could afford were only too willing to let him run roughshod over their copy. To look through the first five years of the *Mercury* is to reap a bumper crop of forgettable pieces by forgotten writers, many of them bearing such shamelessly Menckenesque titles as "Satan in the Dance-Hall," "Babbitt Emeritus," "The Ordeal of Prohibition," "The Christian Statesman," "The Wowsers Tackle the Movies," and "The Uplift Hits the Army." Forced to search for new faces, he occasionally hit pay dirt, as with James Cain, and necessity also inspired him to look in places where other editors feared to tread. He reveled, for instance, in printing the work of black writers. (George Schuyler of

the *Pittsburgh Courier* actually became a *Mercury* regular.) In addition, several authors of importance wrote for him on occasion,
among them Dreiser, Fitzgerald, Sherwood Anderson, Max Eastman, Joseph Wood Krutch, Langston Hughes, Dorothy Parker, and
Carl Van Vechten, and sometimes they sent him their best. Outside
of Fitzgerald's "Absolution" and W. J. Cash's "The Mind of the
South," though, it is hard to find anything published in the *Mercury*
between 1924 and 1929 that has proved to be of permanent interest,
save for the steady stream of essays, articles, and reviews contributed
each month by Mencken himself.

To understand what the *American Mercury* lacked, it helps to look at
another magazine launched a year later. At first glance the *New Yorker*
suggested a cross between the *Mercury* and the *Smart Set*. It was written for educated urbanites, not "the old lady in Dubuque" so famously
written off in Harold Ross's 1924 prospectus. It touched on some of
the *Mercury*'s favorite subjects (one of its most widely discussed articles
was a report on the Scopes trial), though devoting far more space to
fiction and the arts. Like the *Mercury*, it was run on the cheap—Janet
Flanner got forty dollars for her first "Letter from Paris"—meaning
that Ross had to make his own stars instead of buying them off the
rack. But unlike Mencken, he let them develop their own voices. Over
time, he built up an extraordinarily diverse stable of regular contributors (including Mencken himself), and while their work was edited
heavily, even excessively, their individual styles remained intact. "Try
to preserve an author's style if he is an author and has a style," Wolcott
Gibbs wrote in a 1937 manual called "Theory and Practice of Editing
New Yorker Articles." It was a distinction Mencken chose not to make.
One can open issues of the *Mercury* almost at random and find pieces
that are smeared with his inky fingerprints:

> Observe closely the common middle-class American who has
> neither means nor any special talent, and you will find his life
> dominated by two desires: for security and for gentility. He
> yearns to be sure of bed, board and clothing, and he yearns to

work sitting down, to wear a white collar, to feel superior to the men who sweat. . . . I presume that country grand juries are about the same in all our Imperial States. I have no reason to doubt that the average one in New York or California is composed of about the same grade of morons that engage in snooping into their neighbor's affairs in Texas. . . . Censorship is the Cuckoo Klux Klan of Art. All censors are curious birds, but the motion picture censor is the choicest of them all.

It speaks poorly for Mencken that he was willing to offer the public such would-be imitations of his own tangy prose. Yet he also made sure to give them the real thing, and his editorials and book reviews were enough to keep them coming back for more. By 1926 nearly eighty thousand people were buying the *Mercury* each month. H. L. Mencken had gotten what he wanted: he was putting on a one-man show, and it was a hit.

EDITORS WITHOUT MONEY ARE ALWAYS ON THE LOOKOUT FOR NEW writers, but it seems unlikely that Mencken expected to run across any when he agreed to speak to Harry T. Baker's English class at Goucher College on May 8, 1923. Afterward, though, he went out to dinner with Baker and two Alabama maidens with literary aspirations. Sara Mayfield was the current winner of Goucher's annual Freshman Short Story Contest, while twenty-four-year-old Sara Haardt, the Baltimore college's youngest instructor and Miss Mayfield's chaperone for the evening, had actually sold an eight-line poem to the *Smart Set* in 1919. Mencken didn't remember it, probably because there was nothing memorable about it ("Love is a wild thing—/It frightens white flowers,/It warms cold hearts,/Red roses quiver near its fierceness"), but he invited Miss Haardt to send him some of her stories, and two weeks later she and Miss Mayfield were lunching with him. This time it was the student who played chaperone, for Mencken had taken a shine to her teacher.

Unlike Marion Bloom, Sara Powell Haardt came from respectable circumstances. Born in Montgomery in 1898, she grew up playing on the steps of the state capitol with Zelda Sayre—F. Scott Fitzgerald's Zelda—and Tallulah Bankhead. They were hellions, but she was a charmer. "Sara Haardt's as suave as silk," Sara Mayfield's father said approvingly. "She never crosses anybody." She went to Goucher in 1916 meaning to major in history, but her teachers encouraged her to try her hand at writing, and soon she was publishing in the student magazines and collecting *Smart Set* rejection slips. She knew Mencken from the *Sunpapers*, and when she met Fitzgerald in Montgomery in the summer of 1918, he took time out from courting Zelda to inform her that Mencken and Nathan were "the greatest living writers in America." Inspired by his words, she finally made it into the *Smart Set* a year and a half later (six months after Fitzgerald), and the following June she was graduated from Goucher, whose yearbook described her as a "soulful highbrow."

For a time Mencken was close to Scott and Zelda, about whose escapades he wrote in *My Life as Author and Editor* with a mixture of affection and dismay, but he would not meet Sara Haardt until after she returned to Goucher in 1922, this time as an instructor of English. By then her work was appearing in the *Reviewer*, a much-admired little magazine published in Virginia. Mencken had taken an interest in the *Reviewer*—he was friendly with Emily Clark, one of its founding editors—so he was similarly inclined to take an interest in Sara when their paths crossed in 1923, around the time that Marion was finally making up her mind to marry Lou Maritzer. He wrote to her in Montgomery that summer, and both sides of their correspondence survive. His letters are encouraging, hers admiring:

"Before I get stuck in Alabama for the rest of the summer I want you to get me straight on this short-story business. . . . [T]he old notions I had are squirming like a bucket full of bait. How does one do it? I mean, see you." (Sara Haardt to Mencken, May 20, 1923)

"I shall be at Domenique's on Thursday at twelve-thirty unless I hear that that hour is not suited to your schedule. These have been dog days at college and I am looking forward to the meeting as the only inspiring thing that could happen." (Sara to Mencken, May 26)

"Let us have another palaver when you get back. Meanwhile, let me see whatever you do. I think you have a good novel in your head." (Mencken to Sara, June 5)

"Let me see the novel, by all means. I have paved the way with Knopf. I haven't the slightest doubt that you can do a very good job. If you insert any plea for Service into it, however cunningly disguised, you will hear some colossal swearing." (Mencken to Sara, July 26)

"I am working day and night on the new review. We hope to get out the first number in December. Knopf is to be the publisher, and it will be a very slightly, and I hope, dignified monthly. . . . Why not send me some ideas for it? Anything that interests you: the South, the American university, the Anglo-Saxon, anything." (Mencken to Sara, August 17)

"The review I am simply wild about. If Knopf is to publish it it will be a knock-out. I still propose to go up in the air at the issuance of the first copy. . . . As for the university—that needs primarily to be scalded out from the inside. My pet notion is that it is a small town on wheels but minus even the usual small-town thrills. . . . I am returning to Baltimore about the twentieth of September. I have a grand excuse to see you, too. I am saving what there is of the novel to mail to you after I arrive." (Sara to Mencken, August 22)

"Since I am looking forward so very much to seeing you I fully expect to hear that you have been buried in New York for the rest of my natural life or else struck dumb with hay-fever." (Sara to Mencken, September 13)

Once Sara got back to town, she and Mencken lunched together once or twice a week, usually at Marconi's and always with Sara Mayfield in attendance. She was dumped that fall by a suitor from Montgomery, and while she said nothing about it to Mencken, he sensed that something was amiss. Busy though he was with the impending launch of the *Mercury*, he continued to make time to see her in Baltimore and write to her from New York. Letter by letter their friendship grew warmer, but it remained decorous, for Mencken was aware that he was dealing with a lady, albeit one with a taste for irony and a good head on her shoulders. He invariably opened his letters with "Dear Miss Haardt," save on one occasion when he addressed her as "Wohledle, Hochehr- und Tugenbelobte und auch Hochgelehrte Fräulein," German for "Very Noble, Highly Honorable and Virtuous and Highly Learned Miss." He would never have said anything like that to Marion.

In December, Sara was laid low with bronchitis, and Mencken, who had a connoisseur's appreciation of disease, kept close tabs on her. He referred her to Max Kahn, a Baltimore radiologist and fellow member of the Saturday Night Club, and in that same letter, written on January 20, he addressed her for the first time as "Dear Sara," closing with "I kiss your hand." Kahn diagnosed tuberculosis and slapped her into a local sanitorium, and Mencken started seeing her there as soon as he was permitted to do so, sending letters between visits in which he made clear that his interest in her was no longer avuncular: "I refuse to believe that you are ill. Today you looked superb. I suspect that I am mashed on you; nevertheless, my eyes are still reliable."

Was Mencken in love again? If so, it was with a woman who in many ways could not have been more different from her predecessor. Sara was already a published writer, if only a fledgling. Though she had no money, she came from good German stock, and there is no reason to think that she was sexually active in 1924, much less promiscuous. Very likely she was an honest-to-goodness virgin, and now that she was cooped up in Maple Heights Sanitorium, it was improbable that she would soon embark on a career as a temptress. All these things would have been attractive to a man of Victorian tastes, especially one who had spent the past decade involved with a thoroughly modern woman like Marion. But Sara was neither a milksop nor a prude. "For a woman she had a good sense of humor," August Mencken would recall in tones of bemusement. "Normally women don't, you know. . . . She also enjoyed sometimes a little of what you might call a lower kind of humor—more robust, you might say." She may have played the part of the admiring acolyte in her letters to August's brother, but his Virtuous and Highly Learned Miss was in fact both smart and witty—and brave. Then and later she bore her sufferings with fortitude, and nothing could have impressed him more.

Sara stayed at the sanitorium for most of the rest of 1924, writing articles and stories and working on her novel as often as the doctors allowed it. Mencken continued to see her there, and their friendship grew closer. His letters reveal that he even confided in her about his troubles with Nathan. "The Mercury boil, I fear, is about to bust," he wrote on October 19, the same day he sent Nathan the second of the two fateful letters that put an end to their partnership. "A palaver is called for Wednesday, and there will be a great row." He kept Sara abreast of the gory details, both in person and in his letters. In December she went back to Alabama to recuperate, and he kept on writing to her all through 1925. To call their correspondence an epistolary romance would be a forgivable exaggeration. For all his jaunty reassurances ("You will live to be the oldest woman in Alabama!"), he must have taken it for granted that she would never be well again, at least not enough to consider marrying. Nor is there any indication that he had considered proposing to her; he saw other women, and even wrote to some of them. Still, this one meant something special to him, and in his arch and careful way he told her so: "Friend Sary, I miss you like hell. . . . If you were at hand I should probably risk your yells by trying to neck you. I have been practicing on a fat woman. When we meet you will see some technique. So beware again! Ich kuss die Hand!"

An epistolary romance might well have been the best he could do in 1925. His appetite for work had never been greater. Apparently finding Monday Articles and *Mercury* editorials insufficient to keep his typewriter hot, he had agreed the preceding October to write another weekly column, this one published in the *Chicago Tribune* and syndicated throughout the United States. "As H. L. M. Sees It" began that November 9 and ran each week for a bit more than three years.

"First and last, many papers subscribed for it," Mencken recalled in *Thirty-five Years of Newspaper Work*, "but most of them dropped out the first time I took a hack at the Methodists, the American Legion, or any of the other inhabitants of my menagerie of hobgoblins." Even so it added to his national audience, and also gave him more material on which to draw for the fifth and sixth volumes of *Prejudices*, the fourth one having appeared just as the column was getting underway. (The revealing title of the *New York Times* review of *Prejudices: Fourth Series* was " 'Tough Guy' of American Criticism.")

In the meantime he had found a more pliant replacement for Nathan, a twenty-three-year-old Harvard graduate turned newspaperman named Charles Angoff, who became the *Mercury*'s assistant editor and office manager. Unlike Nathan, Angoff was permitted only to reject manuscripts, not accept them. He culled the slush pile in New York and sent his picks to Baltimore for Mencken to read. He also copyedited certain articles and proofread the magazine, and proved to be "a diligent and reasonably competent young man." So Mencken spoke of him in his sealed memoirs; to others, including Sara, he habitually referred to Angoff as "my slave." A longtime reader of the *Smart Set*, Angoff was thrilled to be slaving for H. L. Mencken, but his limited experience had not prepared him to deal with a famous writer who was also a hard-shelled newspaperman. From the day of their first meeting in Mencken's room at the Algonquin Hotel, the editor of the *American Mercury* went out of his way to make his new assistant squirm, and Angoff always obliged. In addition to being diligent and competent, he was humorless, which would have made him the perfect straight man except for one thing: Angoff took notes.

[Mencken] got up, opened the door of the bathroom, and urinated without bothering to apologize. My first reaction was one of shock, then of pleasant astonishment. I could not imagine President Lowell of Harvard urinating in front of anyone. . . . Mencken's act somehow made him an even greater man in my eyes, but I was still a little doubtful about the propriety of his public performance. I noticed that he did not wash his hands, and he must have seen the surprise on my face.

"I never wash my hands after taking a leak," he said. "That's the cleanest part of me."

For the next nine years Mencken told Angoff dirty jokes, mocked his refined tastes in literature, used every ethnic slur he could think of, and generally carried on as though he were a police reporter covering a raid at a bordello. Among other things, he ribbed Angoff ceaselessly about the supposed inadequacies of Harvard (raising the far from remote possibility that having an Ivy Leaguer as his assistant may have caused some awkwardness on his part). But unlike most Harvard men of the twenties, Angoff was Jewish, and found his employer's attitudes toward Jews puzzling. On the one hand Mencken published Jewish writers in the *Mercury* regularly, socialized with them constantly, and gave every indication of being fascinated by them. "I can't understand how anybody can be an anti-Semite," he once told Angoff. "I have never been to a Jewish home that didn't serve good grub, and I have never known a Jew who was a Prohibitionist." On the other hand Angoff added, "he sometimes wrote about [Jews] as though some of them at least had personally harmed or offended him." Most of all he was *aware* of them. In writing his diary and memoirs, it was his reflexive custom to remark on the Jewishness of any Jew—sometimes matter-of-factly, sometimes invidiously, but always. Whenever Alfred Knopf did something he found irritating, Angoff claimed, Mencken would snap, "Goddamn it! He burns me up. But, then, what can you expect from a pants-presser's son?" Sometimes this was part of the great game of Angoff-baiting in which he took such delight; at other times it seemed that Mencken was simply giving Jews the same backhanded treatment he gave any ethnic or religious group. It wasn't until Hitler came to power that Angoff began to wonder if Mencken's peculiarly mixed feelings about the Jews might amount to something more than a quirk.

Isaac Goldberg, a *Smart Set* contributor who made the leap to the *Mercury*, entertained no such doubts, and decided in 1925 to write a biography of his hero. (He might have changed his mind if he had known that Mencken would describe him in *My Life as Author and Editor* as "a Jew . . . of the better sort.") Another book about Mencken, a critical monograph by Ernest Boyd, was already in the works, but Goldberg had grander intentions. "G. plans an elaborate work, unlike anything ever heard of," Mencken told Sara in March. "It will take me several weeks to exhume the necessary early stuff. . . . Inasmuch as he knows no more about me than Coolidge, I must dig up the materials myself." In the end he was forced to dictate what ultimately became a two-hundred-page autobiographical typescript based on his private papers. Goldberg drew liberally on it to write *The Man Mencken*, in which he credited his subject with "a piercing vision that is in itself a form of flight,—often an Icarian flight, often a delusive vision, yet unmistakably, if deviatingly, sunward." Mencken spoke well of the book to Goldberg. To others he was more honest: "God help you if you try to read it. It is a shameless thing." Still, he knew it was no small thing, shameless or otherwise, to be the subject of two books in a single year, and he was pleased by the resulting attention, even though it scarcely equaled the furor that would be stirred up by his brief involvement in a trial held that summer in a tiny Tennessee town called Dayton.

On March 21 the Tennessee senate passed a bill making it a misdemeanor for any public school teacher in the state to teach "any theory that denies the story of the Divine Creation of man as taught in the Bible, and to teach instead that man had descended from a lower order of animal," punishable by a maximum fine of five hundred dollars. Six weeks later, John T. Scopes, a science teacher and part-time football coach at the Rhea County High School in Dayton, was served with a warrant that charged him with violating the Anti-Evolution Law. Similar statutes were being debated in the legislatures of other states, but Dayton was the first venue where anyone had been prosecuted for breaking one of them. Clarence Darrow, the best-known defense lawyer in the country, represented Scopes at the trial; William Jennings Bryan, a two-term congressman from Nebraska and three-time Democratic presidential candidate who served as Woodrow Wilson's first secretary of state, appeared for the prosecution. The jury found Scopes guilty as charged, and Judge John T. Raulston fined him one hundred dollars. But history is not always written by the victors, and the losing side in *Tennessee* v. *Scopes* wrote the version of the case that made it into most of the history books—with more than a little help from H. L. Mencken.

Comparatively few Americans had even heard of Charles Darwin until William Jennings Bryan took an interest in him. To be sure, the theory of evolution was not uncontroversial, but once the warfare over *The Origin of Species* died down, many religious leaders seemed open to the possibility that there might be something to it. One contributor to *The Fundamentals*, the series of conservative Biblical commentaries after which Protestant fundamentalism was named, went so far as to declare that evolution was "coming to be recognized as but a new name for 'creation.'" Then Bryan got into the act, and all hell broke loose. He was called the Great Commoner, and to a generation of banker-hating farmers and working men who bought phonograph records of the stem-winding campaign speeches he delivered in a penny-plain midwestern accent, he was a symbol of hope. In politics he was a big-government, antibusiness progressive, in religion a fire-breathing traditionalist—a combination less exotic in 1925 than it is today, and almost the exact inverse of Mencken's combination of antiutopian libertarianism and religious skepticism. Passionately opposed to American involvement in World War I, he resigned from the Wilson administration in 1915, making him one of the very few cabinet members in American history to quit over a matter of principle. That put an end to his political career, but not to his public life. He set up shop as an itinerant orator, giving speeches throughout the hinterlands in which he advocated the passage of the

Eighteenth Amendment and opposed the teaching of evolution in the public schools, and his words fell on receptive ears. More Americans were going to school in the twenties than ever before, especially in the South (the number of Tennesseans attending high school jumped from fewer than ten thousand in 1910 to more than fifty thousand in 1925), and they were being taught out of state-approved textbooks that took evolution for granted. By alerting their unsuspecting parents to this fact, Bryan galvanized the newborn fundamentalist movement, helping to turn it into a major force in American culture even as Mencken was mustering an opposing army that sought to smash everything Bryan's troops held dear.

Bryan may not have known much about biology, but he instinctively grasped that Darwinism was more than just a scientific theory. It was also an ideology, an intellectual package deal that went well beyond anything Darwin envisioned. Social Darwinism, the sink-or-swim philosophy to which Mencken avidly adhered, was part of the package, as was eugenics. George William Hunter's *A Civic Biology*, the state-approved textbook that John Scopes used as a substitute biology teacher, made short work of epileptics, the mentally ill, and other unwanted types: "If such people were lower animals, we would probably kill them off to prevent them from spreading. Humanity will not allow this, but we do have the remedy of . . . preventing intermarriage and the possibility of perpetuating such a low and degenerate race." Such talk was common among American intellectuals in the twenties. Two years after the Scopes trial, Justice Holmes, writing for the Supreme Court in *Buck* v. *Bell*, would unhesitatingly sanction the court-ordered sterilization of a "feeble minded white woman," explaining that "society can prevent those who are manifestly unfit from continuing their kind. . . . Three generations of imbeciles are enough."

Mencken's essays are similarly peppered with metaphors lifted straight from the eugenicists. Even some of his favored terms of abuse (such as "moron") were part of the newly coined vocabulary of the intelligence tests being used to screen out the genetically unfit, and though he gave them a comic spin, he wasn't kidding: "I repeat that it eases and soothes me to see [chiropractors] so prosperous, for they counteract the vile work of the so-called science of public hygiene, which now seeks to make imbeciles immortal. . . . Every time a bottle of cancer oil goes through the mails *Homo americanus* is improved to that extent. And every time a chiropractor spits on his hands and proceeds to treat a gastric ulcer by stretching the backbone the same high end is achieved."*

The Great Commoner was quick to catch the scent of class warfare in the rhetoric of social Darwinism. As early as 1904 Bryan argued that the theory of evolution "represents man as reaching his present perfection by the operation of the law of hate—the merciless law by which the strong crowd out and kill off the weak." A militant majoritarian, he had no intention of letting such a thing happen on his watch, and starting in 1921, he hit the Chautauqua circuit with a series of speeches that caused tumult in the Bible Belt. Though his goals were modest, even nuanced—he sought only to ban the teaching of human evolution "as true or as a proven fact," and specifically opposed levying criminal penalties against those who might break such a law—his words were apocalyptic. With some justice he portrayed the proevolution lobby as a cadre of Eastern elitists who were seeking unilaterally to impose their worldview on the majority of Americans, in the process ignoring the principle of local control of public education: "A scientific soviet is attempting to dictate what is taught in our schools. It is the smallest, the most impudent, and the most tyrannical oligarchy that ever attempted to exercise arbitrary power. . . . If it is con-

---

*Mencken also freely acknowledged having been influenced by such Victorian liberals as Herbert Spencer, whose political philosophy had a distinctly eugenic tinge. As Spencer wrote in *Social Statics* (1850), the once-celebrated libertarian treatise cited by Justice Holmes in *Lochner* v. *New York*, "The forces which are working out the great scheme of perfect happiness, taking no account of incidental suffering, exterminate such sections of mankind as stand in their way. . . . Be he human or be he brute, the hindrance must be got rid of."

tended that an instructor has a right to teach anything he likes, I reply that the parents who pay the salary have a right to decide what shall be taught."

Up to a point Mencken agreed with him. Not long before *Tennessee v. Scopes* came to trial, he wrote a piece for the *Nation* in which he argued that the Tennessee legislature, having been duly elected by the people of the state, was thereby empowered to run their schools in whatever way it liked:

> No principle is at stake in Dayton save the principle that school teachers, like plumbers, should stick to the job that is set for them. . . . The issue of free speech is quite irrelevant. When a pedagogue takes his oath of office he renounces the right to free speech quite as certainly as a bishop does, or a colonel in the army, or an editorial writer on a newspaper. He becomes a paid propagandist of certain definite doctrines and attitudes, mainly determined specifically and in advance, and every time he departs from them deliberately he deliberately swindles his employers.

Anyone puzzled by this argument does not understand how Mencken's mind worked. Like Justice Holmes, who insisted time and again that "state sovereignty is a question of fact. . . . [W]hat it sees fit to order it will make you obey," he was a hard-nosed *Realpolitiker* who accepted power as a given. That was why he wanted to place the strictest possible constitutional limits on state power—he believed that no mere elected official could be trusted to use it wisely. Supermen were, of course, different, but they couldn't get elected. That was in the nature of democracy, which he defined as "the theory that the common people know what they want, and deserve to get it good and hard." Since the United States was a democracy, John Scopes was a fool to pit himself against the Tennessee legislature. If the people of Tennessee were stupid enough to elect representatives who in turn were stupid enough to pass an antievolution law, that was their business.

Such was the theory. In practice—and in private—he took a sharply different tack. The American Civil Liberties Union had agreed to handle Scopes's defense, and one of its first decisions was to get in touch with Mencken to learn more about "the political and theological pathology of the Bible country," about which he was thought to be an expert. He was briefed on the background of the case at a discreet meeting held in his Hollins Street sitting room. When the ACLU had announced its willingness to defend anyone prepared to challenge the Tennessee law, the city fathers of Dayton, a severely depressed mining town whose population had fallen to eighteen hundred, responded by offering to indict one of their very own schoolteachers in order to garner free publicity for the community. Bryan then volunteered to assist in the prosecution, and the ACLU persuaded Clarence Darrow, an outspoken atheist, to balance the scales by joining the defense's legal team. Recognizing that the presence of both men in Dayton would turn a backwoods trial into a national media circus, Mencken suggested that the ACLU deliberately "sacrifice" Scopes and "use the case to make Tennessee forever infamous" by luring Bryan onto the witness stand "to make him state his barbaric credo in plain English, and to make a monkey of him before the world."

We have only Mencken's word to show that it was his idea to put Bryan on the stand, and it may be that his version of events was self-aggrandizing. It definitely varied in the telling: He also claimed at one time to have persuaded Darrow to join the defense team, a detail missing from the account of the Scopes trial included in the posthumously published *Thirty-five Years of Newspaper Work*. But there is no question that he provided pretrial advice to the ACLU. He made no secret of it. "I have got myself involved in the Tennessee evolutionist trial, as a consulting Man of Vision to Darrow and Dudley Field Malone, both good friends of mine," he wrote to Sara Haardt on May 27. He then talked the *Sun* into sending him down to Dayton to report on the trial, not as an interested party but as a reporter—though it was no secret which side he was backing. In all of America, there was no more voluble opponent of religion, organized or otherwise, and like Bryan, he understood that Darwinism was more than just a scientific theory: It was a hammer with which to beat in the heads of the yokels. Neither abstract theory nor journalistic ethics would stop him from wielding it.

Mencken was not the only reporter to visit Dayton that July. Every major newspaper in the country sent correspondents to cover the Scopes trial; the proceedings were even broadcast live on radio. But his dispatches to the *Evening Sun* were widely syndicated, and they were the ones that would be remembered the longest. The headlines sum up their tone: MENCKEN LIKENS TRIAL TO RELIGIOUS ORGY. DARROW'S ELOQUENT APPEAL WASTED ON EARS THAT HEED ONLY BRYAN, SAYS MENCKEN. LAW AND FREEDOM, MENCKEN DISCOVERS, YIELD PLACE TO HOLY WRIT IN RHEA COUNTY. In addition he paid a visit to an outdoor meeting of the local Holy Rollers, describing what he saw there in a *Sun* story later expanded into one of his best essays, "The Hills of Zion":

> Far off in a dark, romantic glade a flickering light was visible, and out of the silence came the rumble of exhortation. . . . The leader kneeled facing us, his head alternately thrown back dramatically or buried in his hands. Words spouted from his lips like bullets from a machine-gun—appeals to God to pull the penitent back out of Hell, defiances of the demons of the air, a vast impassioned jargon of apocalyptic texts. Suddenly he rose to his feet, threw back his head and began to speak in the tongues—blub-blub-blub, gurgle-gurgle-gurgle. His voice rose to a higher register. The climax was a shrill, inarticulate squawk, like that of a man throttled. He fell headlong across the pyramid of supplicants.

Part of the power of "The Hills of Zion" comes from the fact that Mencken had never before witnessed at firsthand the rituals of religious enthusiasm. Indeed, his trip to Dayton was his first visit to the rural South, and once he finally saw it for himself, he felt something like pity: "A comic scene? Somehow, no. The poor half wits were too horribly in earnest. It was like peeping through a knothole at the writhings of people in pain." But this passage, and all other traces of sympathy, were carefully excised from the version he prepared in the forties for *A Mencken Chrestomathy*, leaving only the keenly observed reportage of a man who had grown too hard to pity anyone foolish enough to believe in anything.

Tennessee was appallingly hot that July, and Mencken wrote "The Hills of Zion" stripped to the waist in his hotel room. He was enjoying himself in spite of the unrelenting heat, but he had a magazine to put out, and after two weeks he went back to Baltimore by prior agreement with the *Sun*. As a result he missed the climax of the trial, which followed his specifications exactly. When Judge Raulston refused to allow the ACLU to present scientific witnesses prepared to testify that Darwin was right, Arthur Garfield Hays, the leader of the ACLU defense team, played his hidden ace and called Bryan to the stand as an expert witness on the Bible. The Great Commoner was happy to oblige: "They came here to try revealed religion. I am here to defend it, and they can ask me any questions they please." Darrow spent two cruel hours putting him through the hoops of biblical inerrancy, asking at one point if the story of Jonah and the whale should be "literally interpreted," to which Bryan replied that "one miracle is just as easy to believe as another." He came off looking like a rube, just as Mencken had planned, and the next day the judge struck his testimony from the record and ruled that the only matter at issue was "whether or not Mr. Scopes taught that man descended from a lower order of animals." After deliberating for nine minutes, the jury decided that he did.

Bryan died in his sleep in Dayton that Sunday. The news reached Baltimore in time for Mencken to write about it in his Monday Article. He knew he had dropped the ball by going home too soon—a faint note of defensiveness can be heard when he talks about it in

Thirty-five Years of Newspaper Work—and he must have been determined to make up for letting the *Sun* down. Indeed, "Bryan" is a good first try, but it was "In Memoriam: W.J.B.," the reworked and expanded version published in the October *Mercury*, that entered the annals of American journalism:

> It was hard to believe, watching him at Dayton, that he had traveled, that he had been received in civilized societies, that he had been a high officer of state. He seemed only a poor clod like those around him, deluded by a childish theology, full of an almost pathological hatred of all learning, all human dignity, all beauty, all fine and noble things. He was a peasant come home to the barnyard. Imagine a gentleman, and you have imagined everything that he was not. What animated him from end to end of his grotesque career was simply ambition—the ambition of a common man to get his hand upon the collar of his superiors, or, failing that, to get his thumb into their eyes.

Edmund Wilson speaks of the "eighteenth-century qualities of lucidity, order and force" to be found in Mencken's prose. His special gift was to combine this bright clarity with a contemporary tone of voice, and the result, so distinctive as to elude all imitators, is everywhere on display in what has become one of the most frequently reprinted of his essays. "In Memoriam: W. J. B." ranks among the great masterpieces of invective in the English language— and it reveals its author to have been something of a poseur. Ever since World War I he had insisted that he was amused, not angered, by the spectacle of democracy in action, and that his own work had no moral purpose. "I have never consciously tried to convert anyone to anything," he claimed. Perhaps not, but the friends to whom he wrote upon his return from Dayton knew he had been stunned by what he saw there. "I had never been on close terms with country people before," he told one of them. "I set out laughing and returned shivering." The last sentence of "Bryan" was actually meant to serve as a call to arms in the crusade against the numskulls: "The job before democracy is to get rid of such canaille. If it fails, they will devour it." He snipped it out of "In Memoriam: W. J. B."—it did not suit the image of detachment he sought to cultivate—but he meant it all the same. "We killed the son-of-a-bitch," he told friends after the trial, a sign of the same ill-concealed fury that inspired him to write of Bryan's "vague, unpleasant manginess." That phrase may have pleased the readers of the *Mercury*, but the reader who first encounters it in *A Mencken Chrestomathy* is more likely to be struck by the fact that "In Memoriam: W. J. B." is immediately followed by an essay in which Mencken calls Woodrow Wilson "a typical Puritan— of the better sort, perhaps, for he at least toyed with the ambition to appear as a gentleman, but nevertheless a true Puritan. Magnanimity was simply beyond him." He never understood that there was more than one way to be a puritan.

Magnanimity, of course, had been the last thing on his mind when he agreed to advise the ACLU defense team. "Getting Scopes acquitted would be worth a day's headlines in the newspapers, and then no more," he told them, "but smearing Bryan would be good for a long while." He was right. Frederick Lewis Allen took Mencken's lurid portrait of Bryan at face value in 1931 when writing *Only Yesterday*. So did Jerome Lawrence and Robert E. Lee, the authors of *Inherit the Wind*, the equally popular and influential 1955 play about the Scopes trial, in which Bryan is turned into an unprincipled rabble-rouser, Scopes and Darrow into secular saints, and the people of Dayton into a slack-jawed mob. (Also figuring prominently in the play is "E. K. Hornbeck," a straw-hatted, wisecracking reporter for the "Baltimore *Herald*." Gene Kelly played Hornbeck in the film, surely the only time that a famous American writer has been portrayed on screen by a famous American dancer.) Mencken may not have succeeded in snuffing out the fundamentalists with his typewriter, but he left his mark on the town of Dayton and the corpse of William Jennings Bryan, and it remains there to this day.

Scopes's conviction was eventually reversed on a technicality, and the law under which he was prosecuted was repealed by the Tennessee legislature in 1967. Thirty-three years later, the president of the chamber of commerce of Dayton, which has tripled in size since 1925, informed a visiting wire-service reporter that it is still "pretty much a creationism town."* The brick courthouse where *Tennessee* v. *Scopes* was tried is still in use, and a reenactment of the trial is held there each year. The Web site of Bryan College, a small evangelical Christian school founded in Dayton in 1930, includes a page that catalogs the inaccuracies of *Inherit the Wind:* "The people of Dayton in general and fundamentalist Christians in particular were not the ignorant, frenzied, uncouth persons the play pictures them as being. . . . The trial record discloses that Bryan handled himself well and when put on the stand unexpectedly by Darrow, defined terms carefully, stuck to the facts, made distinctions between literal and figurative language when interpreting the Bible, and questioned the reliability of scientific evidence when it contradicted the Bible." As for the author of "In Memoriam: W. J. B.," he is mentioned only in passing.

IF MENCKEN HAD LEFT HIS MARK ON DAYTON, THE REVERSE WAS ALSO true. He told Sara Mayfield that the trial had given him a new insight into the nature of democracy, which was that it was based "not so much on any rational theory as upon the organized hatred of the lower orders." This, he added, would be the subject of his next book. The insight wasn't quite new—the phrase appears word for word in a letter he wrote five months before the Scopes trial—but his adventures in the hills of Zion must have sharpened his thinking on the subject of democracy, though his attempts to get it down on paper were slowed by the sufferings of his mother, who had been felled by arteriosclerosis and was now fading fast. "As I worked in my third-

---

*Early in 2002 a federal district judge ruled that Rhea County, the Tennessee county in which Dayton is located, had to stop holding weekly Bible classes in its elementary schools—as it had been doing for the previous fifty-one years.

floor office in Hollins street," he remembered, "I would drop down of an evening to see her in her sitting-room on the floor below.... [T]he sight and thought of her made work almost impossible." The circulation in her hands had become so poor that she could no longer hold a darning needle or pick up a book, and Mencken feared that they might have to be amputated, a prospect that filled him with "horrible imaginings."

She was spared the worst. Anna Mencken died peacefully on December 13 after a minor operation. At sixty-seven, she went under the knife for the first time, and the strain seems to have finished her off. Her son was at once relieved and devastated, recognizing that he owed her a "stupendous debt":

> My father's death in 1899 was really a stroke of luck for me, for it liberated me from the tobacco business and enabled me to attempt journalism without his probable doubts and disapproval to hamper me, but the loss of my mother was pure disaster, for she had always stood by me loyally, despite the uneasiness that some of my ventures must have aroused in her.... Her death filled me with a sense of futility and desolation. It was weeks before I was fit for any work beyond routine: that routine, happily, was so heavy that it rescued me from myself.

Two days before Anna died, Mencken learned that his uncle had gone broke. The family firm of Aug. Mencken & Bro., which had been going downhill for years, would have to be closed as a result. The day after that Dreiser paid an unexpected visit to Hollins Street. He was driving south, he said, and wondered if Mencken could provide him with a bottle of Scotch for the road. He offered repeatedly to pay for the liquor, irritating Mencken, but they sat down in front of the fire—the fireplace was usually lit at 1524 Hollins Street—and chatted. In the course of the conversation, it emerged that Dreiser had left his girlfriend in the car, and Mencken immediately went to fetch her. As they conversed he mentioned that his mother was in the hospital and gravely ill. "I told Dreiser that I feared she might die," he wrote in his diary five years later. "But though he knew her and had been her guest in the house he seemed uninterested, and offered no word of hope that she would recover." Their friendship had already been severely tested by Dreiser's boorishness and Mencken's frank reviews of his books, and that put an end to it. After a cursory exchange of letters the following month, they would have no more contact for nine years.

Mencken was probably relieved to be rid of his old friend. "His customary attitude to the world," he would write in *My Life as Author and Editor*, "was that of any other yahoo, say an Allegheny hill-billy or a low-caste Jew." Nor had he ever balked at breaking with friends out of mere sentiment. "A man of active and resilient mind outwears his friendships just as certainly as he outwears his love affairs, his politics and his epistemology," he had declared in the *Smart Set*. Still, it must have occurred to him that the ties that bound him to the world of his youth were being severed one by one: first Marion, then Nathan, now Dreiser, the old family business, and his beloved Anna. It is striking that he broke a long silence to send Marion a brief letter telling of his mother's death ("What a woman is gone! I feel like a boy of 6"), and that they thereupon resumed their correspondence. Even though he knew it would lead nowhere, he felt a middle-aged man's desire to go looking for lost time.

The editor of a monthly magazine cannot dally long in the past, though, and Mencken was soon so fully reimmersed in the affairs of the *Mercury* and *Sun* that he became capable of taking a giant step away from days gone by: He transferred his office to what had been his mother's spacious second-floor bedroom. He installed his cluttered mahogany desk, flanked by two wastebaskets and a spittoon, facing away from the window that looked out on Union Square. He moved in the sofa on which he read and napped each day, strewing it with his collection of sampler-covered cushions, one of which bore the inspiring motto "Why Worry When You Can Pray?" He installed enough bookshelves to hold, among other things, five different encyclopedias and the complete *Oxford English Dictionary* in twenty volumes, the latter placed at arm's length from his swivel chair. Then he carried in his trusty Corona typewriter and went back to writing about democracy, only to meet with another interruption—one that threatened briefly to put him behind bars.

Herbert Asbury, a reporter for the *New York Herald Tribune*, had sent the *Mercury* a reminiscence of his Missouri youth in which he described the activities of a small-town cleaning lady who moonlighted as a prostitute on her days off. The lady in question, nicknamed "Hatrack" in tribute to her angular figure, went to church every Sunday morning and received clients in the Masonic and Catholic cemeteries every Sunday night. It being impossible to keep a secret in a small town, she was duly shunned by her fellow churchgoers: "From the Christians and their God she got nothing but scorn. . . . I have seen her sit alone and miserably unhappy while the preacher bellowed a sermon about forgiveness, with the whole church rocking to a chorus of 'amens' as he told the stories of various Biblical harlots, and how God had forgiven them." Mencken liked printing true-life stories that made small-town churchgoers look bad, and though he and Angoff both thought this one second-rate, they decided it could be whipped into adequate shape. Angoff did the necessary editing, Asbury was paid eighty dollars, and when a slot of suitable length opened up in the April issue, "Hatrack" went into the *Mercury*.

Neither man suspected that "Hatrack" might bring them trouble. The piece contained no descriptions of sexual activity, and the only use of profanity was the word "damned," which appeared in the last sentence. But Mencken had been pulling the tails of the puritans, not just with his coverage of the Scopes trial but in other *Mercury* articles, and he was especially vexed with Rev. J. Frank Chase, an ex-Methodist preacher who served as secretary of the New England Watch and Ward Society, the agency through whose efforts books and periodicals were "banned in Boston." Mencken had already run an article, "Keeping the Puritans Pure," that was sharply critical of Chase, and in addition to "Hatrack," the April issue contained a piece about Methodism that referred to him in passing as "a Methodist vice-hunter of long practice and great native talent." Three days after the issue appeared on Boston newsstands, Chase banned it, explaining that "Hatrack" contained "filthy and degrading descriptions" of conditions in Farmington, Missouri, and he had arrested a dealer who sold a copy at his Harvard Square newsstand.

At first Mencken was inclined to take the matter lightly. A practical man, he knew that few Bostonians bought the *Mercury*. But the more he thought about it, the madder he got. He saw that "if Chase were permitted to get away with this minor assault he would be encouraged to plan worse ones, and, what is more, other wowsers elsewhere would imitate him." So he called up Arthur Garfield Hays and asked for advice, and Hays recommended that he go to Boston, sell a copy of the April issue in public, and dare Chase to arrest him. The attorney warned Mencken that though he would probably win the case on appeal, it was conceivable that he might spend up to two years behind bars. For Mencken to run such a risk in defense of a principle was wildly uncharacteristic. In 1915 he had tried to persuade Dreiser to compromise with the censors. "Few doctrines seem to me to be worth fighting for," he told Burton Rascoe five years later. But the Scopes trial had emboldened him, and his mother's death meant that he no longer needed to worry about going too far: "I was now free to take chances, and in fact somewhat eager for adventure." He tipped off the wire services, took the train to Boston with Hays on April 5, and went to Boston Common, where Chase was waiting for him, accompanied by the chief of the Boston Vice Squad and another plainclothes officer. Chase gave Mencken a silver half-dollar. He bit the coin to see if it was counterfeit—a nice touch—then handed Chase a copy of the April *Mercury*. "I order this man's arrest," Chase told the police chief, and the five men walked four blocks to a police station, followed by reporters, cameramen, and a riot-size crowd of onlookers, most of them Harvard students who had gathered to watch their hero go up against the puritans.

Mencken was released on his own recognizance, and Hays filed for an injunction restraining Chase from banning the *Mercury*. The parties reassembled in the courtroom of Judge James P. Parmenter the next day. Mencken, Chase, Herbert Asbury, and the vice chief all testified, and the judge announced that he would read the issue himself and render judgment immediately. Mencken returned to his hotel to pass a sleepless night, at the end of which he returned to court to hear Judge Parmenter read a crisp five-minute statement rejecting all of Chase's claims. "I find that no offense has been committed, and therefore dismiss the complaint," the judge concluded, banging the gavel and setting the editor of the *American Mercury* free. Mencken attended a celebration at the Harvard Union, then posed for newsreel cameras and boarded a train for New York, downing a half-pint of whiskey before checking in to the Algonquin to catch up on his sleep. The following morning, he went to the office, opened a newspaper, and discovered that while he had been celebrating in Boston, Chase had gone straight to New York, pulled a few strings, and succeeded in having the postmaster general ban the April issue of the *Mercury* from the U.S. mails.

This was pure harassment, the issue already having been sent out, but it was also desperately serious business. Mencken knew from his censor-dodging days at the *Smart Set* that if the May issue were banned as well, the *Mercury* could have its second-class mailing privilege revoked "on the ground that we had missed two successive numbers, and were thus not 'of continuous publication.' This trick had been worked against various radical magazines during the Red hunt following the World War, and with disastrous effects upon them. . . . If our second-class privilege were taken away from us we'd be wrecked, for we'd then have to pay the full rate of ordinary postage on every copy of the magazine, however innocent its contents." Knowing he could no longer afford to take chances, Mencken decided to pull the May issue, of which several thousand copies had already been printed. He killed an article by Bernard De Voto called "Sex and the Co-Ed," replaced it with an innocuous essay called "On Learning to Play the Cello," sent the issue back to the printer's, and hurled himself into a welter of legal activity that lasted for months.

At the end of it the *Mercury* was safe and Mencken a celebrity—he received more press coverage as a result of the "Hatrack" case than he had ever gotten before—but the price had been high. All told the case cost Knopf $20,000, the equivalent of $190,000 today, and Mencken was shocked that so few of his fellow writers and newspaper colleagues had bothered to defend him aganst the charge that he was running "a sensational and pornographic magazine." One Catholic priest in Denver had preached a sermon calling "Hatrack" "unredeemed dirt, an open sewer, without even the iridescent scum that sometimes half invites and half excuses a glance . . . sheer lubricity, to be silently shunned."* Mencken's disillusion must have inspired him to open *Prejudices: Sixth Series*, published the following year, with a long essay on "Journalism in America" in which he condemned "the general smugness and lack of intellectual enterprise that pervades American journalism . . . the preposterous anointing of Coolidge, the craven yielding to such sinister forces as the Ku Klux Klan and the Anti-Saloon League, the incessant, humorless, degrading hymning of all sorts of rogues and charlatans."

It was with the "Hatrack" case fresh in his mind that he finished *Notes on Democracy*, which he had started writing in earnest only a few weeks before his mother died. "It remains, to me at least, the most unsatisfactory of my books," he later recalled, "and it had the smallest sale. The unhappy circumstances under which it was written are visible on almost every page." When it was published he told his friends he was displeased with the results, and many agreed with him, as did most of the reviewers. But he seems to have changed his mind later in life, for he included several excerpts from it in *A Mencken Chrestomathy*, and though the book is now little read, those

---

*In later years the *Mercury* reprinted "Hatrack" twice, and on the second occasion Asbury received a pithier criticism from a correspondent in Georgia: "I have now read your Hatrack article three times, in 1926, in 1936, and in 1950. It still stinks."

second thoughts were right. For all its limitations—and they are sub-stantial—*Notes on Democracy* is among the most personal of his books.

In it Mencken lays out the idea he said had come to him after watching the Scopes trial. Whatever the nominal theories on which it purports to be based, he argues, democracy in practice arises from the envy felt by inferior men for their superiors. The peasant craves security, the gentleman fights for liberty; since there are by defini-tion more peasants than gentlemen, the inevitable result of universal suffrage is to make it possible for the former to persecute the latter, thus impeding human progress. Knowing this, politicians deliber-ately pander to the prejudices of the mob in order to wield power over those who are too principled to do the same thing. On the rare occasion when a gentleman runs for office and wins, he cannot but fall victim to the iron law of democracy: "If he retains his rectitude he loses his office, and if he retains his office he has to dilute his rec-titude with the cologne spirits of the trade."

Within the compass of its forty-thousand-odd words, *Notes on Democracy* offers a shapely summary of the acidulous view of democ-racy in America that Mencken had been expounding, in print and in conversation, for the past quarter century. The Scopes trial and the "Hatrack" case may have given him a somewhat clearer perspective, but it was more a refinement than a revision; for the most part he had known more or less what he thought since well before the war. Here his purpose is not to break new ground but to say familiar things in a fresh way, as in this portrait of a low-caste southern congressman:

> Until he got to Washington, and began to meet lobbyists, bootleggers and the correspondents of the newspapers, he had perhaps never met a single intelligent human being. . . . His dream is to be chosen to go on a congressional junket, *i.e.*, on a drunken holiday at government expense. His daily toil is get-ting jobs for relatives and retainers. Sometimes he puts a dummy on the pay-roll and collects the dummy's salary him-

self. In brief, a knavish and preposterous nonentity, half way between a kleagle of the Ku Klux and a grand worthy bow-wow of the Knights of Zoroaster. It is such vermin who make the laws of the United States.

As in "In Memoriam: W. J. B.," he is now driven by a newly per-sonal awareness of the nature of democratic man, and from time to time he makes a point of reminding us that he knows whereof he speaks. When, for example, he talks about the tendency of the puri-tans to ascribe "sordid and degrading motives" to those who oppose them, he alludes briefly to his recent experiences in Boston: "When I protested against their sinister and disonhest censorship of literature, they charged me publicly with being engaged in the circulation of pornography, and actually made a vain and ill-starred attempt to rail-road me to jail on that charge."

Most of those reviewers who disliked *Notes on Democracy* did so simply because they disagreed with it. A few grasped its fundamental defect, which is that nowhere does Mencken offer an alternative to representative democracy. "I am not engaged in therapeutics, but diagnosis," he explains in the last chapter, which would be fair enough were it not that given the force of his indictment, one expects something better by way of conclusion than the mere state-ment that "I enjoy democracy immensely. It is incomparably idiotic, and hence incomparably amusing. Does it exalt dunderheads, cow-ards, trimmers, frauds, cads? Then the pain of seeing them go up is balanced and obliterated by the joy of seeing them come down." Not only is this flippancy unworthy of the moral outrage evident in the rest of the book, it is yet another manifestation of Mencken the poseur, unwilling to admit his own passion. Those who remembered "The Mailed Fist and Its Prophet" knew he had not always been so discreet, but now he was playing it closer to the vest. One reader caught his drift, though: Kaiser Wilhelm II, living in exile in Hol-land, sent him a framed photograph inscribed, "With thanks for your splendid book on democracy." Cautious as always, Mencken hung it in the back hall on the third floor of Hollins Street.

The most perceptive reviewers sensed that Mencken, as he himself acknowledged, was "far more an artist than a metaphysician." Walter Lippmann, for example, saw that "the man is bigger than his ideas. . . . It is no crime not to be a philosopher. What Mr. Mencken has created is a personal force in American life which has an extraordinarily cleansing and vitalizing effect." Edmund Wilson, always his most comprehending critic, went even further, grasping that Mencken's "democratic man" was "an ideal monster, exactly like the Yahoo of Swift, and it has almost the same dreadful reality." This tendency to turn his enemies into boldly painted abstractions is part of what gives his political commentary its lasting value—at its best it has something of the poetic quality of the imaginative literature he now disdained—but it also prevented him from acknowledging the human complexity of the politicians about whom he wrote. His "Dr. Coolidge" is a symbol, a 100 percent American sunk so deep in the mire of philistinism that "there is no evidence that he is acquainted with a single intelligent man." Whether Mencken knew it or not, his broad-brushed caricature had little in common with the *cum laude* graduate of Amherst who read Cicero in the original and gave laconic speeches that set forth a philosophy of governance to which his most prominent critic could have taken no exception whatsoever: "The people cannot look to legislation generally for success. Industry, thrift, character, are not conferred by act or resolve. Government cannot relieve from toil." The author of *Notes on Democracy* had no interest in such pesky details. Perhaps he had forgotten a quatrain by the favorite poet of his adolescence: "Ah! What avails the classic bent/And what the cultured word,/Against the undoctored incident/That actually occurred?"

MENCKEN HAD TWO OTHER BOOKS IN THE WORKS AT THE SAME TIME AS *Notes on Democracy*, a second volume of selections from "Americana" and a fifth series of *Prejudices*. No man could shoulder such a load indefinitely, and when Paul Patterson invited him to take part in a "grand tour of the South, to meet editors and look over the ground," he jumped at the opportunity. The two men spent a week storming through Virginia, North Carolina, Georgia, Alabama, and Louisiana, after which Mencken went on alone to Los Angeles for two weeks of "pure holiday." At every stop he told gullible local reporters that "some native bigwig" was "prime presidential material," laughing to himself as they took down his every word. Patterson later wrote to say that he had been "bowled over" by the amount of press coverage they had received, and Mencken's own recollections of the trip suggest the extent of his burgeoning celebrity: "Everywhere we went we were bathed in publicity, and I was interviewed four or five times a day." By then, though, he was being bowled over in a different way, for his purpose in going to the West Coast was to see a petite movie star with whom he was sleeping.

Aileen Pringle is no longer remembered, but in her time she was a very big name indeed. She married a titled Englishman, then proceeded to scandalize him by going on the stage, subsequently making the leap to screen stardom in 1922. Her colleagues thought her a snob, an impression she did nothing to dispel, and she liked to hobnob with visiting intellectuals, in due course coming to be known as Hollywood's "darling of the intelligentsia." Though the coming of sound would reduce her to bit parts, she was still big in 1926, when she met Mencken in June at a weekend house party in Pennsylvania, and she appears to have had a taste for famous writers (she would later marry James Cain), for she got the publisher of "Hatrack" into bed in record time. At first Mencken steadfastly denied that anything was up between him and Aileen. "I am innocent, mon colonel," he told Ernest Boyd. "I made no attempt upon that beautiful artiste." But that was the gentleman speaking. The daily letters and telegrams she sent him after her return to Hollywood tell a different tale: "I love you loving me because I know you have not loved in vain. . . . In spite of the mosquitoes, wilting collars, drooping underwear and the preventions for pregnancy I think we did awfully well."

Mencken's arrival in Hollywood was a field day for the local press, and he added to the noise level by being photographed with such luminaries as Norma Shearer, Lionel Barrymore, and Louis B. Mayer, as well as attending a revival meeting held by the sexy evangelist Aimee Semple McPherson at her Angelus Temple of the Foursquare Gospel, about which he discoursed wittily in the *Sun* and naughtily to his friends back east ("I sat under Aimee yesterday, and had 2 1/2 spells of tumescence"). The town itself he found disagreeable, as he made explicit when he finally got back to Baltimore and wrote up his sour impressions for *Photoplay*. He had no feel for the movies, lacking as he did any strong visual sense, and what comes through above all is his distaste for the Jews who dominated the film business: "Sooner or later the movies will have to split into two halves. There will be movies for the present mob, and there will be movies for the relatively enlightened minority. . . . [T]he present movie-folk, I fear, will never quite solve the problem, save by some act of God. They are too much under the heel of the East Side gorillas who own them. They think too much about money."

More interesting was the Monday Article he wrote about Rudolph Valentino, with whom he and Aileen had dined in New York a few days before the matinee idol's unexpected death. Valentino had sought Mencken out to ask his advice about how to deal with the press, leaving the impression of "a man of relatively civilized feelings thrown into a situation of intolerable vulgarity." Mencken himself had only just gotten free of the "Hatrack" hoopla, and the sensitivity of his response to Valentino's plight is striking: "Here was a young man who was living daily the dream of millions of other young men. Here was one who was catnip to women. Here was one who had wealth and fame. And here was one who was very unhappy."

That was an aberration. A more recognizable-sounding Mencken would later write in *Photoplay* that a star is "simply a performer who pleases the generality of morons better than the average." It would be interesting to know what Aileen Pringle thought of that comment. His attraction to her was much more than casual (I AM MASHED ON YOU WHAT IS TO BE DONE ABOUT IT, he telegraphed on New Year's Eve), but if Marion had been an unsuitable mate, how much more so could a movie actress be, especially one encumbered by an estranged husband? Nor did it help that their association was fast becoming the subject of published speculation. From Hollywood he had bragged to Marion that "two eminent female stars, known everywhere in this wide, wide world, have got mashed on me," but once he and Aileen made it into the gossip columns, his tune changed completely: "The public prints have probably informed you of my marriage in Hollywood. Like most of the other things they report of me, it never happened. The lady mentioned was as sore as hell, and justly so, for she has a husband." If there was one thing he hated, it was reading about his private life in the papers, especially since he was continuing to see other women in New York and Washington while carrying on his bicoastal affair with Aileen, who had made the mistake of mentioning marriage. One of the few letters to her that he made a point of preserving—he kept a carbon—was a note in which he firmly reasserted the lifelong caution he had temporarily laid aside: "You are still young, and beautiful, and still eager for life, and the best of it is ahead of you. But I am beginning to crack . . . I am tied here by the leg and can't get near you. And you are fastened too. Let us at least admire the irony of it!"

Sara Haardt was no less displeased to read of his romance with Aileen, and she made her pique known to Sara Mayfield, who passed it on to Mencken. Though he was now publishing her work in the *Mercury*, their correspondence had grown more reserved, and many months would pass before he moved decisively to break away from Aileen. Even so he had already started to think twice. "I see Sara Haardt here very often: she is better looking than ever. . . . Aileen writes that she is working her head off, making movies that grow more and more idiotic," he wrote in July to a friend who knew both women. Two months later, Sara went to Hollywood to try her hand at writing scripts—she needed the money to pay her medical bills—

and from there she sent letters to Baltimore in which she confessed that the necessity of doing business with the "insufferable" Jews who made movies was giving her "a Nordic complex that is mounting to a phobia." That being Mencken's own view of the men who ran Hollywood, it must have further confirmed him in his growing suspicion that he was chasing the wrong girl. No wonder he spoke so dismissively to a reporter that winter about the institution of marriage: "A man in full possession of the modest faculties that nature commonly apportions him, is at least far enough above idiocy to realize that marriage is a bargain in which he gets the worst of it, even when in some detail or other, he makes a visible gain."

Whatever the state of his private life, Mencken the public figure had never been in a more solid position. *Elmer Gantry*, Sinclair Lewis's bestselling novel about an itinerant and unscrupulous evangelist, was dedicated to him in 1927 "with profound admiration." The *Mercury*'s average circulation that year was 77,000, five times higher than that of its first issue, and everybody who was anybody was aware of its editor. Walter Lippmann readily acknowledged the influence he exerted on "this whole generation of educated people." He even had an admirer on the Supreme Court, for Justice Holmes, undeceived by his boisterous manner, saw in him a kindred spirit: "He I suspect would prove more or less a Philistine at bottom, but Lord with what malevolent joy do I see him smash round in the china shop. . . . With various foibles he has a sense of reality and most of his prejudices I share."

It was around this time that Mencken sat for a picture by Nikol Schattenstein, a once-popular artist whose work is now little known.* Schattenstein's Mencken, a youthful-looking man with melancholy china-blue eyes, is casually dressed in open collar and rumpled, rolled-up shirtsleeves, holding a half-smoked cigar in his

---

*Schattenstein's portraits of Lotte Lehmann, Lauritz Melchior, and Risë Stevens still hang in the Metropolitan Opera House, the only of his surviving canvases currently on public view.

right hand and wearing a pair of striped suspenders given to him by Valentino. The thoughtful expression on his oddly smooth face suggests that he is reflecting on the unending follies of the fundamentalists, or perhaps on the fact that he has become famous enough to pose for a fashionable artist. It is a subtly but unmistakably idealized portrait of the critic as culture hero—the god of the undergraduates—and Marion, not surprisingly, saw right through it. "Old Hank's portrait is a blight on his escutcheon, however you spell it," she testily told her sister. "It's disgusting. I can't make Hank out. Surely you remember the days when he believed that a man of dignity could not afford to fraternize with every one—and now a painting resembling a cub studying art!"

What did the man in the picture make of it all? On the first day of the year, he sat down at his desk and pecked out a memorandum in which he took stock of his professional life to date:

> I have done a great deal less than I wanted to do and a great deal less than I might have done if my equipment had been better, but this, at least, I have accomplished, and it is one of the principal desires of man: I have delivered myself from anonymity. . . . Unless I collapse physically, I think I am good for five or six more books. They will be far better than anything I have done in the past. My existing books, in fact, are all bad. I am at my best in articles, written in heat and printed at once. Unfortunately, they are dead in a few weeks, and so cannot be reprinted. Thus my books, which are made up almost wholly of reworked articles, represent mainly my least effective work. But hereafter I shall write better ones. . . . Here I get back to another source of consolation: that of feeling quite distinct and separate from the masses of men. It is very comforting on blue days. I belong to no party: I am my own party.

This mixed verdict points to the task that would occupy him for much of the next two years. On January 29, 1928, he wrote his last column for the *Chicago Tribune*. "I am going on furlough to write a book," he told his readers. "How long it will take I don't know. But I shall be grateful to any persons who may be moved to assist the business with their prayers." In that book, he would set down his definitive thoughts on the subject of religion, and he would do it from scratch, not by pasting up a stack of newspaper and magazine pieces. (*Prejudices: Sixth Series* was the last of his self-anthologies prior to the *Chrestomathy* of 1948.) At the same time he made a Herculean effort to put his love life in order, and soon it was understood among his closest friends that Sara Haardt was his best girl. She sifted through his scrapbooks to put together *Menckeniana: A Schimpflexicon*, a 132-page compilation of unfriendly comments about Mencken that Knopf published in 1928 as an ironic advertisement for his most controversial author. Though he spent much of the year covering the presidential campaign for the *Evening Sun*, they saw each other as often as possible and were seen together frequently in public. They even began to discuss marriage, and for the first time he did not back and fill.

The *Mercury* reached its all-time circulation peak early in 1928, selling nearly 84,000 copies. Three years later, Frederick Lewis Allen would recall the niche its editor had occupied in American culture:

Up to this time the intellectuals had generally been on the defensive. But now, with Mencken's noisy tub-thumping to give them assurance, they changed their tone. . . . Slowly the volume of protest grew, until by 1926 or 1927 anybody who uttered a good word for Rotary or Bryan in any house upon whose walls hung a reproduction of Picasso or Marie Laurencin, or upon whose shelves stood *The Sun Also Rises* or *Notes on Democracy*, was likely to be set down as an incurable moron.

But the *Mercury* was a countercultural magazine edited by a bourgeois, and its popularity, being based on a contradiction, was bound to end in disillusion. Allen himself misunderstood the magazine completely, supposing it to have been "addressed to the intellectual left wing." Not only did Mencken and Nathan have no such thing in mind, they said so in the first issue and at frequent intervals thereafter, and the time finally came when they were taken at their word. The failure of the *Mercury* is customarily attributed to the change of political scenery caused by the Great Depression, but its circulation first started to decline in mid-1928, and kept on drifting downward thereafter. The left had figured out that Mencken was not one of its own. At best he was a revolutionary who, like so many other revolutionaries, had turned against his own children, in art as well as politics. Nineteen twenty-eight was the year of Evelyn Waugh's *Decline and Fall*, Louis Armstrong's "West End Blues," the Weill-Brecht *Dreigroschenoper*, and the Stravinsky-Balanchine *Apollo*. Sound had come to the movies and cubism to the museums: Chick Austin, the daring new curator of Connecticut's Wadsworth Atheneum, hung his first show of modern art that April. None of it meant a thing to Mencken, who was busy filling his book column with science, sociology, and politics, though he made room in May for a single paragraph in which he disposed briskly of Ernest Hemingway's *Men Without Women* and Thornton Wilder's *The Bridge of San Luis Rey:* "I gather from both of them the feeling that they are as yet somewhat uncertain about their characters—that after their most surprising bravura passages they remain in some doubt as to what it is all about." One might just as well have asked whether Mencken knew what the twenties had been all about, other than the Scopes trial. Sometimes it seemed as if the *Mercury* had missed out on everything else.

Yet it was from a right-wing intellectual that the most astute and farseeing attack on Mencken would come. Harvard's Irving Babbitt preached a post-Christian, democracy-distrusting "New Humanism" that made a deep impression on his best-known pupils, T. S. Eliot and Walter Lippmann. Mencken, who should have known better, thought him a prig, but Babbitt, unlike most American academics of his generation, did not make the reciprocal mistake of dismissing the editor of the *American Mercury* as a pornographer or an ape. He took the full measure of the man and found it impressive, though he also found the "total effect" of Mencken's writing "nearer to intellectual vaudeville than to serious criticism." More important, though, Babbitt was among the first of Mencken's critics to suggest that his noisy war against the booboisie had reached the point of diminishing returns: "One is reminded in particular of Flaubert, who showed a diligence in collecting bourgeois imbecilities comparable to that displayed by Mr. Mencken in his *Americana*. . . . Another discovery of Flaubert's may seem to him more worthy of consideration. 'By dint of railing at idiots,' Flaubert reports, 'one runs the risk of becoming idiotic oneself.'"

7

"I DISCERN NO TREMORS"

Marriage and the Crash, 1929–1935

It was inevitable that Mencken would get around to writing a book about religion, and since he had been dissatisfied with *Notes on Democracy*, he went out of his way to do the job right this time. Insofar as he was capable of concentrating on a single task—which wasn't very far—he immersed himself completely in *Treatise on the Gods*. Outside of his Monday Articles and his contributions to the *Mercury*, he did little other journalistic work in 1929. Angoff had relieved him of the purely mechanical drudgery of bringing out the *Mercury* each month, and once Herbert Hoover defeated Al Smith, he put aside presidential politics as well. Even the sheaf of letters (sometimes as many as fifty) that he dictated each morning to Rosalind Lohrfinck, his new stenographer, reflected his absorption in the job at hand. To Monsignor J. B. Dudek, a Czech-born Oklahoma prelate with whom he corresponded about linguistic matters, went this droll request for help: "By the way, what is the best current book of Catholic doctrine? I want to find out precisely what the church teaches on all the salient points of the faith. . . . No, I am not preparing to disgrace the church as a convert. I simply want to get at some facts for an article I am planning."

The only thing capable of distracting him was Sara's uncertain health. A persistent and increasingly debilitating urinary-tract infection put her in Baltimore's Union Memorial Hospital in June, and Mencken and her doctors did not tell her that it might require the removal of a kidney—a serious operation under any circumstances in 1929, and especially so for a chronically ill patient who might have to undergo surgery in the middle of an "infernal" heat wave without benefit of air conditioning. "She is bearing the thing very bravely," he wrote to Sara Mayfield, "but it is beginning to wear her out." A month later the decision to operate was made. By then she had come to the conclusion that she was going to die, and openly talked about her funeral; the doctors were not much more encouraging, warning Mencken that her prognosis was grim.

Sara's condition forced him to make a decision that he had been dodging for years. As late as November 1928, he had lunched in Baltimore with Marion Bloom. Years later she recalled it as an angry meeting in which he said that her belief in Christian Science meant she was "no longer in his intellectual class." A few days later he sent a coquettish letter of apology: "I have a notion that we called the coroner's inquest before the patient was actually dead. . . . I hesitate to puff you up with flattery, but God roast me in hell if you ain't getting better looking all the time." Aileen Pringle was continuing to pelt him with letters and telegrams, and he was also carrying on an elaborately flirtatious correspondence with an opera singer from Washington, D.C., named Gretchen Hood (she took it seriously, he didn't). Though we don't know what he thought of being on the receiving end of so much feminine attention—his relationship with Sara was the only one about which he wrote—he would scarcely have been the first bachelor to recognize that flirting with several women made it easier to avoid committing himself to any one of them.

Now, with Sara on the verge of death, evasion was no longer possible, and as he waited outside the operating room, mopping his sweaty, tear-streaked face with a handkerchief, he admitted as much to Sara Mayfield. He said he had been told that even if she survived the operation, she would only live for another two or three years at most. "If she pulls through," he added, "I've promised myself that I'll make them the happiest years of her life." As soon as she came out of the anesthetic, he told her of his decision. The obstacle course was not yet completely run—her convalescence was slow and arduous, and at one point her death appeared certain, a piece of news that, according to Mencken, "wrecked me for a week"—but once it was clear that she would recover, they began making plans to marry as soon as possible. On the eve of his forty-ninth birthday, H. L. Mencken had made up his mind.

"If I ever marry," he had told Marion in 1927, "it will be on a sudden impulse, as a man shoots himself." Some find it puzzling that a bachelor with no shortage of romantic alternatives should have yielded to the impulse to marry an invalid whom he had already courted fitfully for six years. Mencken himself had foretold one possible reason in a 1922 "Répétition Générale" item that he chose not to incorporate in the revised version of *In Defense of Women*: "The fact that marriage between older persons—a man over forty, say, and a woman over thirty-five—seldom turns out unhappily, or at least endures till death the twain parts, is one of the best possible proofs, it seems to me, of my long-held contention that nothing is more greatly inimical to a happy marriage than a sense of romance. . . . Romance is for passionate love; and passionate love has no more enduring place in marriage than the moon has in the broad, hard light of day."

He had known passionate love with Marion and found it unsatisfactory, and the same thing would happen with Aileen Pringle. No such thing could happen with Sara. She was a different kind of woman, shrewd and detached—and sickly. Whatever the extent of their physical relationship, it cannot possibly have been enough to bind them together, and her once-delicate features had been rendered "anything but photogenic" (in Mencken's own frank phrase) by the effects of prolonged illness. Yet it is beyond question that they were in love, for all who knew them testified to it, as did Mencken in his diary: "I can recall no single moment during our years together when I ever had the slightest doubt of our marriage, or wished that it had never been." He admired her courage, appreciated her wit, found her satisfactory as a conversational partner, and warmed to her "indefinable pleasant thoughtfulness which passes commonly under the banal name of Southern charm." Like him she was a skeptic, and his alienation from the puritan ways of his native land found a responding echo in the deceptively cool ambivalence with which she described the Deep South that she had left behind but only partly escaped: "I could have latched on to the sweet moulding decay that surrounded me, and convinced myself that it was brave, romantic. In essence all Southerners are bad poets. I loved the old war songs and the perfume of the magnolias, like any other; and I could have lived my days talking about them and repining."

Sara wrote those words in 1934, by which time she had turned herself into a writer capable of being published in the *Mercury* without raising eyebrows among Mencken's friends or enemies. (She also wrote for such magazines as the *Atlantic Monthly, Harper's Bazaar, Liberty*, and *Scribner's*, and one of her stories, "Little White Girl," was included in *Best Short Stories of 1935*.) He had longed in vain to make Marion a professional writer, but Sara became one on her own, leaving behind a body of work impressive enough that the editor of a posthumous collection of her stories and essays would one day describe her as "an important contributor to the literary tradition of the American South." That is an exaggeration, but though she died too soon to realize her potential, Sara Haardt deserves to be remembered as more than just Mrs. H. L. Mencken, not least for "Dear Life," her piercing reminiscence of how it felt to face death in a stiflingly hot Baltimore operating room in the summer of 1929:

The ether poured out of the cone like a blessing. Beyond the rim of my steadily diminishing consciousness I could feel the heat hazes rolling over the Maryland hills. They had reminded me, as I lay there on the stretcher waiting for the surgeons to wash up, of the mists that used to roll over the river down in Alabama. . . . Oh, no use talking, the South was sweet. But it was a sweetness tinged with the melancholy of death. It was because beauty, somehow, is shorter lived in the South than in the North, or in the West; and beauty, more than mere survival, is the most poignant proof of life.

Far from disapproving of her success, Mencken would brag about it, informing Philip Goodman in 1933 that she had "sold more stuff within the last month than she had ever sold before in a whole year. Moreover, she is beset with orders. I hope she'll earn enough by 1934 to support me in reasonable decency." But of course he had not married her merely because she was a good writer or a charming conversationalist, or even because she bore her sufferings without whining, any more than she had married him because he was a famous man with enough money to pay her doctors' bills. His decision to take an ailing bride might have been in some measure a self-aware gesture of gallantry—the sort of thing a Victorian gentleman would have done—but if so, then it paid off handsomely for both of them, and perhaps that is all that needs to be said about their motives, beyond the perfectly obvious fact that they loved each other.

Anyone who doubts it should note that Mencken was unable to work on *Treatise on the Gods* so long as it was possible that Sara might die. As soon as he knew she would recover, though, he started writing again, and he applied the finishing touches on Thanksgiving night, adding the Latin tag *Soli Deo gloria!* (to God alone the glory) as an ironic flourish at the bottom of the last page. He called Western Union and fired off a jubilant telegram addressed to her hospital room: THE BOOK WAS FINISHED AT NINE FIFTEEN TONIGHT GLORY HALLELUJAH I WILL BE READY TO CELEBRATE SUNDAY YOUR PRAYERS MUST

GET THE CREDIT. Then he wrapped up the manuscript and sent it off to Knopf, certain that he had produced a masterpiece.

At first his readers seemed to concur. *Treatise on the Gods* went through seven printings and sold thirteen thousand copies by the end of 1930. But it soon slipped from view, and by 1941 even Mencken was forced to admit that "it is seldom discussed, or even alluded to." Nevertheless he insisted that it was

> my best book, and by far. It is smooth, good-tempered, and adroitly written, and during all the years since it was pub-lished no one has ever successfully challenged a statement of fact in it. It is a model of condensation; I can think of no other American book in which such a large body of material is brought into such small compass without any sacrifice of form or style. . . . It gets closer to fundamental realities than I have ever got elsewhere, and in it, in all probability, is my best writing.

Some reviewers agreed with him, among them the left-wing liter-ary critic Granville Hicks, who called it "the best popular account we have of the origin and nature of religion," but those who found Mencken's incorrigible skepticism unsympathetic were less inclined to admire *Treatise on the Gods*. "The gleam of fanaticism is in Mr. Mencken's eye while he inveighs against the bigotry of the priests and the stupidity of their followers," wrote the Protestant theologian Reinhold Niebuhr in the *Atlantic Monthly*. "It is only when dealing with moral and social issues that he achieves the heights of complete detachment, and in this case the detachment is that of the cynic rather than that of the scientist."

Even allowing for his understandable reluctance to praise a book about religion by a man who freely confessed to being "quite devoid of the religious impulse," Niebuhr was closer to the mark. *Treatise on the Gods*, like *Notes on Democracy* before it, is less a rigorously argued "treatise" than yet another Punch-and-Judy show in which the enlightened superman does battle with the mob. To the oft-repeated argument that religious faith is a mere exercise in wishful thinking, motivated chiefly by fear, Mencken could add nothing new save the brilliance and verve of his mature prose style—and the surprising thing about *Treatise on the Gods* is its failure to scintillate. It is revealing that he spoke of it as "smooth" and "good-tempered," and that he now thought these worthwhile things for a book to be, when what had made *Notes on Democracy* so readable was its unabashedly foul temper.

After Mencken died Charles Angoff claimed to have read *Treatise on the Gods* "chapter by chapter, as it was written," and to have warned Mencken that it was not up to par:

> I hesitated. "Oh, I guess what I want to say is that it's a little out of your line."
>
> His face became red. "Why?"
>
> "Frankly, I'm afraid of what the professors will do to it. The newspaper and magazine reviewers will do fine by it. But what do they know about the history of religion?"
>
> "To hell with the professors," said Mencken. "I beat them at their game in *The American Language* and I'll beat them at the religious game."
>
> "I hope so," I said. "I hope so."

This is among the least plausible of Angoff's party pieces, but like so much of what he said about his old boss, it contains a nub of truth. "I have been torn all my writing life," Mencken admitted, "by two conflicting desires—the journalistic desire to say it at once and have it done, and the more scholarly desire to say it carefully and with some regard to fundamental ideas and permanent values." The flap copy for *Treatise on the Gods* indicates that he thought he had written a book of the second kind, or at least hoped he had:

"Nothing in this book has been printed before, either in THE AMER-
ICAN MERCURY or elsewhere. It is the first book by Mr. Mencken
since 1927." He was slightly self-conscious about his homemade
scholarship—most autodidacts are—and while he had no need to
feel uncomfortable about the breadth of his preparatory reading,
which took in everything from the "higher criticism" of German
biblical scholars to the then-current work of such psychologists and
anthropologists as Jean Piaget and Bronislaw Malinowski, the learning he wore so lightly in *The American Language* is too frequently
brandished like a pikestaff in *Treatise on the Gods*. He was incapable
of being altogether dull, and there are times when he rises effortlessly to the rhetorical occasion, as in his description of how
modern-day believers are no less prone than their animistic ancestors to personalize their gods: "The God of the Episcopalians is
an elderly British peer, courtly in manner, somewhat beefy, and, in
New York, vaguely Jewish. . . . The God of the Methodists is an
agent provocateur, forever fingering His pad of blank warrants." Just
as often, though, he sounds as if he wanted to make absolutely sure
that his readers knew how much he knew, and *Treatise on the Gods*
suffers from this seeming need, as well as from Mencken's complete
inability to conceive of a truly intelligent man who could look upon
religion as anything other than a poultice for the fear of death:
"Men may live decently without it and they may die courageously
without it. But not, of course, *all* men. The capacity for that proud
imperturbability is rare in the race—maybe as rare as the capacity
for honour. For the rest there must be faith, as there must be
morals. It is their fate to live absurdly, flogged by categorical imperatives of their own shallow imagining, and to die insanely, grasping
for hands that are not there."

In one chapter, though, Mencken wrote with his customary directness. Praising the Bible for its "lush and lovely poetry," he remarks
that it is "astounding" that the Jews should have been responsible for
"nearly all of it," since they could be "very plausibly" described as

"the most unpleasant race ever heard of." Nor did he let it go at that. In a passage full of faint echoes of Nietzsche, he went on to explain:

> As commonly encountered, they lack many of the qualities that mark the civilized man: courage, dignity, incorruptibility, ease, confidence. They have vanity without pride, voluptuousness without taste, and learning without wisdom. Their fortitude, such as it is, is wasted upon puerile objects and their charity is mainly only a form of display. Yet these same Jews, from time immemorial, have been the chief dreamers of the human race, and beyond all comparison its greatest poets. . . . All this, of course, may prove either one of two things: that the Jews, in their heyday, were actually superior to all the great peoples who disdained them, or that poetry is only a minor art. My private inclination is to embrace the latter hypothesis, but I do not pause to argue the point.

Mencken later called this passage "a chance reference," which is absurd. He was too experienced a controversialist not to have known it would draw fire, though the intensity with which the Jewish press responded—his scrapbooks are full of articles, pro and con alike, published in Jewish magazines and newspapers—may have surprised him. For whatever reason, he took the unusual step of agreeing to be interviewed by a Jewish paper about *Treatise on the Gods.* "Don't forget that my book is about religion, not about individuals," he told Joseph Brainin. "I don't like religious Jews. I don't like religious Catholics and Protestants." His objection was to "professional Jews" who "go about like sandwich men carrying big signs: 'I am a Jew.' They parade it in front of you. They shout it into your face—not as an answer to a question, but aggressively, without solicitation. . . . It's just like those Catholics clamoring that they are the only people who will go to heaven." Instead of separatism he recommended assimilation, assuring the readers of the *Pittsburgh Jewish Chronicle* that it would not lead to "losing your identity. Just try not to be unpleasant and stop thinking that you're 'it.' You live in this country; try to behave as others do."

The headline of the piece, "Is H. L. Mencken an Anti-Semite?," must have taken him aback. As even Angoff acknowledged, Mencken did not regard himself as a Jew hater, and he was entirely serious when he claimed to be no rougher on "professional Jews" than any other body of believers, a defense his admirers would echo ever after. Possibly he failed to see what many of his other readers found embarrassingly obvious, which is that while "religious Catholics and Methodists" were members of a church, Jews were—to use his word—a race. To be sure, many of his Jewish friends (including Alfred Knopf) had similar feelings toward non-German Jews. Yet the relevant passage from *Treatise on the Gods* appears unambiguous: He was speaking of Jews "as commonly encountered," not just Eastern Jews or religiously observant Jews. While his explanation quelled some of the uproar, a seed of doubt had been planted, and everything he wrote about Jews would henceforth be subjected to close scrutiny. After 1930 he reserved his bluntest comments about them for his private papers. With their publication long after his death, it became clear that, as usual, he had meant exactly what he said.

For the moment the uproar proved short-lived, in part because most Americans were preoccupied with the dire state of the economy. "No Congress of the United States ever assembled, on surveying the state of the Union, has met with a more pleasing prospect than that which appears at the present time," Calvin Coolidge had proclaimed in his final State of the Union message. "In the domestic field there is tranquility and contentment, harmonious relations between management and wage earner, freedom from industrial strife, and the highest record of years of prosperity." Then he handed over the reins of power to Herbert Hoover, and seven months later, as Mencken was writing the last pages of *Treatise on the Gods*, the bottom fell out of the stock market, leaving some three million men looking for work. Prices rallied early in 1930, but soon they started slipping downward again. In May the governor of the Federal

Reserve Board acknowledged that the economy had entered "what appears to be a business depression"; by year's end the number of unemployed had risen to six million.

Mencken wrote nothing about the crash, and two years would go by before he took note of it in a *Mercury* editorial, though he later claimed to have foreseen the Great Depression "ever since the rise of the Coolidge Prosperity, and had been at pains to keep my own money out of speculative investments." With Sara on the mend and his latest book about to be published, he had more amusing things on his mind. Three days after Christmas, he sailed for Europe to cover the London Naval Conference for the *Evening Sun* and visit the Continent, having filed a tepid review of Ernest Hemingway's *A Farewell to Arms* for the January *Mercury* ("The virtue of the story lies in its brilliant evocation of the horrible squalor and confusion of war. . . . But Henry and Catherine, it seems to me, are always a shade less real than the rest"). He got tipsy with Fritz Kreisler, the great Austrian violinist, on the cruise to Cherbourg, received the first copy of *Treatise on the Gods* from Blanche Knopf, Alfred's wife, in Paris, and took the Orient Express to Vienna, where he visited the graves of Beethoven, Mozart, Schubert, Brahms, and Johann Strauss and heard a "superbly sung" performance of *Così fan tutte* at the Vienna Staatsoper. After a side trip to Budapest, he went to London to earn his keep, sharing a "lordly suite" with Paul Patterson at the Savoy. He wrote to Sara from every stop, knocking down the walls of reticence and speaking openly of his love: "How lovely you are, and how I'll miss you! You will never know how much I think of you, and depend on you, and love you. . . . Your letter was lovely, as you are, and always will be. I am horribly homesick for you. But the next time you will be aboard, and everything will be perfect. . . . I miss you dreadfully, and love you completely."

He returned in February, and Sara went home to Montgomery two months later to tell her family that she had become engaged. She wrote to her new fiancé nearly every day, sounding more like a giddy schoolgirl ("I am so happy I am dizzy. You're the most perfect person in the world") than the accomplished author whose first novel, *The Making of a Lady*, had just been signed by Doubleday. In her absence Mencken rented a high-ceilinged duplex apartment on the third floor of a brownstone at 704 Cathedral Street, a few blocks from the downtown branch of the Enoch Pratt Free Library. Sara moved in and started decorating shortly after she came back to Baltimore in May. The old-fashioned furnishings were mostly of her choosing—flowery satin wallpaper, antique rosewood furniture, and several hundred wax flowers and pin boxes—though Mencken supplied a few homey touches of his own, including the baby grand piano from the Hollins Street parlor, one of his inscribed portraits of Kaiser Wilhelm II, an abstract painting by Thomas Hart Benton that he hung as a joke, and a jumbo lithograph of the Pabst breweries in Milwaukee. (This last loomed large over a mahogany sideboard in the dining room, where Mencken and Clarence Darrow can be seen admiring it in a 1930 photograph.) Soon he began sending jovial letters to his friends and colleagues, telling them the news. "Barring acts of God and the public enemy," he wrote to Edith Lustgarten, his *Mercury* secretary, "I am to be married on August 27th. The bride-elect is Sara Haardt, a talented girl of great courage. Tear up her card in the index: if she ever works for the magazine hereafter it will have to be for nothing!"

Mencken knew that the official announcement of his engagement would cause a tremendous stir. Paul Patterson offered to handle the press arrangements, and he accepted with relief. "My fear was that the newspapers would play it up as sensational news," he recalled, "for I had long been one of their principal examples of a confirmed and incorrigible bachelor. It was Patterson's job to get something approaching decorum into their reports, and this he achieved with great skill." The announcement was made in Montgomery on August 3, and all requests for interviews with the bride and groom were directed to Patterson, who released a one-line statement from Mencken: "I formerly was not as wise as I am now." Three weeks later he and Sara answered a series of written questions submitted by United Press. Asked if he had "any of the traditional fluster of a bridegroom-to-be," he replied, "I discern no tremors. Getting married, like getting hanged, is probably a great deal less dreadful than it has been made out. . . . It seems to me that the best rule for marriage is the best rule for all human relations: 'Be polite.' I am marrying one of the politest of women, and she is getting a husband whose politeness has the high polish of a mirror."

As he expected, his fellow journalists had a field day with the news that America's cynic in chief had yielded to the delights of middle-aged love, and some were even clever enough to quote from the twenty-fifth chapter of *In Defense of Women*, "Late Marriages": "The marriage of a first-rate man, when it takes place at all, commonly takes place relatively late. He may succumb in the end, but he is almost always able to postpone the disaster a good deal longer than the average poor clodpate, or normal man." Their stories filled seven fat scrapbooks and parts of three more, with the best headline of all, BELLICOSE MENCKEN WILL TRIP TO ALTAR JUST LIKE OTHER BABBITTS, appearing in a Mississippi paper. Mencken took the teasing in good humor, perhaps because he had a card up his sleeve. He secretly moved the wedding date up a week in order to prevent photographers from "busting into the church and shooting off flashlights."

An industrious *New York Times* reporter soon broke the story, but Patterson talked his fellow editors into settling for the pictures of a single, handpicked *Sun* photographer, and Mencken and Sara were married without incident on the afternoon of August 27 at the Church of St. Stephen the Martyr. It was an Episcopal ceremony—the state of Maryland did not recognize civil marriages—with August as best man and eight other family members in attendance. The only outsiders were Patterson, Hamilton Owens, and the photographer, whose work Mencken found unlovely: "No Hearst photographer, trained upon divorcees and murderesses, could have done worse. In every picture Sara showed an expression of extreme foreboding, almost of alarm, and her complexion was four or five shades darker than it was in fact." That morning, he had discreetly bundled up Aileen Pringle's letters and telegrams and sent them back to her in Hollywood, closing the books on his bachelor days. He had already written to Aileen and Gretchen Hood to tell them of his impending marriage, but no such letter to Marion has survived.

The newlyweds returned from their Canadian honeymoon in October and settled into their seven-room apartment, the fifty-year-old groom's first new home since 1883. Since Sara was as serious about her writing as her husband was about his, they soon established a firm routine. They breakfasted separately, he at eight and she at nine, after which Sara spent the rest of the morning tending to household chores while Mencken answered letters and read *Mercury* manuscripts (by then he was doing all his serious writing at night). Though she had never kept a house, she took to it like the well-bred lady she was. The table was always set for four at noon—Mencken often brought home guests without warning—and the cook and maid were dressed in spotlessly turned-out black uniforms. Following lunch and a nap, the Menckens worked all afternoon. They dined at six, went back to work at seven, then joined each other at ten for a drink and a chat in front of the fire before retiring to their separate bedrooms for the night. "I don't think that we ever bored each other," Mencken wrote in his diary five years after her death. "I know that, for my part, the last days of that gabbling were as stimulating as the first. I never heard her say a downright foolish thing."

Mencken affected to be flabbergasted by the ease with which he adjusted to married life. As he told Philip Goodman, "I expected to make rather heavy weather of the first year. I feared I'd be homesick for Hollins Street, and that it would be more difficult to work in new surroundings, eating purely Southern cooking—hams and greens, corn-pone, hot biscuits, etc. Nothing of the sort ensued. I am far more comfortable than I was in Hollins Street. Sara takes all telephone calls. . . . She never uses coercion on me—that is, never in an obnoxious manner." In fact she had tailored her own habits to fit her husband's long-settled routine. "Marriage hasn't changed Henry a bit," she told a reporter in 1932, adding that "I never think of [him] as a radical. He has always seemed a conservative and very conventional."

Insofar as possible she let him do what he wanted, including brewing his own beer and going out each week to carouse with the Saturday Night Club, which Sara unbent enough to permit to meet at their apartment on infrequent occasions.* He responded by turning himself into the very model of a doting, hand-holding husband, a change at which his friends marveled. (He even replaced his clattery Corona typewriter with a brand-new noiseless Underwood, so as not to disturb her sleep.) He knew that her fragile health might give way suddenly and permanently—she was hospitalized with pleurisy twice in the four and a half years of their marriage—but the knowledge only made him all the more determined to wring every moment of pleasure from whatever time they were destined to spend together. If Mencken's long, uncomfortable romance with Marion Bloom shows him at his worst, then his poignantly brief marriage to Sara shows him at his very best.

Preoccupied as he was with his new wife and home, he could have been forgiven for failing to notice that the wild times of normalcy and fiesta—of Scopes and Gatsby, Al Capone and Lucky Lindy, Silent Cal and Rudolph Valentino and the *American Mercury*—were gone for good. Yet one unignorable sign of change had greeted him on his return from Europe that February: the last issue of the *Mercury* in which George Jean Nathan's name appeared as contributing editor. As of March "Clinical Notes" and "The Theatre" were dropped, and Nathan vanished from the masthead. No formal

---

*"My wife can always tell when I've been out drinking a lot of beer," he told a radio interviewer in 1932, "because she says I come home and am very complimentary and bland."

announcement of the changes was made, but they were widely noticed nonetheless, and newspapers from coast to coast carried an Associated Press story reporting the departure of the magazine's cofounder. To the *New York Times* the tactful Nathan said only, "My friend Mencken is the *Mercury*'s ideal editor." Mencken said nothing in public, and many of his friends were unaware of the extent of their estrangement. The two men continued to see each other socially from time to time, and Nathan gave Mencken and Sara lunch when they came to Manhattan after their honeymoon. Sara Mayfield, who was present, left with the impression that "they were certainly on the best of terms." Mencken's private thoughts can be inferred more accurately from a letter he sent to Blanche Knopf earlier that year: "George tells me that he wants to use the title, 'Clinical Notes,' on a book to be published by some other publisher. I have asked him to hold up the business until Alfred gets back. The title is my invention, and I regard it as the property of The American Mercury. I may want to revive it later on, with someone else doing the stuff. . . . I see no reason why we should give George anything for nothing." With those cold words he said farewell to the most significant friendship and the most eventful decade of his life. He knew true happiness and wrote his best books in the eighteen years that remained of his career, but once the twenties were over, the road ran downhill.

THOUGH MENCKEN HAD PASSED THE PEAK OF HIS POPULARITY, THE enthusiasm with which the press covered his doings serves to remind us that he remained the prewar equivalent of a modern-day media figure. In November, Sinclair Lewis won the Nobel Prize for Literature—the first American author to do so—and went out of his way to make mention of Mencken in his acceptance speech, large parts of which could have been drafted by the critic who had banged the drum for *Main Street* and *Babbitt*. As late as 1932 he was considered worthy of an admiring profile in *Vanity Fair*, in which Ernest Boyd spoke of him as having "wielded an influence in this country comparable to that of Shaw in England. . . . [H]e is *homo americanus*, raised to the Nth power, the American citizen quintessential, as expounded and dreamed of in the Jeffersonian philosophy."

Yet in many ways he was strangely isolated. At fifty he no longer kept in touch with new developments in literature and the arts, and not since the Scopes trial had he done serious firsthand reporting on any aspect of American life except for party politics. Absorbed by the challenge of making a life with Sara, he had become a pundit, writing off clips and gossip instead of doing his own legwork. Hence the increasing detachment of his post-1930 writings, a quality emphasized by his unwillingness to confront the darkening realities of the depression. At first he thought it simply another twist in the business cycle, one whose ultimate effect would be to purge the economy of fraud and waste, though Angoff believed (or claimed to believe) that its persistence gave him "inner doubts about his own grasp of events." True or not, Mencken was no economist, and while he never wavered in his loathing for socialism, he might well have joined his fellow countrymen in wondering why the United States was failing to right itself.

The *Mercury* expressed no such bafflement. It proceeded on its normal course, nicely summarized in the 1930 edition of *Writer's Market*, the annual publication that describes the contents of magazines for the benefit of freelancers looking for new outlets: "Strongly opinionated articles written more or less in the style used by the editor. Political sketches summing up the career of some demagogue. Short stories of exceptional merit. Very little verse. Articles on some social problem treated from liberal [i.e., libertarian] viewpoint. Shorts, written by men of wide experience in their field, in Arts & Science Department." Though Nathan had been dropped in March and the cover had been given a newer look in July, the rest of the magazine remained unchanged. One could open any issue and find articles that might have been written in 1924: "The Word of God came to the United States Military Academy in the twenty-third year of its existence, on a certain Winter morning, late in 1825. It arrived with a spiritual detonation as thunderous as if one of the campus cannon had suddenly fired off and kicked itself over the cliffs into the Hudson."

Mencken's involvement in the magazine's day-to-day affairs was as vigorous as ever, and he continued to wield the blue pencil with every sign of enthusiasm. (One regular contributor who sent in a piece about safe-cracking "yeggmen" received a long letter full of queries and suggestions, prefaced by this engaging exhortation: "This article is so good that I hesitate to ask you to give it another lick, but I see some holes in it that you can't see yourself, and when it goes into The American Mercury *I want it to be perfect.*") That was the problem. He had launched the *Mercury* in order to deal with a wider range of topics than the *Smart Set* had allowed itself to address, but his own declining interest in imaginative writing had cut down on the sheer number of *voices* in the magazine, making it duller and narrower. From the outset it had mirrored his style and opinions; now it reflected little else. James Cain claimed that of all the magazine editors with whom he worked, Mencken was the worst, explaining that "you couldn't argue with [him]. You had to make up your mind to take it and like it, as I decided to do, or not write for him at all." Even such minor tics as his insistence on referring to politicians as "Dr. Coolidge" and "Dr. Wilson" had begun to grate. "Can't Henry see that this silly gag is what excludes most professional writers from his magazine and means he gets it out with an endless succession of pieces by amateurs?" Arthur Krock of the *New York Times* griped to Cain. "Because the pro is not going to submit to a rule that makes it appear to the reader . . . that he is one more guy trying to write like Mencken."

Mencken was unapologetic, claiming that many of his contributors "needed so much help that my struggles with their manuscripts spread the legend that I was trying to make all contributors to the *American Mercury* write alike. I had no desire to make them write alike, but my own way of writing was the only way I knew, and when theirs turned out to be impossible I had to substitute mine." Yet other editors—above all Harold Ross, who had found his footing and was turning the *New Yorker* into the genuinely wide-ranging magazine the *Mercury* should have been—understood that the purpose of good editing is not to impose a uniform style on writers but to help them realize their own individual styles more completely. But Ross was not a writer, much less a famous writer with a highly distinctive style of his own. Mencken was, and such writers rarely make good editors, any more than virtuoso performers make good teachers. Like Jascha Heifetz showing an intimidated young student how to toss off a Paganini caprice, he knew only one way to write: his own.

He also only knew one way of looking at the world, and as the raucous twenties gave way to the anxious thirties, fewer Americans were disposed to agree with it. Cain thought that the *Mercury* "was utterly unable, once the times grew darker, to meet the new challenge, as howls of laughter don't cover much ground at the expense of the bread-line." Perhaps it wasn't so much that Americans no longer longed to laugh as that they were tired of laughing at the things Mencken found funny. It is noteworthy that his unwillingness to engage with new ideas was not limited to the collectivist ideology of the Left. A case in point is his response to the publication in 1931 of *I'll Take My Stand*, the book-length manifesto in which a group of southern intellectuals, among them John Crowe Ransom, Allen Tate, and Robert Penn Warren, argued for a spiritually conscious agrarian conservatism that sought to revive the virtues of the Old South. Like them or not, they were serious thinkers who were taken seriously, but Mencken's response to their critique of modern American life bordered on outright anti-intellectualism: "The factitious, drug-storish 'superiority' of the professional pedagogue hangs about it. It has little more bearing upon life as men and women must now live it in the world than the Presbyterian metaphysics of Paul Elmer More or the amusing but falsetto vituperations of Irving Babbitt." For him the South was still the Sahara of the Bozart, the land of

Coca-Cola, hookworm, and Bryan-loving Holy Rollers, and no amount of evidence to the contrary—not even from Sara—could make him change his mind.

Illness prevented Mencken from writing the same kind of extended narrative history of his involvement with the *Mercury* that he supplied for his years at the *Smart Set*, but what he did record, in *Thirty-Five Years of Newspaper Work* and the diary he started to keep late in 1930, suggests that he blamed the magazine's decline on the bad management of Samuel Knopf, Alfred's father, who served as business manager for both Alfred A. Knopf, Inc., and the *American Mercury*. In the long run, Mencken said, the elder Knopf's wrong-headed emphasis on "a large circulation and heavy general advertising" doomed it to long-term difficulties by artificially inflating its circulation beyond the natural level to which it had started to fall after 1928. He would never acknowledge that the magazine's contents might also have been at fault. Whatever the reason, the *Mercury* was in hot water, and by 1931 its founding editor had started to "make plans for clearing out. . . . I must have told Alfred Knopf of my desire to withdraw well before the end of 1931, but he begged me to hang on, and his father's death in June, 1932, left things in such confusion that I had to do so until the end of 1933."

That was his version, written a decade after the fact, and it may have been true, though there is no suggestion in his diary entries for 1931 or 1932 that he had told Knopf of his desire to quit the *Mercury*. But he had more than enough other projects in mind to explain his loss of interest. Early in 1931 he started a sequel to *Treatise on the Gods*, this one dealing with ethics, and it had proved to be painfully hard slogging: "I must consult authorities at every step, and there is a great deal of rewriting." After that he wanted to revise *The American Language* and *In Defense of Women*, prepare a two-volume collection of *Prejudices*, and write "a frank discussion of moral problems" to be called *Advice to Young Men*. He also envisioned "a large treatise on the human race" called *Homo Sapiens*; a book about government, "larger

and better than 'Notes on Democracy' "; and "a psychological autobiography, describing the origin and growth of my ideas." It was an
ambitious program, and he must have known that editing other men's
articles would keep him from finishing more than a fraction of it.

Why he now decided to keep a diary is impossible to explain—
possibly he saw it as a resource for some future autobiography—but
he began dictating regular entries to Rosalind Lohrfinck in November, and the results, brisk, impersonal, and gossipy, told much about
his crowded life in the thirties when they were published sixty years
later. We see him dining with Scott and Zelda Fitzgerald, drinking
home-brewed beer with the taciturn T. S. Eliot and chatting about
the difficulties of getting out a monthly magazine, going to Carnegie
Hall with Alfred and Blanche Knopf to hear Serge Koussevitzky
conduct Mahler's Ninth Symphony. (Willa Cather was at the same
concert, and Mencken was dismayed to learn that she preferred
another work on the program, a "cheap" piece by Ravel, to Mahler's
"very fine writing.")

Once in a while he betrays a glimmer of sentiment, as when he
reports that Gertrude and August are thinking of leaving Hollins
Street to share an apartment in the suburbs: "My inclination is to
take over the house, lock it up, and keep it substantially as it is. I
couldn't bear to think of strangers living in it." Yet he wrote little
about Sara, and nothing about the depression. According to Angoff,
he was insisting as late as mid-1931 that "the depression was nonexistent, that it was the invention of 'charity racketeers,'" but it is
unlikely that he was quite so unaware as that. Between 1929 and
1932, the gross national product dropped from $104.4 billion to
$74.2 billion, and unemployment rose from 3.1 to 24 percent.
Mencken was too well informed not to have noticed such things, but
he preferred to keep them out of his diary—and, at least for now, his
magazine. In November 1931 the *Mercury* published "The Tragedy
of the Sioux," an essay by Chief Sitting Bull; an annotated collection
of suicide notes and a glossary of "Circus Words"; a piece by the

avant-garde composer Henry Cowell on "The Basis of Musical Pleasure"; articles on college sororities and American doctors in Vienna; an editorial on "calendar reform"; reviews by Mencken of *Oklahoma City's Younger Leaders*, three books on Prohibition, and a biography of Mme. Blavatsky, the founder of theosophy; and the concluding installment of "The Worst American State," a three-part series by Mencken and Angoff. (The prize went to Arkansas, whose Ku Klux Klan chapter responded by declaring the editor of the *Mercury* to be a "moral pervert.")

Angoff claimed to have picked a fight with Mencken over this issue, arguing that the *Mercury* had become "an antiquarian magazine" at the very moment when the United States was in "the throes of a historic crisis." A few months later Edmund Wilson, long one of his greatest admirers, wrote a piece for the *New Republic* in which he contended that the man he had once praised as "the civilized consciousness of America" was now reduced to "having the same emotions, stuck in the same intellectual jam, content to rage and complain without hope." Around the same time the *Omaha World Herald* felt moved to publish "Mellow Mencken," an editorial whose sugary praise would have irked its subject far more than the sharp words of a left-wing literary critic who would soon join Theodore Dreiser, Sherwood Anderson, and John Dos Passos in endorsing William Z. Foster, the Communist Party's presidential candidate: "Life grows sweeter to him now, with a happy marriage, with a plumper middle age, with a more sedentary existence."

The *Mercury* finally deigned to acknowledge the Great Depression in March 1932. Mencken's monthly editorial, "What Is Going On in the World," summed up his view of the crash and its aftermath with typical straightforwardness:

> It is only the story of those Americans who yielded irrationally
> to professional seers and visionaries, as yokels yield to travelling
> corn-doctors and evangelists. . . . Are the rest of us in the same

boat? I doubt it. The boat we are in is getting some unpleasant rocking from the foundering of the other, but it is tighter of seam and will survive. We have all lost something, but not many have really lost everything. In actual values the country is still rich, and any man who owns any honest part of it still has that part, and will see it making money for him when the clouds roll by. . . . It seems to me that the depression will be well worth its cost if it brings Americans back to their senses. Once they rediscover the massive fact that hard thrift and not gambler's luck is the only true basis of national wealth, they will discover simultaneously that a perfectly civilized and contented life is possible without the old fuss and display.

His own boat was being rocked, too. "My total income for 1932," he later told a German correspondent, "was less than forty percent of my income for 1931, and in 1933 there will be an even further drop. My book royalties in one year actually dropped eighty percent. These figures should show you that the Depression on this side is very real. . . . The number of people, once well-to-do, who are quite without means is really appalling." But though he knew that something was wrong, he remained incapable of admitting its extent. In May he gave an interview to a United Press reporter in which he proclaimed that the effects of the crash had been "greatly exaggerated, mainly by interested parties. Some of them are communistic sympathizers who try to make it appear that communism would save us. Others are charity mongers who naturally howl calamity whenever there is a chance, and yet others are simply idiots of the kind who yell 'fire!' in crowded theaters." These remarks, like his *Mercury* piece, were widely and unfavorably noted by editorial writers and columnists, and with good reason. Before he had merely been out of touch; now the Fool Killer of the twenties was willfully refusing to face the verities of the thirties. Hubris was courting Nemesis.

The twain met face to face when Mencken went to Chicago in June to cover the Republican and Democratic conventions for the

Sunpapers, there to write in tepid praise of the man he had dubbed "Roosevelt II" in 1920, describing him as "a pale and somewhat pathetic caricature of his late relative, Theodore . . . resembling the Rough Rider much as a wart resembles the Matterhorn." When Franklin Roosevelt reemerged twelve years later as the leading candidate for the Democratic presidential nomination, Mencken echoed those words in a Monday Article: "He is one of the most charming of men, but like many another very charming man he leaves on the beholder the impression that he is also somewhat shallow and futile. It is hard to say precisely how that impression is produced: maybe his Christian Science smile is to blame, or the tenor overtones in his voice. Whatever the cause, the fact is patent that he fails somehow to measure up to the common concept of a first-rate man." Yet he had even less use for "the pathetic mud-turtle, Lord Hoover," and once FDR came out in favor of legalizing the sale of beer, he had Mencken's support, such as it was: "I believe that, taking everything into consideration, the Democrats are wetter than the Republicans—yes, quite a little wetter. I think, taking one day with another, that the Democratic Party is a little bit better than the Republican Party. They're a shade decenter people." The high-flying rhetoric of Roosevelt's acceptance speech, in which he demanded "a new deal for the American people," meant nothing to Mencken and little to anyone else. Walter Lippmann had already summed FDR up as "a pleasant man, without any important qualifications for the office," and few were disposed to disagree. Roosevelt was elected not because of anything he said or did in the 1932 campaign, but because his name was not Herbert Hoover.

Knopf encouraged Mencken to spin his convention coverage into one of the first "instant" books, *Making a President: A Footnote to the Saga of Democracy*, published in September to good reviews and minuscule sales. Knopf sent an advance copy to Roosevelt, who acknowledged it in a letter in which he declared that "I am particularly happy to have it because I happen to know Mr. Mencken very well." Mencken was amused by the note, and even more amused that

Knopf fell for FDR's flattery. That was the extent of his interest in the presidential campaign of 1932. Unexcited by either candidate, he chose not to go on the road for the *Sun*, content to pay a visit to Roosevelt's October 25 speech at Baltimore's Fifth Regiment Armory, which left him resolutely unimpressed: "He was built for lyrical work, not for whooping and howling. . . . His voice shows heavy strain, and he begins to force his tones like a wornout opera tenor."

Two months after Roosevelt's election, he told Knopf that he wanted out of the *Mercury*. According to Angoff, the two men had been "bickering" over the magazine's contents for some time, and Knopf suggested that what his friend needed was "a good long vacation." Yet another version of the story comes from August Mencken: "He had decided—aided and abetted by his wife . . . to quit the *Mercury*, take a year, and do nothing but serious work." Whatever the truth of the matter—and none of these explanations should be taken at face value—Mencken and Knopf agreed that it was time for a change. After months of waffling by Knopf, they invited the conservative journalist Henry Hazlitt to take over as of the first issue of 1934, the tenth anniversary of the *Mercury*'s founding. It should have been a landmark moment: Mencken would be out of the magazine business for the first time since 1915. But by the time he cleaned out his desk, the world had changed so dramatically that his departure would end up being little more than a footnote to the activities of a pair of master politicians whose rise to power caught him flat-footed.

NINETEEN THIRTY-THREE WAS THE YEAR THE WEST BLEW UP. ON March 5 the Nazis won a plurality of seats in the German Reichstag, and Adolf Hitler, the country's new chancellor, promptly moved to establish one-man rule, turning his armed thugs loose on the Jews and opening a concentration camp at Dachau. Meanwhile Franklin Roosevelt, having assured the American people that the only thing they had to fear was fear itself, started bombarding the incoming Congress with a cornucopia of legislation aimed at curing the depression by putting the federal government in charge of the American economy.*

It was a time of maximal confusion, and Mencken was briefly thrown off his bearings. "I had a low view of Roosevelt," he wrote in *Thirty-Five Years of Newspaper Work*, "but he was at least better than Hoover, and so I was inclined to grant him every reasonable assumption when he was sworn in. But when, after calling Congress in extra session, he began reaching out for a vast congeries of extraordinary and often extra-constitutional powers, and, what is worse, filling Washington with a horde of bogus 'experts' of a hundred varieties, I found myself on the opposition bench once again—my natural, most comfortable and apparently inevitable place." That was the way he remembered it, but what he wrote in his first Monday Article about the Roosevelt administration, "A Time to Be Wary," was different in tone:

> What I here suggest is by no means a general walk-out on Dr. Roosevelt. He is tackling the problem before him in a very vigorous manner, and he seems to be moving toward its solution. He is obviously not infallible, but it is certainly refreshing to behold his courage and enterprise . . . There has probably been no President since Cleveland who was better worthy of the immense and almost singular powers now put in his hands. But even the best dictatorship ought to have clearly defined limits, and its end ought to be kept in sight from its beginning. If the American people really tire of democracy and want to make a trial of Fascism, I shall be the last person to object. But if that is their mood, then they had better proced toward their aim by changing the Constitution and not by forgetting it.

---

*Among his first acts as president was to legalize beer, and Mencken celebrated at midnight on April 6, 1933, by ordering the first stein of post-Prohibition 3.2 percent beer to be served at the bar of Baltimore's Rennert Hotel, with a *Sun* photographer on hand to snap a picture of his goggle-eyed colleague pouring the contents down his throat. "Pretty good. Not bad at all," he declared in a story published in the next day's *Sun* under the headline "Mencken, High Priest of Brew, Gives Sanction to Beer."

He found Hitler no less confusing at first. "The German news is probably at least nine-tenths bogus," he wrote to a friend on March 24. "Certainly the Germans are not beating up Jews, as such. It simply happens that a good many Communists are Jews." Not until May, two days after the Nazis held their first book burning, did he decide that what was happening in Germany went well beyond anything he was prepared to support—and even then, he refused to attack Hitler in print. It wasn't that he admired the Nazis. As he explained to a pro-German correspondent, he thought them boobs, but nothing worse than that: "I am keenly conscious that this is the time, above all times since 1914, for those who know and respect the country to come to its defense, but I see no way to make that defense convincing to fair men so long as the chief officer of the German state continues to make speeches worthy of an Imperial Wizard of the Ku Klux Klan, and his followers imitate, plainly with his connivance, the monkey-shines of the American Legion at its worst."

Mencken intended this letter as a definitive statement of his position on the new regime in Germany. (He sent a copy of it to Paul Patterson in order to explain why he had said nothing about the Nazis in his column.) So it was, though not in quite the way he meant. What it shows is that he misunderstood Adolf Hitler as completely as he had misinterpreted Nietzsche and Conrad, and for much the same reason: He had no feeling for the darkness in the heart of man. He looked at evil and saw ignorance. To him Hitler was Babbitt run amok, and he thought it inconceivable that such a buffoon could long pull the wool over the eyes of the most civilized people on earth. Sooner or later they would have to catch on. In the meantime Americans needed to keep their wits about them, and above all to avoid being inflamed by the "professional kikes" whose hotly partisan rhetoric he feared would only strengthen Hitler's hand: "It was plain from the start that the campaign against the Jews would set their brethren in the United States to howling mightily, and I well knew Roosevelt's friendliness toward them and his obligation to them. . . . The only consequence I could see in the rise of the Nazis was a revival of anti-German agitation all over the world. But it was one thing to feel thus, and quite another thing to help the Jews foment that agitation." That was why he kept silent about Hitler in public, though he sent letters to German friends telling them that their *Führer* was a boob.

In the wake of the Holocaust it is impossible to see things the way Mencken saw them in 1933. He had no idea that Hitler would undertake to murder every Jew in Europe, successfully doing away with six million of them before the Allies put an end to his "final solution." Had he known it, he would have been sickened—and he would have said so, as loudly and eloquently as possible. Two years earlier, when a black man was lynched in Maryland, he had written the first in a series of powerful Monday Articles that brought him death threats: "Not many observant Marylanders, I take it, were surprised by the news of last Friday's extraordinarily savage and revolting lynching at Salisbury. Something of the sort had been plainly hatching down in that forlorn corner of the State for a long while. . . . The Ku Klux Klan, which was laughed at in all the more civilized parts of Maryland, got a firm lodgment in the lower counties of the Shore, and the brutish imbecilities that it propagated are still accepted gravely by large numbers of the people, including not a few who should know better."*

---

*The ferocious scorn with which Mencken wrote about lynching, though entirely sincere, is at the same time inseparable from his contempt for the culture of the rural South. Consider, for example, his 1932 review of Ernest Hemingway's *Death in the Afternoon*:

> I emerge cherishing a hope that bullfighting will be introduced at Harvard and Yale, or, if not at Harvard and Yale, then at least in the Lynching Belt of the South, where it would offer stiff and perhaps ruinous competition to the frying of poor blackamoors. Years ago I proposed that brass bands be set up down there for that purpose, but bullfights would be better. Imagine the moral stimulation in rural Georgia if an evangelist came to town offering to fight the local bulls by day and baptize the local damned by night!

Yet he also believed that "the white man is actually superior to a Negro and on almost all counts," just as he believed that the Jews were "the most unpleasant race ever heard of," and that their unpleasantness was in large part responsible for the mass resentment that put the Nazis in power. He stated this latter conviction most explicitly in a private memorandum written around 1945:

> There was very little anti-Semitism in [Germany] before and during World War I—in fact, I have heard argument that there was less than in England or the United States—and under the democratic constitution adopted after the war [Jews] had equal rights with all other citizens. But when this opening showed itself, they began to push into the government in large numbers, and in a few years they were in a fair way to dominate it, and many Germans became alarmed. There was a Jew in, or close to every key position, and Jewish ideas began to have the call on German ideas. . . . The result was an enormous growth of anti-Semitism. Hitler organized it, and aroused the mob to support it, mainly on unsound grounds, but he did not have to create it.

Such views were not rare in the United States in 1933, even among well-to-do Jews who recoiled from their greenhorn brethren. If anything Mencken's opinions were moderate, at least by comparison with the vicious Jew-baiting so widely accepted in the thirties that it was possible for Father Charles E. Coughlin to broadcast openly anti-Semitic talks over his own nationwide radio network. It was a decade when "restricted" hotels and Jewish quotas at Ivy League universities were taken for granted. (Otto Kahn, chairman of the board of New York's Metropolitan Opera Company from 1911 to 1931, was never allowed to rent a box seat in the Metropolitan Opera House. Even after he was baptized an Episcopalian, he had to watch the company's performances from the orchestra.) That Mencken had not changed his mind about German anti-Semitism by

1945 is another, more disturbing matter, though even in 1933 some of his friends were unable to stomach his position on the Nazis. Philip Goodman broke with him at that time, a development that Mencken described tersely in a footnote to *Thirty-Five Years of Newspaper Work*, explaining that "the troubles of the Jews in Germany reduced [Goodman] to such a state of hysteria that he announced that he was a Communist, and my friendship with him, begun in 1917, terminated." But most of the other Jews who treasured their own friendships with Mencken were no more prepared to part company with him over Hitler than they had been over *Treatise on the Gods*. Knopf, for one, did not boggle at publishing that book, any more than he boggled at accepting an ad from the German Tourist Information Service, which ran in the *Mercury* in the summer of 1933: "Germany is the appeasing refuge from weariness of daily routine, the struggle for gain, the stress of social activities. It offers genuine refreshment of the soul, rehabilitation of health; and sends you back with new ideas and broader visions."

As for Mencken, he had his long-deferred say about Hitler in his last *Mercury* book review, published that December. His retirement had been announced two months earlier, inspiring a rash of obituary notices, the most perceptive of which appeared in the *New Yorker*: "He let a good deal of wind out of the country's fat stomach. Today the national flatulence is of another order—less the acute pains from eating too much at Rotary luncheons, more the large retchings of a persistent emptiness." No less telling, though, was a full-page ad that Knopf placed in *Esquire* to announce the magazine's "new editorial direction": "THE AMERICAN MERCURY will be seriously concerned with the tremendous social, political and economic problems raised by the world crisis. . . . [It] will be less concerned with the now comparatively harmless forms of imbecility displayed by Rotarians, Babbitts, and the lower inhabitants of the Bible Belt."

Mencken's own valedictory editorial, in which he explained that he now intended to devote himself to writing more ambitious books requiring "hard study and continuous application . . . hardly compatible with the daily duties of a magazine editor," was overshadowed by "Hitlerismus," a review of *Mein Kampf* and four other books on Hitler and Nazi Germany in which he sought to provide an even-handed description of how Hitler seized power: "The disadvantage of the Jew is that, to simple men, he always seems a kind of foreigner. . . . Thus he is an easy mark for demagogues when the common people are uneasy, and it is useful to find a goat." He went on to oppose a "general boycott" of Nazi Germany, explaining that "here in the United States it might easily launch a formidable anti-Semitic movement, especially in view of the fact that many of the current Jewish leaders in this country are very loud and brassy fellows, with an unhappy talent for making the *Goyim* Jew-conscious." When Knopf saw the galleys he wrote at once to warn that the review would "leave in the minds of a great many readers the idea that you are, yourself, at best lukewarm in your feelings" about Hitler. Mencken thought otherwise. All he was trying to do, he replied, was to "disentangle the facts from the blah of both sides," though he was willing to cut the last paragraph, in which he had suggested that Hitler had moderated his views since writing *Mein Kampf.* Knopf declared the change sufficient, and the piece as published ended by recommending a policy of benign neglect: "That a majority of Germans will go on yielding to Hitler *ad infinitum* is hard to believe, and in fact downright incredible. Either he will have to change his programme so that it comes into reasonable accord with German tradition and the hard-won principia of modern civilization, or they will rise against him and turn him out."

With that prediction, the founder of the *American Mercury* left his creation in the hands of Alfred Knopf and Henry Hazlitt. Knopf sold the magazine in 1935, and it survived until 1980, serving for some years as one of the earliest journalistic voices of the postwar conservative intellectual movement before passing into the hands of a series of anti-Semitic cranks, a horrendous irony that went largely unnoticed solely because the *Mercury* itself was long forgotten. For all practical purposes it died when Mencken quit, and in the minds of most Americans it had died well before that, for reasons best summed up by an anonymous staffer at *Time:* "Its angry crusadings against provincialism, the Bible Belt, Rotarians, evangelists, puritanism, Prohibition and Babbittry in general became sufficiently successful to seem a little antiquated and unnecessary."

HAVING BEEN OFFICIALLY DECLARED A BACK NUMBER, MENCKEN polished off his book on ethics, then took Sara on a Mediterranean cruise in February, a two-month-long trip he later remembered as "our happiest time together," during which they visited Greece, Italy, North Africa, Turkey, and—perhaps surprisingly—Palestine.* No sooner had he announced that he was leaving the *Mercury* than he was inundated with offers from "virtually all the important American newspaper syndicates," but the only query that piqued his interest came from the *New Yorker*'s Katharine White (whose husband, E. B. White, had almost certainly written the magazine's editorial about his departure). Mencken told her that while he was about to go abroad and thus could make no promises, he would keep the *New Yorker* in mind. Out of this brief exchange grew the most important professional relationship of his post-*Mercury* days, though five years would pass before it came to full flower.

He came back to Baltimore in April to find that Henry Hazlitt had already proved impossible (Mencken told a Saturday Night Club

---

*Like many other people of his generation who had mixed feelings about Jews, Mencken became something of a Zionist, though his visit to Palestine left him wondering whether the Jews would be able to make a go of it over the long haul: "On the one hand, it is being planted intelligently and shows every sign of developing in a healthy manner. But on the other hand there are the Arabs—and across the Jordan is a vast reservoir of them, all hungry, all full of enlightened self-interest. Let some catastrophe in world politics take the British cops away, and the Jews who now fatten on so many lovely farms will have to fight desperately for their property and their lives."

member that Hazlitt had been "competent in all qualities except in managerial ability") and had been replaced by Angoff, whose ill-starred tenure as editor would last for exactly as long as it took Knopf to sell the *Mercury*. It was Knopf's problem now, not Mencken's. He cared only about his new book, *Treatise on Right and Wrong*, and its unfavorable reception was a bitter blow. Once again it was the *New Yorker* that hit the bull's-eye: "One is puzzled to note that the author refers to his treatise as this 'present ribald book.' I wager it would not bring even the mildest flush to the cheeks of a Radcliffe junior." In fact his attempt to write a popular history of what he called "moral science" was no more or less successful than *Treatise on the Gods*, with which it is consistent in tone and approach. The difference was that his reputation had lost its luster in the four years that separated the two books, and reviewers who once might have treated him with the respect due a distinguished elder were no longer inclined to pull their punches. Irwin Edman's *Saturday Review* notice was especially harsh: "Mr. Mencken gives the impression at times of a parent telling the facts of life to an eighteen-year-old boy who has long been practically and intellectually bored with them. Surely Mr. Mencken must realize that his pupils, the 'civilized minority' for whom he used to edit a magazine, may have grown up by this time."

Mencken spent the rest of 1934 revising *The American Language*, writing his Monday Articles, and turning out a new weekly column, this one for the Hearst papers. Whatever anger he felt at the reception of *Treatise on Right and Wrong* was soon forgotten, for he had something of vastly greater importance about which to worry: Sara's uncertain health had taken a sharp turn for the worse. She had shown signs of undue fatigue during their Mediterranean vacation, and Benjamin Baker, the Mencken family doctor, put her to bed in September, afraid that some stubborn pocket of tubercular infection had become active. Her temperature returned to normal in December, but she remained too weak to leave the apartment, even when her mother died of a stroke on Christmas Eve.

Mencken watched over her tenderly, drowning his anxieties in work. One of the assignments with which he sought to distract himself was the "loyal opposition" speech he gave before Washington's Gridiron Club on December 8, the occasion of his humiliation at the hands of the hated FDR—further proof, if he needed it, that his days of glory were over. Bad as that night was, though, it could not rival the possibility that the love of his life might be dying. She went into the hospital in March, and Mencken wrote dozens of letters to the Knopfs, reporting on her changing condition and simultaneously recording the ups and downs of his own agitated state of mind: "Sara has had a little setback, but it doesn't seem to be serious. I am howling for the quacks, and no doubt they'll be able to do something about it. . . . She is still showing some temperature, but the chiropractors report that it is not disquieting. . . . Sara's recovery is proceeding with depressing slowness. . . . I begin to suspect that Yahweh is on our trail." Even under the direst of circumstances, Mencken remained himself, characteristically referring to Sara's doctors as "quacks" and "chiropractors." But the time for slapstick was over now, and his detachment would soon be tested as never before.

Sara returned to Cathedral Street in April, and Mencken rented a cottage in the Adirondacks where he planned to take her to spend the summer. Then she developed "alarming symptoms" on May 25, and Mencken pecked out a stoic note to Max Brödel four days later in which he acknowledged the worst: "Sara has meningitis, with t.b. bacilli in the spinal fluid. It is, of course, completely hopeless. She seems comfortable—at least far more comfortable than she was a few days ago. The horrible headache has passed off, and she sleeps peacefully all day long. She may be aroused for half a minute, but hardly for more." He visited her in the hospital for the last time that same day. She recognized him and spoke a few coherent sentences, then lapsed into a coma, dying on May 31. ("I did not look upon her in death," he wrote in his diary a decade later. "It was too dreadful a thing to face.") August, who was recovering from a sinus operation, offered to move to Cathedral Street to keep him company; Mencken suggested that they take a trip to England after the funeral, "if only to get away, for a little while, from the endless reminders of Sara." As she had requested, she was cremated on June 3 after a brief ceremony, and her ashes buried in Baltimore's Loudon Park Cemetery, at the foot of the grave of the mother-in-law she never met. Mencken spent the next few days answering letters of condolence, and the Saturday Night Club met at his apartment that week, after which he and August left for England.

He had never been one for fond farewells. Dreiser, Marion, Nathan, the *Smart Set*, the *Mercury* that he had so carefully made in his own image: He had walked away from them all without looking back. Even the loss of his father had left him feeling "free at last." Only his mother's death had put more than the lightest of marks on him, and that could not compare to the emptiness he felt as he faced the bleak prospect of life without Sara, unable to find solace in the belief in immortality that neither of them possessed. "Now I feel completely dashed and dismayed," he wrote to Ellery Sedgwick. "No sensible work is possible. . . . What a cruel and idiotic world we live in!" He paid tribute to her in 1936 by editing a collection of her best stories, *Southern Album*, published by Doubleday and prefaced by a brief biographical sketch:

> The thing that obsessed her, thinking of the South, was the dramatic contrast between the old dying and the young struggling to live and thrive. It was a picture, in its way, of her own brief life. She was doomed like the Old South, and yet there was a tremendous vitality in her. I have never met a more patient person, or a more gallant. . . . I find it hard, even so soon after her death, to recall her as ill. It is much easier to remember her on those days when things were going well with her, and she was full of projects, and busy with her friends and the house, and merry with her easy laughter.

It was a graceful homage to the woman who had succeeded in pulling him away from the fixed course of his bachelor life, and whose steadfast love must have eased the hurt of his fall from public grace. But he saved the most telling statement of his anguish for his private files. "I was fifty-five years old before I ever envied anyone," he wrote in a 1941 notebook entry, "and then it was not so much for what others had as for what I had lost."

"WE'LL EVEN IT UP"

The Age of Roosevelt II, 1935–1941

Once again Mencken reached for his universal antidote. As soon as he and August came back to Baltimore in July, he resumed his Monday Articles and Hearst columns and began working in desperate earnest on *The American Language*. He had already begun to contribute occasional pieces to the *New Yorker*, and those continued as well. August returned to Hollins Street in November, leaving him to carry on by himself. "The ghosts stalk that dreadful apartment," James Cain said, "and he does nothing but talk, almost mechanically, as though that were the only way to keep them from jumping out and saying boo." He played music with his friends and visited August and Gertrude at Hollins Street each Sunday, but otherwise he spent his days and nights writing—and remembering. "In the weeks after Sara's death I was flat and hopeless; the whole thing seemed preposterous and intolerable," he told a newly widowed friend five years later. "What pulled me out of the hole was simply hard work—the hardest imaginable. I tackled the revision of 'The American Language,' and put in from 10 to 12 hours a day on it. By the time I got to Schellhase's in the evening I was so tired that I was only half alive."

It seemed as though the only things about which he cared to write were the American language and the perfidies of Franklin Roosevelt. The measured appraisals of his early columns on the New Deal gave way to trench warfare: "His infallibility has dropped from him as one after another of his vast and cocksure schemes for bringing in the millennium has blown up, and the assumption of his lofty altruism has been considerably corroded by his constant and transparent politicking." He even went so far as to make a guest appearance in the pages of the *Mercury*, declaring the politician he had once thought a gentleman to be nothing more than a common fraud:

> He is directly and solely responsible for every dollar that has been wasted, for every piece of highfalutin rubbish that has been put upon the statute books, and for the operations of every mountebank on the public payroll, from the highest to the lowest. He was elected to the Presidency on his solemn promise to carry on the government in a careful and sensible manner, and to put only competent men in office. He has repudiated that promise openly, deliberately, and in the most cynical manner.

Mencken's distrust of FDR went well beyond his economic policies. By 1935 he also thought that some kind of international armed conflict might be in the offing—possibly fomented by England, a country of whose devious power-balancing diplomacy he always took a dim view—and that Roosevelt "would try to take the United States into it, if only to conceal the colossal failure of the New Deal." This concern was yet another reason for his continuing silence over Hitler's persecution of the Jews. In letters to his Jewish friends he explained that "the uproar being made by the Jews in this country is doing them far more harm than good," inflaming anti-Semites in Germany and America alike. "A very definite anti-Semitic movement is gathering force behind the door, and whenever a convenient opportunity offers it will bust out," he told Benjamin De Casseres, one of the *Mercury*'s Jewish contributors. "At that time you may trust me to mount the battlements and holler for the Chosen." But he took it for granted that any European conflict into which the United States might be drawn would pit America against Germany. For

Mencken, who remembered with blinding clarity what it felt like to be spied upon by his own government, such a prospect was infinitely more repugnant than whatever Hitler might or might not now be doing to the Jews of Germany, and he believed that anti-German agitation by their American brethren made it that much more likely.

As fantastic as his point of view now sounds, it was well within the boundaries of normal political discourse in the mid-thirties. To be sure, the United States was full of people who felt (in the words of a Jewish writer remembering her thirties childhood) that "even more than a hero, which he certainly and most grandly was, Roosevelt was a fact of nature." Yet the tuxedoed socialites of Peter Arno's famous *New Yorker* cartoon who went to the Trans-Lux to "hiss Roosevelt" were not the only ones who had their doubts about him. As Mencken hastened to point out in his postelection autopsy, nearly twenty-eight million Americans may have voted for FDR, but another sixteen and a half million voted for his Republican opponent, and more than a few of them were thoughtful people with well-considered reasons for questioning the New Deal. World War I was not remembered as a "good war," and polls showed that as late as 1939 most Americans wanted to remain neutral in any future European conflagration. (A Roper poll taken shortly after World War II began showed that 37 percent of Americans wanted to stay out of the war completely, while another 30 percent were unwilling to do business with any warring nation, even on a cash-and-carry basis.) Even Mencken's half-admiring Anglophobia was common enough at the time.*

Because the pro-Roosevelt interventionists finally prevailed, and because most American writers and intellectuals took their side, those who felt differently now tend to be portrayed as more margin-

---

*"The only really intelligent people on earth are the English," he told De Casseres in a 1935 letter. "They have no graces and their selfishness is brutal and undisguised, but they can think at least four times as fast as the French and ten times as accurately."

alized than they were. FDR himself would give a speech in the summer of 1936 declaring that the United States should "shun political commitments which might entangle us in foreign wars." Two years later a constitutional amendment was proposed that would have made it illegal for Congress to declare war save in the event of an armed invasion, and otherwise only subject to the approval of "a majority of votes cast in a national referendum." In the House of Representatives 209 members voted to bring it to the floor—nine short of the number needed, and an actual majority.

To all this should be appended a well-known fact about which a great many American intellectuals were reluctant to speak in the thirties and forties, though it was never far from Mencken's mind. As he told Upton Sinclair in 1936:

> I am against the violation of civil rights by Hitler and Mussolini as much as you are, and well you know it. But I am also against the wholesale murders, confiscations and other outrages that have gone on in Russia. . . . You protest, and with justice, every time Hitler jails an opponent, but you forget that Stalin and company have jailed and murdered a thousand times as many. It seems to me, and indeed the evidence is plain, that compared to the Moscow brigands and assassins, Hitler is hardly more than a common Ku Kluxer and Mussolini almost a philanthropist.

In such a polarized climate Mencken's difference-splitting position—he was against the Nazis, but unwilling to do anything about it—would not have seemed in any way fantastic, save to those who, unlike him, understood that Adolf Hitler was something more than a Ku Kluxer. But it was hugely unpopular with the liberal intellectuals who had made up much of his audience, and who found it impossible to understand why the idol-smashing hero of their youth had become a Roosevelt-bashing isolationist who perversely insisted on talking about Hitler and Stalin as if they were two sides of the same counterfeit coin. Hence his dilemma: He kept on writing Monday Articles about the dangers of New Deal–style statism, but nobody seemed interested in what he had to say, and when the Nazis ratcheted up their persecution of the Jews still further, his idea of being helpful was to give an interview to a Jewish newspaper in which he pointed out that "getting rid of Hitler will not solve the Jewish question, whether in Germany or elsewhere. It will go on demoralizing the Jews and cursing the world until the Jews learn to go from Saturday to Friday without recalling once that they are Jews, just as the rest of us put in whole weeks without recalling that we are Aryans, or Chinamen, or members of Blood Group No. 4, or what not." He was sounding more and more like what he seemed to be: a crotchety, miserable man whom time had passed by, living by himself in an apartment full of fast-fading memories. He needed a change of scene—and of subject matter.

The first sign that he had found something new to write about came in January, when he sent a typescript called "Ordeal of a Philosopher" to Katharine White at the *New Yorker*, accompanied by a cover letter that sounded almost tentative: "This is the piece I threatened you with last week. It is really not a short story, but what it is I don't know. I had a lot of fun writing it, and so I am passing it on." It was a tale of his Baltimore childhood, and from the first page it was plain to see that he had struck gold:

> That learning and virtue do not always run together I learned early in life from the example of Old Wesley, a man of color liv-ing in the alley behind our house in Hollins street. Wesley dwelt in illicit symbiosis with Lily, the stately madura cook of our next-door neighbors, and once a year his younger brother, who pastored an A.M.E. church down in Calvert county, dropped in on him to remonstrate against his evil ways. But Wesley always won the ensuing bout in moral theology, for he had packed away in his head a complete roster of all the emi-nent Biblical characters who had taken headers through No. 7,

beginning with King David and running down to prophets of such outlandish names that, as I now suspect, many of them were probably invented on the spot.

Mencken had been spinning such yarns for years, but he had never before tried putting one down on paper, and when he did so, he discovered that written reminiscence came naturally to him. The ornately colloquial prose with which he had spent years belaboring the follies of novelists and politicians proved an ideal medium for the telling of shaggy-dog tales about his childhood. And he had found the best of all possible markets: Harold Ross loved to publish the recollections of writers with a sense of humor. From James Thurber's *My Life and Hard Times* and Clarence Day's *Life with Father* to Vladimir Nabokov's *Speak, Memory*, some of the most winning memoirs of the twentieth century first saw print in the pages of the *New Yorker*. The story of Wesley and Lily was right up Ross's own alley, and in April he published the first installment in the long series of autobiographical essays with which Mencken would write himself back into the good graces of the American public.

The next month Knopf brought out the fourth edition of *The American Language*, whose title page declared it to be a "corrected, enlarged and rewritten" version of the 1923 edition (than which it was 278 pages longer). A week after it hit the bookstores, the first printing of five thousand copies was "pretty well exhausted," and once the reviews began to appear, it sold even faster. Liberals who were wondering whether Mencken had lost his mind responded with sighs of relief. "The articles razzing the New Deal which have been appearing in The American Mercury over the name of H. L. Mencken have sounded disconcertingly like the compositions supposed to be communicated from the other world by the spirits of Mark Twain, Oscar Wilde and William James," Edmund Wilson wrote in the *New Republic*. "It is therefore reassuring to have authentically from Mencken's living hand this new edition of 'The American Language'. . . . It is one of the few recent books of which is is really possible to say that it ought to be in every American library and that it ought to be read by every foreigner who wants to understand the United States."

Not long afterward Gertrude decided to move from Hollins Street to the suburbs, and her older brother, at last accepting the inevitable, closed the apartment he had shared with Sara and returned to the house in which he had lived for a half-century, there to keep "bachelor hall" with August, his favorite sibling. Though August, who never married, led an unexciting but apparently satisfying life of his own as an engineer—he even produced a few books, including an illustrated history of the railroad passenger car and an anthology of news stories about hangings called *By the Neck*—he now seemed no less content to spend the rest of his days dwelling in the long shadow of his famous brother. Mencken brought with him Hester and Emma, the cook and maid from Cathedral Street, set up shop in his old second-floor office, and quickly settled into a routine almost indistinguishable from the one he and Sara had established for themselves: up at eight, breakfast at nine, mail at ten, lunch downtown, literary chores in the afternoon, a nap at five, and dinner at six-fifteen, after which he settled in for an evening's writing. "I'd have to watch the clock," August remembered, "because at 10:00 precisely I'd hear him say, 'Let's go down.' No matter what he was doing—bang." He would retire to the downstairs sitting room with August, build a fire, and pour himself a Scotch. Then they chatted, with Mencken doing most of the talking: "All the stories that he told me about people—and he knew a great many, you know. All his plans. If he had a book under way, he'd pretty near drive me nuts just talking about the thing, but sometimes it was very interesting." It was as if he had never been away.*

He was still angry, and his coverage of the 1936 presidential cam-

---

*Unsentimental as always, he would observe in a diary entry six years later that "my own marriage was too short to have much effect on me. . . . [I]t was unhappily over before I was 55, and so there was still enough resilience in me to enable me to get back on my old track."

paign showed it. When Roosevelt rolled over Alf Landon in November, winning election to a second term by a historic landslide that supplied him with "a blank check from and upon the American people, authorizing him to dispose of all their goods and liberties precisely as he listeth," Mencken snarled in the *Sun* that the next step was a full-fledged dictatorship: "It will come, inevitably, or something like it under another name, if the present carnival of demagogy goes on." FDR had robbed him of his ability to laugh at American politics, and his *Sun* columns would never again be as readable as they had been in the days when he was emptying both barrels into the hide of the booboisie. But his private life was back on an even keel, and the *New Yorker* had given him an outlet for a new kind of writing, one that would allow him to escape from the unsatisfactory present and revel in the painless memories of a time when he was "young, goatish and full of an innocent delight in the world." For the first time since Sara's death, he had reason to hope.

ONCE THE ELECTION WAS OVER AND HE HAD FILED HIS POSTMORTEMS, he checked into Johns Hopkins, beset by his usual assortment of low-to-medium-grade woes. His medical file contains the hospital's official record of his condition in December 1936: "Complaint: Tickling in back of throat, present for six days, with malaise, muscle pains and very mild fever. Diagnosis: Acute infectional tracheitis; acute infectional sinusitis; acute epididymitis (right) cause unknown.... One month afterward a cyst was removed from the right epididymus by Dr. Smith, and Steinach operation performed." This was a vasectomylike procedure devised by the Viennese surgeon Eugen Steinach for the purpose of rejuvenating aging men. W. B. Yeats submitted to one around the same time, and was thereafter known in Dublin as "the gland old man." It is unlikely that Mencken did it for the same reasons, though he does not appear to have been altogether celibate after Sara's death. (Aside from his claim to have tried out his vasectomy on an obliging friend in New York, he had at least one extended flirtation in the mid-thirties, with an obscure poet named Jean Allen

Balch, as well as particularly close friendships in the forties with a pair of younger women, the English engraver Clare Leighton and Marcella Du Pont, the soon-to-be-estranged wife of a scion of the Delaware chemical family with whom he was also friendly. Both women were open to the possibility of intimate involvement, but Mencken appears to have drawn back both times.) At fifty-six he felt his energies starting to ebb—his diaries of the time are full of somber entries in which he records his various illnesses and ponders how much longer he will be able to keep working—and it seems likely that he thought the operation might slow down his physical decline.

He produced no new books between 1934 and 1940, though the reception of "Ordeal of a Philosopher" had given him the idea of writing at greater length about his childhood. the *New Yorker* published a second piece in the same vein, "Innocence in a Wicked World," in February of 1937, and he soon broached to both Katharine White and Alfred Knopf the possibility of doing a whole book of autobiographical essays to be serialized in the *New Yorker*. "At the moment, I feel pretty good, and am planning to tackle a new book," he wrote to a friend in March. "In all probability, it will be a rambling account, more or less truthful, of my first twelve years. The Baltimore of the 80's was an odd and amusing place, and I greatly enjoyed my early youth in it." But he was preoccupied with other matters, and by September he was stalling. "My conscience has me down and is torturing me unmercifully," he told White. "I should have written those sketches at least six months ago, but various other jobs intervened and I am still struggling to clean them up. Once they are off my hands, I'll certainly sit down to my Baltimore reminiscences."

Another two years would go by before he resumed work on *Happy Days*. He had been doing a great deal of writing for the *Sunpapers* in addition to his Monday Articles, including multipart series on the University of Maryland and the Johns Hopkins Hospital and a sixty-page chunk of *The Sunpapers of Baltimore*, the official history of the *Sun*, and in January of 1938 he agreed to do something he had sworn never to do again: He went back to newspaper editing. Paul Patterson asked him to take charge of the editorial page of the *Evening Sun* for three months, evaluating the existing staff with an eye to the possible promotion of Philip Wagner, a *Sunpaper* editorial writer, to the post of editorial-page editor. Seeing the proposal as an opportunity to nudge the paper back toward his way of thinking on Roosevelt, he agreed to take on the task. He published his final Monday Article (in which he predicted that FDR would be unable to resist running for a third term, one of his few political forecasts to prove accurate) on the last day of January, then set up shop at the *Sun*, determined to stir up as much controversy as possible in the time available.

No sooner had he showed up for work than he started redesigning the page and building bombs to hurl in the general direction of the White House, including a gigantic full-page editorial about the New Deal and a stunt still cited by historians of American journalism: "I gave over six whole columns of the editorial pages to a mass of small dots—each supposed to represent a jobholder at the Federal trough. A brief editorial in the first column, headed by a hand pointing to the wilderness of dots, explained that their purpose was to give taxpayers a graphic representation of the vast army of parasites devouring their money." Intolerant of all viewpoints save his own, he ended up writing or rewriting most of the editorials that appeared on the page in February and March. Even so he made a favorable impression on his new colleagues, who marveled at his drive and were disarmed by his personal quirks. A confirmed tobacco chewer, he brought his spittoon from Hollins Street and used it all day long. Each morning at ten A.M., he put on a Princeton beer jacket before descending to the composing room to make up the page, whose contents, one *Evening Sun* editorial writer recalled, "varied wildly from day to day, reflecting both Toryism and radical change, pedestrian pomposity and wonderful highjinks." When it was over Philip Wagner remembered Mencken's brief tenure as "something fantastic and altogether delightful."

Mencken kept his memories to himself (save for the passing reference on the last page of *Newspaper Days*, where he spoke of "a brief and unhappy interlude" at the *Sun*), but they were less than benign.

He had determined within days of his arrival that except for Wagner, the editorial-page staff "ranged downward from third-rate" and had grown lazy and slack under Hamilton Owens's leadership. Wagner himself was too liberal to be trusted and incapable of writing "an editorial of genuine force," but Mencken found him serviceable when kept under careful supervision, and recommended that he be given a trial as editorial-page editor.

His duty to the *Sunpapers* done, he checked into Johns Hopkins on May 18, complaining of a "peculiar sensation" around his heart. The doctors told him that he had arteriosclerosis and ordered him to lose weight and take a vacation, and he decided to visit Germany for the seventh and (as he rightly suspected) last time. A well-connected friend there assured him that "the German General Staff regarded another great war as inevitable," in which "the French would be disposed of almost as quickly as in 1870." Another acquaintance, a professor at the University of Berlin, said that though "the roughing of Jews upset him," they had only "their own gigantic folly" to blame for it, adding that "Hitler, whatever his outrages, had found a way to restore hope and self-respect to the country, and what is more, confidence." Mencken recorded these remarks without comment in *Thirty-Five Years of Newspaper Work*, though they were consistent with his oft-stated belief that the German Jews were at least partly responsible for their present plight, of which he claimed to have seen little firsthand evidence save for "one or two smashed windows" in Berlin. After his return to Baltimore in July, he told a *Sun* reporter that Germany "looked like a church to me, it was so quiet."

Other visitors to Germany had reported more alarming sights in 1938, but Mencken was probably inclined to dismiss their accounts as partisan, remembering as he did the made-up atrocity stories of World War I. Five months after he returned to the United States, though, it became impossible for him to keep on ignoring what Hitler was up to, or to remain silent about it any longer. On the night of November 9 the Nazis unleashed a countrywide pogrom in which 91 Jews were killed and 26,000 more packed off to concentration camps.

The streets of Berlin and other German cities were filled with the smashed window glass of such Jewish-owned shops as still existed. Three weeks later Mencken published a column about *Kristallnacht* in the *Morning Sun* (to whose op-ed page he had moved in May), in which he kept his long-deferred promise to mount the barricades and holler for the Chosen. "It is to be hoped," he wrote, "that the poor Jews now being robbed and mauled in Germany will not take too seriously the plans of various foreign politicos to rescue them. Those plans, in all cases, smell pungently of national politics, and in not a few cases they are obviously fraudulent." Dismissing as "next door to murder" the English plan to resettle German Jews in Tanganyika and British Guiana, he pointed out with gleeful precision that the White House was doing no better, since FDR had stated that he had "no intention of proposing a relaxation of the immigration laws" that prevented most Jews from emigrating to the United States. "Such gross and disgusting peck-sniffery," Mencken said, "is precisely what one might expect from the right hon. Gentleman, and I only hope the American Jews who have swallowed so much of his other buncombe will not be fetched by it." The only way to help Jewish refugees, he added, was "to find places for them in a country in which they can really live. Why shouldn't the United States take in a couple of hundred thousand of them, or even all of them?"

This little-known *Sun* column was exhumed in the eighties when the posthumous publication of Mencken's diaries led to renewed accusations that he was an anti-Semite. "Help for the Jews" was offered as proof to the contrary, though those who quoted from it almost always neglected to mention the last two paragraphs:

> I am here speaking, of course, of German Jews, and of German Jews only. . . . The question of the Eastern Jews remains, and it should be faced candidly. Many German Jews dislike them, and were trying to get them out of Germany before the present universal disasters came down. There is a faction of them that tends to be troublesome wherever they settle, and there is

apparently ground for the general belief that in this country
they incline toward the more infantile kinds of radicalism.

Fortunately, they are not numerous in German Jewry. The
really large accumulations of them are in Poland and Rumania.
It would be obviously impossible, even if it were prudent, for the
United States to take them all in. But there is still plenty of room
for them, and in a land where there is no prejudice against them,
and their opportunities are immensely better than in Tanganyika or Guiana, or even Palestine. That land is Russia.

Having made so outrageously cynical a suggestion, Mencken
affected to be dismayed when the *Sun* received "floods of letters of
protest and abuse" from Jews who declined to go along with his solution to the problem of Eastern Jewry. He then decided that since
nothing he wrote short of "the most preposterous flattery" would
satisfy the Jews, he would write no more about them, "no matter
how poignant their sufferings." Except for a follow-up column in
January and a brief mention of "the calamity of World War II" in the
essay about his 1934 visit to Palestine that he published in *Heathen
Days*, he kept his word. When *Treatise on the Gods* was reprinted in
1946, it was in a revised version from which he had silently excised
the passage about the "unpleasantness" of the Jews that had brought
him grief sixteen years before. Nowhere in his diary, his published
writings, or his private correspondence does he make any direct reference to the Holocaust, which is also conspicuously missing from a
list of baleful consequences of World War II that he gave a reporter
for *Life* in 1946. He did not deny that it took place: He simply had
nothing to say about it. He would never again have anything of substance to say about the Jews—in print.

From 1938 on Mencken began to devote an increasing amount of
time and energy to writings not intended for immediate publication.
In addition to his diary he had already accumulated thousands of
random notes, many of them "rough jottings" set down in a terse,

shorthandlike style. (Here is one of the few surviving ones, written in 1932: "The movement toward realism and sense, following World War I, was over. It became evident that Am. people could not stand the truth. Had to reach out for some religion. It came as New Deal, a despairing reaction f om [*sic*] the Hoover religion. I saw what was afoot on Sun. Obvious Am. Mercury was out.") A few had been polished and published in "Répétition Générale" and "Clinical Notes," but most were simply tossed into a bin and forgotten until he started shining them up with an eye toward turning them into a book. He would continue to make such notes for the rest of his working life. Some of the unpublished ones deal with personal matters, such as his previously mentioned penchant for dropping old friends "who begin to bore me." In others he reflects on race in America: "The Negro is held back by the fact that every time a superior member of his race obtains a right a dozen imbeciles take a free ride on his shoulders." More than once he observes that America is on its way to becoming a totalitarian state, and that another world war will finish the job already started by FDR, whom he now thought was scheming to bring the United States into the next world war in order to win a third term. But the subject that comes up most often is the Jews:

> The sharp, unyielding separateness of the Jews, based on their assertive racial egoism, marks them off as strangers everywhere. . . . The chief whooping for what is called racial tolerance comes from the Jews, who are the most intolerant people on earth. In the United States, as in all countries that they inhabit, they interpret tolerance to mean only an active support of their own special interests. . . . I hate to have to ask a man what he is—Jew or Christian, German or Englishman, European or Asiatic. There are so few persons in the world who are really worth knowing that it is wasteful and absurd to sort them into categories. But the average Jew leaves one no alternative. He is Jewish before he is a man, and presses the fact home with

relentless lack of tact. This habit, I suspect, is one of the chief causes of Jewish unpopularity, even among those who are not rationally to be called anti-Semitic.

Mencken may have felt that he was "not rationally to be called anti-Semitic," but few who are fully acquainted with all the available evidence are likely to agree with him, though by the looser standards of his own time and place he might well have been acquitted of the charge. He was, after all, no Nazi—but he was willing to leave the Nazis to their own malevolent devices rather than see the United States go to war with Germany. He loathed Adolf Hitler—but no more than he loathed FDR. He thought America should open its doors to Jewish refugees—but only if they were his kind of Jews.* He sympathized with their distress—but thought they had brought it on themselves. Perhaps most telling is the language he used to criticize Jewish friends such as Philip Goodman, who decided that they could no longer keep silent about the plight of the Jews under Hitler: Their crime, he said, was to have "turned Jewish on me."

In the end one grows weary of hair-splitting, and recalls the words of Old Wesley, the back-alley philosopher of Hollins Street: "Truth is something that only damned fools deny." To admit the truth about H. L. Mencken is not to dismiss the self-evident fact that there was far more to him than his deeply equivocal feelings about the Jews. It is not his anti-Semitism for which he will be remembered—but that he was an anti-Semite cannot now reasonably be denied.

MENCKEN HAD LONG SUSPECTED THAT CHARLES ANGOFF WAS PLAN-ning to write something nasty about him, and in the winter of 1938 Angoff finally came through with an article for the *North American*

---

*He also went out of his way to assist a number of Jewish friends (and their relatives) who were attempting to get out of Germany, signing immigration affidavits and offering other forms of concrete assistance, in one case going to the State Department to inter-vene personally on behalf of the granddaughter of an old *American Mercury* contributor.

Review in which he declared his old boss to be a forgotten man: "Three years ago, even two years ago people used to ask what has happened to Mencken. Now almost nobody asks. It is a pity, for he can still be interesting, gracious, polite, and he still can write very good English when he forgets his juvenile stylisms." His essay was a shameful attempt to condescend to the man who had pulled him out of a suburban Boston newspaper and given him a taste of the more abundant literary life, and it must have bitten that much harder because there was, as usual with Angoff, a touch of truth to it. *The American Language* had been a great success—a special Book-of-the-Month Club edition had sold a hundred thousand copies—but it was a rewritten version of a twenty-year-old book. None of Mencken's more recent writings had attracted anything like the same favorable attention.

Then, in January, he returned to writing about his Baltimore childhood, and the twilight began almost immediately to grow brighter. He sent the *New Yorker* "The Ruin of an Artist," a piquant description of his youthful musical studies, and Ross bought it on the spot by return telegram. The piece ran in May, and for the rest of the summer the magazine published additional installments as quickly as Mencken could mail them in. By September the manuscript was complete, and Knopf and Ross were begging for a sequel, this one about his years at the *Herald*, which Mencken declared himself happy to supply "if the holy saints spare me so long." For once he was serious. He had suffered the mildest of strokes in July, and though the doctors at Johns Hopkins assured him that he was in no imminent danger, he nonetheless took it as a sign that "I was growing old at last, and that the end of my sojourn in this vale was looming into view."

Happy Days was published in January to a chorus of critical praise, promptly showing up on bestseller lists around the country. Virtually without exception, the reviewers declared that Mencken had outdone himself. "A book to be read twice a year by young and old, as long as life lasts," proclaimed the *Atlantic Monthly*, and no one seemed moved to differ. Even the august *Times Literary Supplement* of London, which rarely went out of its way to speak well of contemporary American authors, praised *Happy Days* in Olympian language that must have gone straight to its author's head: "In one chapter he confesses his abiding admiration for 'Huckleberry Finn,' his childish discovery of which he terms 'probably the most stupendous event of my whole life.' Perhaps it is then no accident that this book, too, should exude a Mark-Twainish robustness, even earthiness. . . . This book lacks the final magic of the best of 'Huckleberry Finn,' but it may well stand on the same shelf."

Happy Days is no longer widely read, but connoisseurs of Mencken's writing know it as one of his most completely realized achievements—arguably the best of his many books, though *Newspaper Days* is at least as good. It is unsentimentally affectionate, a mixture rarely encountered in childhood memoirs, and it is also a masterpiece of pure style, offering those readers who recoil from Mencken's politics an opportunity to wander without guilt along the long, looping arcs of his cunningly poised sentences:

> At the instant I first became aware of the cosmos we all infest I was sitting in my mother's lap and blinking at a great burst of lights, some of them red and others green, but most of them only the bright yellow of flaring gas. The time: the evening of Thursday, September 13, 1883, which was the day after my third birthday. The place: a ledge outside the second-story front windows of my father's cigar factory at 368 Baltimore street, Baltimore, Maryland, U.S.A., fenced off from space and disaster by a sign bearing the majestic legend: AUG. MENCKEN & BRO. The occasion: the third and last annual Summer Nights' Carnival of the Order of Orioles, a society that adjourned *sine die*, with a thumping deficit, the very next morning, and has since been forgotten by the whole human race.

The success of *Happy Days* was the best possible answer to Angoff: No one capable of writing such a book could possibly be forgettable.

But success had not mellowed Mencken to any noticeable extent. He was still writing about Roosevelt with all the savagery at his command—and still haunted by memories of Sara. On the fifth anniversary of her death, he wrote about her in his diary, aware as he did so that his own end might also be drawing near: "We have parted forever, though my ashes will soon be mingling with hers. I'll have her in mind until thought and memory adjourn, but that is all. . . . We were happy together, but all beautiful things must end." Then he quoted with heartbreaking appositeness from Goethe's *Faust*, a play that must have been much on his mind as Germany hurtled heedlessly toward its own rendezvous with Nemesis: *Deny yourself! You must deny yourself!*

Six weeks later, he looked on from the press gallery as the Democratic National Convention, meeting in "almost incandescent" heat in Chicago, nominated Franklin Roosevelt to run for a third term, wondering as he watched whether this was the last presidential convention America would ever witness: "This one had gone through the traditional forms, but in substance it was actually no more democratic than one of the outpourings staged by Hitler in Nurnberg or Berlin." He was in deadly earnest, though he made a joke of it, shaking hands with his fellow reporters and telling them, "Goodby, old chap. No, we won't see each other in 1944. This is the last political convention that will ever be held in this country." As he clumped heavily out of the hall, chewing on a cheap cigar, a Democratic publicist said to a reporter, "That's Mencken. He's a card, isn't he?" He went on the road with Wendell Willkie in August, filing dispatches to the *Sun* that shine with his old verve, but he feared that at sixty his reporting days were behind him: "I was still able to cover a story competently, but I was tiring more and more quickly, and it was obvious that a really exhausting job of work might easily floor me."

He also knew it was time to leave the *Sunpapers*, which had resumed their slow drift toward Roosevelt and away from isolationism as soon as he turned the editorial page of the *Evening Sun* over to Philip Wagner. The idealistic goals he had outlined in his long-ago memorandum calling for "a great newspaper that seeks to lead rather than to follow" now seemed to him more unattainable than ever. By the end of the year both papers were "supporting Roosevelt II's effort to horn into World War II in a frantic and highly unintelligent manner," and he no longer wanted anything to do with them. One night in January he sat down with Paul Patterson in his Hollins Street parlor and delivered an hour-long monologue full of bad news: "I was resolved, I said, to quit writing for the editorial page of the *Sun*, at once and forever. I was resolved, further, to undertake no more writing for its news pages. Yet further, I refused to take any more responsibility, whether direct or indirect, for its editorial conduct, and would not undertake to offer him any more advice on the subject." Patterson begged him not to sever his ties to the institution with which he was so closely identified, and Mencken agreed to remain on the board of directors of the *Sun*'s parent company (to which he had been appointed in 1934) and to continue to represent the paper in its negotiations with the Newspaper Guild (as he had been doing since 1938), but that was as far as he would bend.

Once he had made up his mind to leave, he wasted no time in going. He published a warm reminiscence of his early days at the paper in the January 26 issue, followed the next Sunday by a piece called "Progress of the Great Crusade" in which he forecast a grim future for America: "Win or lose, the United States is doomed to suffer some appalling headaches during the next dozen years. If the cause of humanity triumphs we will have the bills to pay, and if Beelzebub gets the better of it we'll have the damages. In either case half of the human race will hate us very earnestly, and the rest will hold us in a kind of esteem hard to distinguish from contempt." Without further ceremony he then ended his thirty-five-year run on the editorial pages of the *Sunpapers*, fully expecting never to return. He would write nothing more about public affairs—for the *Sun* or any other publication—until 1948.

"Yesterday was the first Sunday of my release from the *Sun* editorial page," he wrote to Monsignor Dudek a week later, "and I feel almost like a prisoner liberated from the hoosegow." His situation was more like that of a man who had decided not to commit suicide. He knew that if and when the United States went to war with the Axis, his opinions would overnight become as unpublishable as they had in 1917.* By quitting well in advance of that dread event, he ensured that his reputation, which the favorable reception of *Happy Days* and *The American Language* had done so much to salvage, would remain unsullied. Yet it must have been unbelievably hurtful for him to have walked away from the *Sun* twice in a lifetime, whether or not he cared to admit as much. August, who had come to know Mencken better than anyone, was aware of what the *Sunpapers* meant to him: "In my brother's life, the *Sun* came first. It's hard to believe, but I think he would have dropped anything he was doing to attend to the *Sun's* business. Nothing gave him greater pleasure or greater pride than being [a] director on the *Sun*. He always looked on himself as basically a newspaperman, you see, and he always looked on the *Sun* as his prime responsibility. I don't know of anything that really interested him more than that newspaper. In every way." Perhaps that was why he now undertook to set down his memories of the paper he had served so well. Once he finished writing *Newspaper Days* (which had begun appearing in the *New Yorker* in February) and editing a book of quotations on which he had been working for years, he began dictating a detailed chronicle of his years at the *Sunpapers*. It came quickly—there were days when he turned out as many as three thousand words—and by October, half of it was written.

Newspaper Days was published by Knopf in August. Its reception was, if anything, more enthusiastic than that of *Happy Days*: Mencken the political columnist may have hung up his copy spike, but Mencken the memoirist was more popular than ever. War-conscious Americans who opened their newspapers with fear and trembling each day must have found it a relief to immerse themselves

---

*When asked by a Philadelphia reporter that month whom he wanted to see win the European war, he replied, "I don't care, so long as it is not England."

in the adventures of young Henry Mencken, the cub reporter who at
the tender age of eighteen found himself "at large in a wicked seaport
of half a million people, with a front seat at every public show, as free
of the night as of the day, and getting earfuls and eyefuls of instruc-
tion in a hundred giddy arcana, none of them taught in schools." It is
possible, perhaps even useful, to read *Newspaper Days* for what it tells
us about turn-of-the-century American journalism—no reporter of
Mencken's generation wrote a better professional memoir—but that
would be like reading *The Bonfire of the Vanities* to learn about civic
administration. First and foremost *Newspaper Days* is the story of a
naive young man turned loose in the big city, told with a gusto
unmatched by any other American author. It, too, is a not-so-minor
masterpiece of affectionate reminiscence, one that in a better-
regulated world would be recognized as a modern classic. It is also a
useful corrective after long immersion in Mencken's notes of the
thirties and forties, an experience that can leave even the most sympathetic of readers with the illusion that he was a hardhearted, small-
souled crank. He may have been hard (or at least wanted to be), and
he was without question cranky, but there was nothing small about
him, not even his flaws.

Around the same time that *Newspaper Days* was being shipped to
bookstores, Charles Lindbergh, no less than Mencken a symbol of
America in the twenties, came to Des Moines, Iowa, to appear at an
America First rally. The much-admired pilot had emerged as the
best-known spokesman for the largest and most influential group
opposed to American intervention in the war in Europe, and now
that the Roosevelt administration had all but committed itself to the
Allies, he proceeded to say in public what Mencken had been saying
in private for years:

> Tolerance is a virtue that depends upon peace and strength.
> History shows that it cannot survive war and devastation. A few
> farsighted Jewish people realize this, and stand opposed to
> intervention. But the majority still do not. Their greatest dan-

ger to this country lies in their large ownership and influence in our motion pictures, our press, our radio, and our government. . . . [T]he leaders of both the British and the Jewish races, for reasons which are understandable from their viewpoint as they are inadvisable from ours, for reasons which are not American, wish to involve us in the war. We cannot blame them for looking out for what they believe to be their own interests, but we also must look out for ours. We cannot allow the natural passions and prejudices of other peoples to lead our country to destruction.

Lindbergh's speech, received enthusiastically in the Des Moines Coliseum by a crowd of eight thousand, was torn to shreds in the media. Had he not given up his column for the *Sun* and retired to his office in Hollins Street to write about his youth, Mencken would unhesitatingly have endorsed every word of it.

Three months later the Japanese air force bombed Pearl Harbor, a catastrophe to which Mencken paid no attention in his diary or correspondence. Nor did he write about FDR's message to Congress the following day, or the announcement on December 11 that Germany and Italy had declared war on the United States, or any of the war news with which the *Sunpapers* were thereafter flooded. The retreat to Bataan, the surrender of Hong Kong, the capture of Wake Island: None of it meant a thing to him. Not until December 29 did he make another entry in his diary: "I went out to Loudon Park Cemetery on Christmas morning with August, and we laid a wreath on the family tombstone. It was the forty-first Christmas since my father's death, the seventeenth since my mother's, the seventh since Sara's, and the second since my sister-in-law Mary's. My mother has begun to recede into the shadows, but Sara is still clearly before me, and I think of her constantly. Nevertheless, I am half glad that she is dead, for the future looks dismal, and 1942 is certain to be the worst year I have ever seen."

"IT WILL BECOME INFAMOUS TO DOUBT"

Skeptic in Exile, 1942–1948

Mencken did not ignore World War II for long. He couldn't: It left traces everywhere he looked. Though he was no longer writing for the *Sun*, he cared about what it published and how it prospered, and the war had become its primary concern, as it was of every other newspaper in the United States. Not surprisingly he found its military coverage to be both credulous and partisan, and complained about it in his diary, once going so far as to personally take the paper's editors to task for their refusal to violate "the war-time order against giving weather information that could be of use to the public enemy" by reporting the exact number of inches of snow that had fallen on Baltimore in an unexpected blizzard. He kept up with war-related gossip.* He received and answered letters from soldiers who were reading *Happy Days*, which was published in a special Armed Services paperback edition. "I write to every soldier who writes to me," he told a reporter. "I offer to buy him a beer when he gets home. I'd like to buy him ten beers; in fact, I'd go into debt buying

---

*Among other things, he heard in the winter of 1943 that Dwight Eisenhower was relaxing from his duties as supreme commander of the Allied forces in Europe by having an affair with Kay Summersby, his pretty chauffeur, an indication of how well-connected Mencken remained throughout his exile.

beers. You don't have to pay taxes on debts." He even thought about writing a book called *Notes in Wartime*, though he soon realized that there was no point in pursuing the idea, since it would have been unpublishable: "To me the war, or, at all events, the American share in it, is a wholly dishonorable and ignominious business. I believe that that will be history's verdict upon it. I seem to stand in a very small minority. . . . The human mind is a curious machine, operated mainly by baloney power."

For the most part, though, he paid as little attention to the war as he could. He had always felt distant from his countrymen: "I was born here and so were my father and mother, and I have spent all of my 62 years here, but I still find it impossible to fit myself into the accepted patterns of American life and thought. After all these years, I remain a foreigner." Now he found himself pleasantly surprised by how little effect the war had on him, beyond the arrival in his Union Square neighborhood of a pack of "oakies, lintheads, hill-billies and other anthropoids" who had come north from Appalachia to Baltimore in search of better-paying war work. No one he knew was killed or wounded in combat. He received a comfortable stipend from the *Sunpapers*, not a cent of which he spent on war bonds, though he griped constantly about the high taxes he paid on it. He avoided the rigors of wartime travel by staying home, and food rationing had no noticeable effect on his table. "The best lunches I ever get in Baltimore," he bragged, "are in my own house." The Saturday Night Club was slowly unraveling—Max Brödel, who had shared a piano bench with Mencken for thirty years, had died the preceding October, after which things were never quite the same—but he was content to stay home with August. Though his own health was now erratic, it had never been much better than adequate (at least as far as he was concerned), and he still had enough steam to do "precisely what I want to do," which was to work on his books, answer his mail, and drop by the *Sun* every weekday. The one thing he couldn't do was write about the war, but that was a minor inconvenience: "I am anything but a propagandist and even if I had the utmost freedom to write in print I'd probably make no effort to dissuade the American people from their follies." In return for his silence the government ignored him, and he thought that a reasonable price to pay for being left alone.

Sometimes he would make an unexpectedly jarring remark to a reporter. "I think the war will never end," he told Bob Considine in 1944.

> I think it's a better state than peace. You never hear a man who's been off to war say he wishes he had stayed home. They're sore for a time when they're drafted, but they wouldn't miss it for the world. It's tough only on soldiers who have nine or ten children at home. Roosevelt will keep this war running at least until the end of his fourth term. He knows that if the war stops, he loses his war powers and his jobs. I've changed my mind about Roosevelt. I'm going to vote for him this time. Any country who put him back in office twice, when they had a chance to vote him out, deserves him. And he deserves the country.

But these rare breaks in his silence were dismissed as slapstick from a professional curmudgeon. *Time* called him "the nation's comical, warm-spirited, outstanding village atheist." He might as well have been dubbed the court jester. He had been officially welcomed back into the fold, and no one cared to consider that behind his mask of geniality might lurk a violently heterodox dissenter who still deserved to be taken at his word.

Though he kept busy enough during his exile from the *Sunpapers*, he was aware of the dizzying speed with which the world was passing him by. One sign was the publication of *The American Mercury Reader*, a volume edited by Angoff and Lawrence Spivak, who had in turn succeeded Henry Hazlitt as editors of the magazine that once embodied to perfection its founder's taste. Printed on crumbling wartime pulp, the *Mercury Reader* was a representative but nonetheless odd mixture of Mencken-era articles and pieces that would never have gotten into the magazine in the first place had he been around to reject them. In "Remember Us!," the screenwriter Ben Hecht pointed out that under Hitler, the Germans had "animated the myth of the Jewish menace beyond any of their predecessors and tried to prove their case by presenting the world with a larger pile of Jewish corpses than has ever before been introduced into the ancient argument." About such matters Mencken kept still, for he knew what a firestorm of protest would be ignited were he to speak his mind in public. "The American people are now wholly at the mercy of demagogues, and it would take a revolution to liberate and disillusion them," he wrote in his diary. "I see no sign of any such revolution, either in the immediate future or within the next generation. When the soldiers come home it will become infamous to doubt—and dangerous to life and limb."

In April he came down with a bad cold that got worse, and he went into Johns Hopkins at the end of the month and stayed there for two weeks. (When he moved back to Hollins Street after Sara died, he and August had promised one another that "neither of us would ever be sick in bed in the house. We agreed, in case we ever had to lie up for more than a few hours, to go to hospital. There are no facilities in a bachelor's house for caring for the ill.") The doctors warned him that his blood pressure was dangerously high and that he could survive "only by leading a quiet life, taking as much rest as possible, and avoiding all severe labor, whether physical or mental. . . . I may have a stroke at any minute, or the coronary occlusion may fetch me." The news inspired him to finish preparing his private papers and correspondence file for eventual transfer to the Enoch Pratt Free Library and the New York Public Library, as well as to speculate on the long-term fate of his writings, which he thought might someday have a revival:

> In 1920 I was able to press that revival myself. That will be impossible this time, but the fact does not bother me, for in

the long run every man has to shut down, and if he is remembered thereafter it is by the effort of others. Not many American authors will ever leave a more complete record. It took
me a couple of years to arrange my papers, and the cost of
binding them was very high, but now they are in the Pratt
Library, and if anyone ever wants to investigate them they will
be easily accessible. It is likely, of course, that no one will ever
want to do so, but nothing can be done to change that implacable fact.

That summer Knopf published his *New Dictionary of Quotations on
Historical Principles from Ancient and Modern Sources*, a million-word
monster with which he had been wrestling for a decade. He had long
kept a card file of quotations for his own use, and in 1932 got the idea
to expand it into a full-scale dictionary; Charles Angoff worked on it
with him for two years, after which he carried on alone. Though not
generally recognized as such, it is one of his major achievements,
comparable in scope to *The American Language* and no less personal
in its method. "The Congressman hunting for platitudes to embel-
lish his eulogy upon a fallen colleague will find relatively little to his
purpose," he warned in his preface, and many readers have thus concluded that he compiled the *New Dictionary of Quotations* with tongue
in cheek. Like Dr. Johnson's dictionary, it is wrongly remembered
for its eccentricities, among them an extensive selection of invidious
remarks about the Jews and a sprinkling of unattributed "proverbs"
that sound as though they had been coined by the editor himself. In
fact it contains a vast number of well-chosen, well-organized, accurately attributed and dated quotations on every imaginable subject,
ranging widely among both familiar and arcane sources. The only
important author missing from its 1,347 pages is Mencken, who told
Time that "I thought it would be unseemly to quote myself. I leave
that to the intelligence of posterity." Yet the *New Dictionary* bears
the dark stamp of his skepticism on every page, and at least one critic,
Morton Dauwen Zabel, was quick to grasp the fact: "The impression

soon becomes inescapable that what Mencken has produced as a 'Dictonary of Quotations' is really a transcendent 'Prejudices: Seventh Series,' a 'Notes on Humanity,' or more expressly 'Mencken's Philosophical Dictionary, Written by Others.' "

After that the flow of his public writings dried up. Outside of the *New Yorker* essays that went into *Heathen Days*, the third volume of his reminiscences (which was well received when it came out in 1943, though it was more miscellaneous in subject matter and somewhat less even in quality than its predecessors), he published virtually nothing in 1942 or 1943 save for an article on the origins of the expression "O.K." that appeared in *American Speech*. More and more, he was writing for his desk drawer. He finished the 520,000-word *Thirty-Five Years of Newspaper Work* four days after leaving the hospital, then launched into *My Life as Author and Editor*, in which he would seek to provide a similar account of his work on magazines, though he was only able to carry it as far as 1923, just prior to the launch of the *Mercury*. Neither manuscript was intended for publication. He put them in sealed wooden crates and left them to the Enoch Pratt Free Library with strict instructions that they not be opened by anyone until thirty-five years after his death, when they would be made available to scholarly "resurrection men" seeking to better understand him.*

Just as he intended, *Thirty-Five Years of Newspaper Work* and *My Life as Author as Editor* have become the primary sources for all who now write about the *Sunpapers* and the *Smart Set*. Nor are they devoid of revelations about his personal life and character, though many of these are indirect. We learn as much about him from what he tells us about other people as we do from what he tells us about himself. The frankness with which Mencken discusses what he perceived to be the shortcomings of his associates may give a misleading

---

*This macabre locution, a favorite of Mencken's, is revealed in *The American Language* to refer to men who "robbed graves for the doctors in the days before the Anatomy Acts gave them a lawful supply of cadavers."

impression of how he felt about them as human beings, since he meant for both manuscripts to remain under lock and key until "time has released all confidences and the grave has closed over all tender feelings." Perhaps his desire to tell "the exact truth . . . without equivocation" made him seem crueler than he was. Or perhaps not: It is unnerving to watch him reflect on his friends with a cool detachment of which they had no notion, judging them mercilessly in the privacy of his Hollins Street office and taking care that his judgments would survive him.

Aside from settling old scores, there were a few more books he wanted to write, some of which might even prove suitable for wartime publication. *The American Language* was already in need of a supplemental volume or two, and he continued to polish his unpublished notes in the hopes that Knopf might be willing to print a selection from them, going so far as to draft a preface in the summer of 1943. He also had the idea of cobbling together "a sort of Mencken Encyclopedia, made up of extracts from my writings over many years, arranged by subject and probably with additions." The one thing he would not do was write for the *Sun*. Paul Patterson tried to talk him into covering the 1944 presidential conventions, but Mencken was adamant: the *Sun* "had departed so far from my notions of sound journalism" in its coverage of the war that he considered himself "out for life."

Even if he had been briefly tempted by Patterson's suggestion, the decline of his health might well have led him to reject it. By then he was having nightmares caused by cardiac insufficiency, though their content was a manifestation of his troubled state of mind: "At 3 o'clock this morning I awoke from a dream of pursuit with my heart thumping violently and a feeling of general distress. The dream was of the familiar sort: I had missed a train and was running after it in an effort to get aboard." One day he lost control of his bowels while shopping, an experience that shook him severely: "Such things are disquieting, and also humiliating. . . . [W]hen a man begins to disintegrate all his controls weaken, and he returns to a sort of infancy—physically first, and then mentally." Five days later, he signed a new will, one that would provide for August and Gertrude should they outlive him, as he now expected. Charles, the middle brother, meant less to him—they did not see each other for years at a time—and thus received a flat cash bequest instead of a life income.

As well as leading him to put his affairs in order, advancing age awakened in him a longing to mend fences. He was back in touch with George Jean Nathan, to whom he wrote in April 1942, describing a visit to his draft board: "When I dropped in yesterday to enroll myself in the army of Christian democracy, I found the place full of men even more decrepit than I am. It was somehow consoling, though it was rather upsetting to see three of them drop dead in the course of eight minutes." (The two men continued to see one another in New York, and posed together in 1947 for a double portrait by the fashion photographer Irving Penn in which they suggest *Sunset Boulevard*–like mummies of their youthful selves, poorly preserved yet somehow glamorous.) He had also patched things up with Theodore Dreiser, whose death at the end of 1945 put him in mind of "the long-ago days when hope hopped high in both of us, and we had a grand time. I can hardly imagine grand times hereafter. A man past 65, if he can still work, can manage to be more or less contented, but he has become too old to be gay." Yet he could still give a plausible impersonation of a high-spirited man, especially in the lively letters he dictated each morning to friends old and new. To James Cain, who popped the question to Aileen Pringle in the summer of 1944, he dispatched "[a]ll the congratulations that the Western Union refused to send—and a million more. Of all the women now extant in this great Republic Aileen is probably the most amusing."

He had grown closer to Alfred and Blanche Knopf over the years, though he was capable of writing testily about both of them in his diary, describing them on one occasion as "extreme neurotics, and given to hysterics." Mencken had long offered private counsel to Knopf, recommending that he publish such authors as Thomas

Mann and William Carlos Williams, but it was music as much as the making of books that put the two men on common ground. Until the coming of war brought a temporary halt to the annual Bach Festival in Bethlehem, Pennsylvania, they had been going there together since 1923 to immerse themselves in the "perfect and glorious" sounds of the B Minor Mass, and whenever he visited Knopf at his country home in upstate New York, they divided their time between talking and listening to records. He had no such common interest with Harold Ross, but they, too, had become good friends since he began writing for the *New Yorker*, sharing as they did a robust sense of the absurd, and in between their meetings in Manhattan they exchanged frequent and funny letters in which they swapped stories about the magazine business and complained about their health. "I am going up to Boston to see my lady doctor for a check-up next Monday," Ross wrote in 1943. "She will pump my stomach out and find the usual amount of acid. I have done everything right but work too hard, and smoke, otherwise I have been like Christ in my simplicity and patience."

Above all, he had August, to whom he paid homage in a touching diary entry that bore witness to their closeness, as well as to the shrinking of his own horizons as he moved inexorably toward old age: "Without August I'd be lost indeed. He and I read much the same books, are interested in many of the same things, and think alike about all matters of any importance. I thus enjoy sitting with him in the evenings. He has a very pungent humor, and his comments on the mountebanks now running the world are often excellent." It was a relationship that within a few years was to become more essential to him than he could possibly have imagined.

THE WAR HE SAID WOULD NEVER END BEGAN TO WIND DOWN IN 1945, IN spite of his continued insistence that it would go on "so long as Roosevelt is in office, for if he made peace he would lose all his war powers, and his disintegration would follow quickly. Thus I'll never see any freedom." That prediction, like so many before it, proved false, for a reason that Mencken failed to foresee. FDR suffered a massive cerebral hemorrhage as he sat for a portrait at his retreat in Warm Springs, Georgia, his mistress looking on in horror. He died within minutes, plunging the country he had led for twelve years into a spasm of collective grief. Few were unmoved by his passing save Mencken, who took to his typewriter to hammer out a pair of diary entries in which he rejoiced at having outlived his worst enemy: "His career will greatly engage historians, if any good ones ever appear in America, but it will be of even more interest to psychologists. He was the first American to penetrate to the real depths of vulgar stupidity. He never made the mistake of overestimating the intelligence of the American mob. He was its unparalleled professor." The only thing that gave him pause was that Baltimore's saloons and restaurants closed that Saturday night in a gesture of respect as the funeral train passed through the city. "As a result," he grumbled, "the Saturday Night Club missed its usual post-music beer-party for the first time in forty years."

With Roosevelt's death Mencken started to emerge from the shadows. He had persuaded Knopf to publish a pair of supplemental volumes to *The American Language*, and the appearance of the first one, whose first printing sold out prior to publication, was greeted with jubilation by Edmund Wilson in the *New Yorker*: "Mencken, like Van Wyck Brooks, has been writing a part of our social history, and he has brought to it, as well as indefatigable diligence—to use a favorite word of his—his humor, his excellent writing, his special relish, at once acrid and genial, for the flavor of American life." Stimulated by its reception, he put aside *My Life as Author and Editor* to write Supplement II and "revise and rewrite" *Treatise on the Gods*. In January, Knopf sent him the largest royalty check he had ever received, for $14,547.68. And one day before the first anniversary of the dropping of an atomic bomb on Hiroshima, *Life* published an interview in which he set forth his "views on public questions" for the first time in five years. He broke his silence with a vengeance, taking full advantage of the occasion to sum up what he saw as the consequences of World War II:

First, the Asiatic barbarians, held at bay since 1683, have been let loose in Western Europe. Second, Italy has been returned to the Black Hand. Third, France has become a chattel of England. Fourth, the Japs have been run out of China and Malaysia, and the English philanthropists are back. Fifth, the Chinese highbinders have been supplied with materials for continuing their civil war *ad infinitum*. Sixth, the Finns, Lithuanians and so on are enslaved and the Swedes are menaced. Seventh, the English have taken over the Italian colonies. Eighth, the Jews have been euchred out of what was promised them in Palestine. Ninth, half the human race is starving. Tenth, millions of young Americans, robbed of the most precious years of their lives, are dismally trying to start anew. Eleventh, prices are rising, inflation impends, no American investment is safe, and another huge collapse seems to be certain. Twelfth, irresponsible and unconscionable labor racketeers, having piled up immense slush-funds from the swollen wages of the war-workers and scared to politicians into setting them above the law, are planning to take over the government altogether. Thirteenth, gangs of smart fellows are preparing to organize the bewildered veterans and climb into high office on their backs.

Even more biting was this paean to atomic warfare: "Large numbers of the victims, I was proud to note, were women and children. They were slowly fried or roasted to death like people burned by radium or x-rays. In many cases their agonies were prolonged, and they suffered worse than any bishop will ever suffer in Hell. . . . I can well imagine the professional exaltation of the physicists who planned this glorious holocaust of infidels. They will live in history alongside Cotton Mather and Torquemada. How they must glow when they look into their shaving-glasses of a morning." No doubt a few readers noted that Mencken had nothing to say about the other, more lethal holocaust in Europe, but except for that singular omission, he had said his piece about the war. *Life* received 313 letters about "Mr. Mencken Sounds Off," 117 positive and 182 negative—a tally that he would presumably have taken as proof that the American people had been well and truly suckered by FDR.

He went straight to work on Supplement II, and when it was finished the following April, he began to think seriously about the "Mencken Encyclopedia" he had proposed to Knopf in 1943. It might not have come to pass, though, had he not suffered another small stroke on July 30, this one more serious than its predecessor. His diary entries for that day and the next, whose original typescripts are full of misspelled words, tell the ominous story: "I awoke this morning in a considerable state of confusion. It was impossible for me to write a line on the typewriter without striking wrong keys and making blunders in spelling, and for [a] while I could scarcely write by hand or sign my name. In talking, too, I found it necessary to grope for words. . . . I fear I am in for it." Several more days went by before his typing returned to normal, and more still before he felt able to write, so he started sifting through his out-of-print books and uncollected writings in search of material suitable for inclusion in an omnibus collection. His stenographer prepared clean typescripts of each selection, which he then edited by hand, and by October she had copied nearly four hundred thousand words. He intended to "pile it up without plan, and then make my selections," assuming that if the first volume sold well, Knopf would want a sequel as soon as possible.

In February he went to Florida with August for his first vacation since before the war, having accumulated another hundred thousand words' worth of *Chrestomathy* material. Supplement II to *The American Language* came out shortly after his return, and he resumed work on *My Life as Author and Editor*, perhaps sensing that he was running out of time. Louis Cheslock had noticed at a Saturday Night Club gathering in March that his "strange errors in rhythm [and] time upset the group," another sign that hardening of the arteries was taking its slow toll on his mental processes. Nevertheless, he was determined to work as hard as he could for as long as he could, and when Paul Patterson suggested in April that he attend that year's presidential conventions, he agreed, though leaving open the question of whether or not he would write anything about them for the *Sunpapers*.

Before he could make up his mind, a dreadful calamity struck his household. Early one May morning, Elsie Burley, the daughter of Hester Denby, his cook, went mad and killed her mother with a knife. Then she headed for Hollins Street, crying "Now for the boss!" Her pounding on the front door woke August, who looked out the window, saw that she was deranged, and called the police. Mencken spent the next several days settling Hester's affairs, attending her funeral with August; Emma Ball, the housekeeper, thereafter doubled as cook. It was, as he wrote in his diary, an "appalling business," and its effects on his state of mind are impossible to imagine, though one assumes it must have made him still more conscious of his own mortality.

Possibly Hester's death supplied the final push needed for him to resume full diplomatic relations with the *Sunpapers*. He packed his typewriter when he went to Philadelphia with Patterson for the Republican convention, and on June 19 the readers of the *Sun* opened their morning papers to discover that its most famous contributor was back after a seven-year absence. He filed four dispatches before succumbing to an attack of tracheitis, the most memorable of which described the first network telecast of a presidential convention: "I was sitting quietly in the almost empty press-stand when the first 10,000-watt lamp was turned on. The initial sensation was rather pleasant than otherwise, for it was a good deal like that of lolling on a Florida beach in midsummer. But in a few minutes I began to wilt and go blind, so the rest of my observations had to be made from a distance and through a brown beer bottle." Three weeks later he went back to watch as Harry Truman strode onto the platform in a gleaming white linen suit and told his fellow Democrats that he planned to "win this election and make these Republicans like it—don't you forget that." Mencken, who was seated near the rostrum, was unimpressed: "Despite his braggadocio, it was plain that he was not sure of himself." Like every other reporter in town, he expected Thomas Dewey to win in a walk.

He was far more interested in his third trip to Philadelphia, this time to cover the Progressive Party convention. Henry A. Wallace, the hapless nominee, was little more than a stooge of the Communists who were running the show, and Mencken pounded away with glee as Wallace's "psychopaths" handed him on a platter the last convention story he would ever write:

Such types persist, and they do not improve as year chases year. They were born with believing minds, and when they are cut off by death from believing in an F.D.R. they turn inevitably to such Rosicrucians as poor Henry.

The more extreme varieties, I have no doubt, would not have been surprised if a flock of angels had swarmed down from Heaven to help whoop him up, accompanied by the red dragon with seven heads and ten horns described in Revelation XII, 3. Alongside these feeble-minded folk were gangs of dubious labor leaders, slick Communists, obfuscators, sore veterans, Bible-belt evangelists, mischievous college students, and suchlike old residents of the Cave of Adullum.

Understandably pleased to find that "my old facility for rapid writing in the face of a moving news story had not been lost," he started turning out occasional editorial-page columns about the campaign. "I quit in disgust when [the *Sun*] began to support Roosevelt's war schemes," he wrote to a friend. "But now it is disillusioned, and I can go back with something resembling a smirk." At the same time he launched a new series of *New Yorker* articles called "Postscripts to the American Language" and put the final touches on the self-anthology he was now calling *A Mencken Chrestomathy*. He had culled twice as much material as he needed, and by the end of September, eleven days after his sixty-eighth birthday, he shipped off a

185,000-word revised draft to Knopf in New York, leaving the rest to shape into a sequel if the sales of the first volume justified it; Knopf had also agreed to publish a volume of notebook entries, for which Mencken had already devised a title, *Minority Report*.

The *Chrestomathy* was a kind of super-*Prejudices*, a fat volume enshrining his thoughts on everything from the music of Beethoven (good) to the presidency of FDR (bad). He claimed, somewhat disingenuously, that his purpose in editing it was "simply to present a selection from my out-of-print writings, many of them now almost unobtainable." In fact, the text was the climax of a process of meticulous serial revision that in some cases had lasted as long as three decades, during which Monday Articles had been transformed into *Smart Set* reviews, polished yet again for inclusion in one of the *Prejudices*, then put through one last revision to create a definitive version for the *Chrestomathy*. This editorial process is of particular relevance because Mencken's output consisted mainly of essays. Few of his books were, to coin a Menckenism, *durchkomponiert*. By choosing the best of these essays, revising them extensively, and collecting them in one carefully arranged volume, he produced a book that is at once an anthology and a deliberate act of literary and intellectual self-definition. *A Mencken Chrestomathy* is not quite as comprehensive as it looks: Much of his work was still in print in 1948 and is therefore not included. But despite the absence of any excerpts from *A Book of Prefaces, Treatise on the Gods, Treatise on Right and Wrong*, or the three *Days* books, it nonetheless contains a broadly representative cross-section of his writings, one from which subsequent generations of readers would acquire a total sense of Mencken as man and artist. It has remained continuously in print since its original publication in 1949, and with the exception of Malcolm Cowley's *Portable Faulkner*, no anthology of a modern American writer's work has done more to shape the reputation of its subject.

Even for a man who lived to write, Mencken's final burst of creative energy is startling. Perhaps he was whistling past the graveyard—he knew perfectly well that he might blow an artery at any moment—but if so, he meant to go out with a bang. In 1948 the state of Israel was founded, Mahatma Gandhi was assassinated, Whittaker Chambers accused Alger Hiss of being a Soviet spy, and Alfred Kinsey published *Sexual Behavior in the Human Male*. (Mencken read it, and didn't believe a word of it.) The bookstores were full of challenging novels, including Truman Capote's *Other Voices, Other Rooms*, James Gould Cozzens's *Guard of Honor*, Graham Greene's *The Heart of the Matter*, Norman Mailer's *The Naked and the Dead*, and Gore Vidal's *The City and the Pillar*. The first mass-produced electric guitar, the Fender Broadcaster—it would be renamed "Telecaster" two years later, after TV had caught on—went on sale in 1948, and the first McDonald's hamburger stand opened in California. It was the year of Stravinsky's *Orpheus* and Richard Strauss's *Four Last Songs*, of bebop and film noir and abstract expressionism. A twenties-like ferment was in the air again, and Mencken must have felt the itch to write about it while he still could.

That he would have had provocative things to say about postwar life was evident from the columns he published in the *Sun* on November 7 and 9. In the first one he noted with amusement how completely the press had been foxed by Truman's upset victory over Dewey: "The super- or ultra-explosion that staved in the firmament of heaven last Tuesday not only blew up all the Gallups of this great free republic; it also shook the bones of all its other smarties. Indeed, I confess as much myself. Sitting among my colleagues of the *Sunpapers* as the returns began to come in, I felt and shared the tremors and tickles that ran up and down their vertebrae." He followed it up with a piece calling for the desegregation of Baltimore's public tennis courts, a stance that may have conflicted with his belief in the inferiority of blacks but was all of a piece with his civil libertarianism: "Certainly it is astounding to find so much of the spirit of the Georgia Cracker surviving in the Maryland Free State, and under official auspices. The public parks are supported by the taxpayer, including the colored taxpayer, for the health and pleasure of the whole people. Why should cops be sent into them to separate those people against their will into separate herds? . . . It is high time that all such relics of Ku Kluxery be wiped out in Maryland."

For the first time in recent memory, he had even been excited by a new novel—and by an Englishman to boot, though it was set in California. *The Loved One*, Evelyn Waugh's scathing satire of life and death at Hollywood's Forest Lawn Memorial Park, impressed him with its grasp of the finer points of American euphemism, and when he learned that Baltimore would be Waugh's first stop on a tour to do research for a *Life* article on the Catholic Church in America, he agreed to a lunch date on November 24. (The arrangements were made by William Manchester, a young World War II veteran turned *Sun* reporter who was then at work on his first book, a biography of Mencken.) It is fascinating to imagine what Waugh, the Yank-baiting, Oxford-schooled Catholic aristocrat manqué, would have made of Mencken, the Anglophobic agnostic who never went to college and whose father owned a cigar factory. Waugh was a stupendously difficult man, but Mencken had succeeded in charming some hard cases in his day, and he was not easily bullied. Besides, the two men shared a profound sense of alienation from their time and place, as well as a similarly equivocal view of World War II. Unlike Mencken, Waugh had been briefly thrilled by the stark moral clarity arising from the Nazi-Soviet pact: "The enemy at last was plain in view, huge and hateful, all disguise cast off. It was the Modern Age in arms." But when Hitler turned on his erstwhile ally, and England and America hastened to make an expedient peace with the Kremlin, he found himself nauseated by the prospect of having to fight on the same side as Stalin, just as he squirmed at the heroic wartime rhetoric of Winston Churchill—sentiments that Mencken would have applauded.

Outside of their radically differing views on religion, the main difference between them was one of attitude. Both were repelled by the age in which history had forced them to live, but Waugh, tortured by the greater power of his imagination, found it all but unbearable. "His strongest tastes were negative," he wrote of Gilbert Pinfold, his fictional alter ego. "He abhorred plastics, Picasso, sunbathing and jazz—everything in fact that had happened in his own lifetime. The tiny kindling of charity which came to him through his religion sufficed only to temper his disgust and change it to boredom." Mencken, no less disgusted by the modern world, was prepared to let it go to hell in its own good time, withdrawing more or less contentedly into his work. He believed that the only way for a man like him to survive was to use his writing as a receptacle for "all his fears and hatreds and prejudices—all his vain regrets and broken hopes—all his sufferings as a man, and all the special sufferings that go with his trade. . . . There is always a sheet of paper. There is always a pen. There is always a way out." Right or wrong, it was a reflex that he could not deny. At sixty-eight, he was far too old to undo the habits of a lifetime.

The meeting, had it taken place, might have been a disaster—or a revelation. Either way, it was not to be. Mencken had another appointment to keep.

"UNTIL THOUGHT
AND MEMORY ADJOURN"

Scourge at Bay, 1948–1956

"As I grow older," Mencken wrote in 1943, "I am unpleasantly impressed by the fact that giving each human being but one life is a bad scheme. He should have two at the lowest—one for observing and studying the world, and the other for formulating and settling down his conclusions about it. Forced, as he is by the present irrational arrangement, to undertake the second function before he has made any substantial progress with the first, he limps along like an athlete only half trained." More than most men he had contrived to live both lives at the same time, but only by spending countless hours at his desk. He had never exercised, never played games, never engaged in any activity much more strenuous than playing the piano every Saturday night. Whenever he checked into Johns Hopkins, his doctors would tell him to lose weight, and from time to time he would embark on a diet, but he never stuck to it. He fully expected to drop dead without warning, and viewed the prospect with equanimity. "Moderation in all things," he wrote in his notebook. "Not too much life. It often lasts too long."

One of the minor miracles of his life was that after a half-century of hypochondria, his youngest brother still took his complaints seriously. In the last months of 1948 August's anxiety about his brother's health was so great that he was unwilling to leave him alone for any great length of time. On the evening of November 23, Mencken had to persuade him to go out and have dinner with friends. After August left his brother went across town to visit the Maryland Club, then stopped by his stenographer's home to pick up some typing and take her to dinner. Rosalind Lohrfinck had worked loyally for him since 1928, typing hundreds of manuscripts and taking down thousands of letters with quiet efficiency. "I have never encountered a more self-effacing person," he wrote of her in his diary. "At work, she some-how manages to bury her personality altogether." (No doubt he had enough for both of them.) Having listened as he dictated *Thirty-Five Years of Newspaper Work* and *My Life as Author and Editor*, she knew more about him than anyone except August, and he would have had to do something extraordinary to startle her—something not unlike what happened that night.

Not long after Mencken arrived at Mrs. Lohrfinck's house, he suddenly began babbling incoherently. Astonished, she called Dr. Benjamin Baker, who arrived within minutes, put Mencken into his car, and took him straight to Johns Hopkins. Since August was out, he telephoned Gertrude to tell her the news, and she kept calling Hollins Street until August returned at eleven. August drove imme-diately to the hospital, but none of the interns on duty had any idea where his brother was. Then he ran into Baker, who brought him to Mencken's room. According to August, the doctor told him, "Your brother has got a massive hemorrhage. He can't conceivably live until morning. He was practically dead on arrival. That's the way he would have wanted it to be." Baker later denied that he had said any such thing to August, but there is no question about what he told Alfred Knopf, who called the next day to ask about Mencken's con-dition: "Mr. Mencken has suffered a stroke, and I am sorry to say he is recovering from it."

Paul Patterson kept the story out of the *Sunpapers* and off the Associated Press wire. William Manchester was not even allowed to tell Evelyn Waugh what had gone wrong, and Waugh departed Bal-timore believing that Mencken had deliberately insulted him. But the news got out anyway, and spread quickly around town. Louis Cheslock heard it in the lobby during the intermission of a concert and called August, who told him that Mencken's speech had been affected by the stroke. Cheslock in turn told the members of the Saturday Night Club, who briefly considered cancelling that week's meeting. Instead, they gathered together to play the G Minor Symphony of Mozart, the tragic masterpiece of a composer whom Mencken had once praised as "beyond critical analysis: he simply happened."

The following week Johns Hopkins announced that Mencken had suffered "a small stroke," adding that his condition was "serious only insofar as the underlying process has serious connotations. His outlook has improved in the last few days and his condition is as satisfactory as can be expected." No other information was forthcoming, and the resulting news story was the first official word that he was ill. Nothing more was said about the extent of his illness, or about the fact—painfully and immediately obvious to all who heard him talk—that his brain had been badly damaged. Within a few days he was up and about and, in August's words, "raising hell to get home." But Baker, who was not a neurologist, believed him to be "violently insane" and had arranged for him to be admitted to what August later described as "a high grade nursing home for alcoholics and narcotic addicts."

Paul Patterson recommended instead that he be brought home and tended by a full-time nurse, and on January 6 he returned to Hollins Street. Cheslock went to see him there the next day, writing down his impressions of the visit as soon as he left:

> I was amazed to find Henry in the doorway of his sitting room, standing and beaming and saying how glad he was to see me. My immediate reaction was a very happy one. However, his first few sentences were among his best. His speech at times became quite confused and irrelevant. . . . His chief complaint

was about his eyes—not being able to read—but frequently referred to his eyes as his "ears" . . . alien words & stray fragmentary phrases kept cropping up despite his intense effort to steer a straight course in his thinking & speech.

His "Postscripts to the American Language" pieces continued to run in the *New Yorker* (the last one, an expanded version of his *American Speech* monograph on the origins of "O.K.," appeared in October). *A Mencken Chrestomathy* was published, and *Time* mentioned in its review that "a severe stroke . . . left the 68-year-old gadfly partially paralyzed and stilled his buzzing." Few outside his circle of family and friends knew the whole story, and those who did were stunned by the grotesque irony of it. Nothing was wrong with Mencken's eyes: The stroke had played havoc with the portion of his brain that allowed him to read and write. All he could do now was sign his name, scrawl an occasional one-sentence note full of misspelled words, and recognize the names of people he knew when he saw them in the paper, though he had trouble remembering them otherwise. He had not even been able to correct the proofs of the *Chrestomathy*. (Nathan did it for him.) Nor could he talk with anything like his accustomed ease. He could carry on a conversation and make himself understood, but often the wrong words came out: *clown* instead of *bound, youse* instead of *yard, notts* instead of *nothing, bugs* instead of *birds.** Once he understood that his career was over, he was devastated, and his doctors briefly feared he might try to commit suicide. That was never in the cards, but for a time his depression was profound. He went to a Saturday Night Club meeting in June and tried to play a Strauss waltz, but gave up after six measures; a full year went by before he would talk on the phone, listen to phonograph records, or permit anyone to read aloud to him.

---

*Sometimes he resorted to paraphrase. Unable to remember the word *scissors,* he asked instead for "the thing to cut with."

Long before that, though, he had come to terms with his situation, though he never stopped complaining about it, sometimes wryly, more often bitterly. "If I could only read," he would say to August. "The rest of it doesn't matter. But if I could just read I'd be the happiest man in the world." It became his custom to speak of himself as if he were already dead: "The trouble is, I didn't live long enough to write all I wanted." But he was still alive, and determined to get whatever he could out of whatever was left to him. He and August exchanged regular visits with the Cheslocks. Television was too moronic to bother with (one shudders to think what he would have made of Milton Berle), but he started going to the movies twice a week and found, very much to his surprise, that though most of them were "trash," some were worth seeing. He had no patience with melodrama of any kind, but big-budget musicals delighted him, especially if they were in color, and he enjoyed such Walt Disney cartoons as *Peter Pan* and *The Lady and the Tramp*, along with certain comedies, among them *Father of the Bride*, *The Quiet Man*, and the film version of George Bernard Shaw's *Pygmalion*, his last encounter with the "Ulster Polonius" who had launched his book-writing career in 1905. He even had a few favorite stars: Claudette Colbert, Olivia de Havilland, Donald O'Connor, Robert Ryan, Mickey Rooney, Clifton Webb, and above all Fred Astaire.

Within a few months he was willing to be seen in public again. In October, he came to the laying of the cornerstone of the *Sun*'s new publishing plant on North Calvert Street—it contained a copy of *The Sunpapers of Baltimore*, the history he had helped to write—and his picture ran in the paper the next day. Four months later he went to his last board meeting, successfully managing to take part in the latest discussion of whether the *Sun* should contribute to such charities as the Red Cross and the Community Fund. (As always, he voted no, and told August when he came home that "it gave him pleasure to have one more chance to vote against such quacks.") After that he started going out to concerts with Louis Cheslock, and before long music had regained its place at the center of his existence. He listened to records and the radio every day, and rarely went by the piano in the parlor without sitting down to play a fragment of some remembered piece, usually saying to himself, "If I could only read again!"

That he would never be able to do, though his speech slowly improved, and he found that he could make a kind of sense out of simple stories in the newspaper. He began sorting his unfiled correspondence with the help of Mrs. Lohrfinck, dictating an occasional letter in which he sounded like a faded but recognizable version of his old self: "Mr. Mencken sends his best thanks for your Easter card. He asks me to tell you that during the present week, of course, he has devoted his whole time to the church. He went to services this morning at 4 A.M. and he tells me that he will remain until 11.35 this evening." Increasingly confident, August let the nurse go and thereafter made do with Emma and an orderly, though the greatest burden of all fell on him: He took it upon himself to keep Mencken as happy as he could.

As August helped him to put the pieces of his broken life back together, the rest of the world began to realize what it had lost, and to remember that he was about to turn seventy, a suitable age for high honors. The *Chrestomathy*, like *The American Language* and the *Days* books before it, had been a bestseller.* In February of 1950 the American Academy and National Institute of Arts and Letters awarded Mencken its gold medal for distinguished achievement in essays and criticism. A few of his friends boggled when they heard the news—he had nothing but contempt for the academy and its members, turned

---

*It made the *New York Times* bestseller list almost immediately following its publication in May of 1949, appearing alongside John P. Marquand's *Point of No Return* (in which the *Mercury* is mentioned), Thomas Merton's *The Seven Storey Mountain*, Nancy Mitford's *Love in a Cold Climate*, George Orwell's *Nineteen Eighty-Four*, and Ayn Rand's *The Fountainhead*.

down awards of any kind, and told his writing friends in no uncertain terms that they should do the same—and wondered if illness had softened his brain. The inner circle knew the whole thing was a mistake. Misunderstanding a conversation with Mencken, Mrs. Lohrfinck had written on his behalf to accept the award. "I don't want that damn medal," he told August when he found out. "Stop it." But there was no graceful way to decline: The medal had already been struck and invitations printed, so he let the ceremony proceed as scheduled. Mark Van Doren received the medal on his behalf in New York, saying that Mencken "has been to his people what Shaw and Nietzsche were to theirs."

Another spurt of media attention came in April when Edgar Kemler published *The Irreverent Mr. Mencken*, the first full-length biography since Isaac Goldberg's *The Man Mencken* in 1925. Kemler had been in touch with Mencken prior to his stroke, and was genially told to do his damnedest: "I certainly hope you don't stay your hand when the time comes to write the book. Such jobs ought to be done realistically. Please bear in mind that I have been on the chopping block for years and have got used to it. No matter how dreadful you make the picture, the truth will always be worse." Kemler obliged, calling Mencken "a skeptic of the first rank—an American Rabelais, Swift or Shaw—who has somehow abused his gifts. As an artist, he might have written a *Gargantua* or a *Gulliver's Travels*. Instead he devoted himself almost wholly to the passing scene . . . after having assaulted and demolished the delusions of one era, he became a spokesman of the delusions of another." Still more problematic was his account of Mencken's isolationism, which was close enough to the truth to be damaging, if not quite enough to be fair:

Since the advent of Hitler in 1933, if not before, he had been quietly flirting with the newfangled doctrines of Nazism and Fascism. . . . Mencken was no Nazi. But he sympathized with Hitler's program for European conquest, and this heresy marked him off from the bulk of his fellow Tories, "swimming and swooning in the British orbit." Thereafter, his descent was swift and inexorable. Within less than a year, indeed within less than six months after the Munich crisis, he had forfeited his brief respectability and found himself in the political wilderness with Colonel Robert McCormick of the Chicago Tribune, Father Coughlin, Colonel Charles A. Lindbergh and other notorious isolationists as his allies.

In his final chapter Kemler wrote of Mencken's postwar visit to Ezra Pound in Washington's St. Elizabeths Hospital, where the poet was committed pending trial for having made wartime propaganda broadcasts from Italy: "Mencken found it a very enlightening experience. Though he had only an occasional correspondence with Pound, and in general disapproved of his poetry, they had once been joined in common hostility to the Puritan tradition in letters. More recently they had been joined again in the intellectual precincts of Fascism; and but for the grace of God, he too might have ended in some such hideous predicament." Again it was an unfair portrayal—he was no more a fascist than a Nazi—but Kemler had touched a nerve. Had Hitler been less of a vulgar rabble-rouser, it is possible that Mencken might have been tempted to a far greater degree of indiscretion in his political writings of the late thirties, though he would never have stooped to anything like baiting Jews on the radio. Unlike Pound, he was not that kind of anti-Semite.

The *Sun* celebrated his birthday in September by running a story in which the state of his health was summed up in decorous evasions: "An elder now himself, gray of hair, leaner since his illness and not up to major tasks, but still an avid connoisseur of long-leaf tobacco, fine wines and the published foibles of the nation, Mr. Mencken has not changed." He attended a small luncheon party held in the board-room, complete with cake and candles. "Well, I'm ready for the angels," he said, and blew them out. A photographer was on hand to record the occasion, and the pictures, unlike the story, leave no doubt of just how much old age and illness had changed him. He was wizened and stooped, and the blue eyes that once had shone as bright as gas jets were now dull and lifeless.

Four months later William Manchester published *Disturber of the Peace*, an infinitely more sympathetic biography than *The Irreverent Mr. Mencken*. Manchester adored Mencken and made the mistake of trying to imitate his style, leading George Jean Nathan to blast him in the *New York Times Book Review* for having stooped to "hero-worship of such lush proportions as haven't been matched since Frank Harris' famous testimonials to himself." Another of Mencken's old associates was similarly critical: Charles Angoff, writing in the *Saturday Review*, called Manchester's prose "so Menckenian in style that it is sometimes embarrassing; carbon-copy menckenese is not easy to take." Both men had a point, though they also had axes to grind. In retrospect the book's main significance is that Manchester was the first person to print the truth about his subject's condition: "For several sentences he might talk quite coherently; then the rest of his message would be senseless. The remembrance of names, dates, and specific events in their entirety was completely beyond him. . . . His condition was hopeless."

Many other tributes were paid to Mencken on and around his seventieth birthday, both fulsome and backhanded, but one in particular stood out, perhaps because it was neither. Clifton Fadiman, reviewing Lionel Trilling's *The Liberal Imagination* for the *New Yorker*, used Mencken as a benchmark in trying to sum up Trilling's particular gift: "Perhaps not since the days before Mr. Mencken mellowed have we had, however, [a critic] who could talk to the generality of educated rather than merely literary Americans." It was an odd compliment—it is hard to think of two writers who had less in common—but still says a great deal about the esteem in which Mencken had come to be held, especially now that illness had rendered him incapable of giving offense to the easily offended. At long last the ironic title by which he had come to be known, the Sage of Baltimore, was starting to sound like the simple truth.

* * *

ONE MONTH TO THE DAY AFTER HIS SEVENTIETH BIRTHDAY, MENCKEN suffered a heart attack severe enough that the doctors at Johns Hopkins told the Associated Press there was "no chance" he would recover from it. The tributes gushed forth yet again (including a mistakenly published obituary in the *Gazette* of Gastonia, North Carolina), but a hospital spokesman was reporting a week later that he was drinking beer and asking for cigars. Louis Cheslock paid him a visit in December and found him in fine fettle: "His speech more coherent than in many a moon. His humor up to par. He talked about his having seen the Metropolitan [Opera]'s Christmas day performance of Hansel & Gretel on television—and having had recourse to the urinal midway in the opera—cigar in mouth, and eggnog with the overture—wondered who could be enjoying the performance in greater comfort than he?!" Another friend who visited him was struck by how his writerly imagination warred with his aphasia, causing him to utter sentences full of verbal twists: "My nurse called the cops in and I got some medicine." (The same friend noted that when he couldn't think of a word, he filled in the blank by cursing.)

He went home in March, unapologetically sorry to be alive. "God, why didn't the quacks let me die when they had me?" he asked Philip Wagner. But since he was, he resolved to make the most of it, working with a speech therapist once a week and keeping up with the news as best he could. William Manchester interviewed him for the *Sun* and reported that for once, he had some not entirely bad things to say about a couple of politicians. He called Dwight Eisenhower a "better than average" president who was doing well "for a general," and added, more surprisingly, that Harry Truman "did some good work." Lest anyone think he had mellowed, though, he went on to call Douglas MacArthur "a dreadful fraud who seems to be fading satisfactorily."

Manchester had remained close to Mencken after the publication of *Disturber of the Peace*, and in 1954 the young author came to

Hollins Street four mornings a week to read to him. He listened with
pleasure to *Huckleberry Finn*, still his favorite American novel, and
Conrad's "Youth," a story he loved so much that he had resorted to
musical metaphor when he wrote about it in the *Smart Set* in 1922:
"Some time ago I put in a blue afternoon re-reading Joseph Conrad's
'Youth.' A *blue* afternoon? What nonsense! The touch of the man is
like the touch of Schubert. One approaches him in various and
unhappy moods: depressed, dubious, despairing; one leaves him in
the clear, yellow sunshine that Nietzsche found in Bizet's music."
They listened to records together (Liberace inspired him to mumble
"something obscene," but he smiled at Gilbert and Sullivan) and
marveled at the Sam Sheppard murder case as it unfolded in the
pages of the *Sun*. Manchester was struck by the undisguised affection
he felt for the boys and girls of the neighborhood: "Children had
become dear to him; unlike their parents they were unembarrassed
by his condition. . . . [E]ach Christmas the old man distributed huge
sacks of candy to all the children who lived around Union Square."*
He had always been good with the children of his friends—a fact that
never failed to astonish those who knew him mainly by his reputation—but now they meant more to him than ever before.

In time Manchester gave way to Robert Allen Durr, a Ph.D. candidate from Johns Hopkins whose reminiscence of Mencken's last
days is the most touching thing ever written about him. Durr noted
with amusement the hand-rubbing anticipation with which
Mencken awaited his daily dose of tidbits from the *Sun*'s police blot-
ter: "What's in the papers? Any good stuff, anything rich? Any mur-
ders or rapes? Any robberies?" In the absence of violent crime, he
liked to hear about airplane crashes, boat sinkings, and the failed

---

*Alistair Cooke came calling one day and noticed the same thing: "We went out into
the dooryard for a breath or two of a beautiful spring morning, and two miniature heads
bobbed up over the wall. They were the neighbor's children, and they piped, as in a nurs-
ery rhyme, 'Mis-ter Men-cken, Mis-ter Men-cken.' He waved at them and shouted, 'Hi,
there, rascals.'"

romances of royalty. But he also liked to eat breakfast (fruit juice, two soft-boiled eggs, and a slice of bread) at his office desk, sipping his coffee as he gazed out the window and watched his neighbors' children walking to school:

> The winters were hardest on Mr. Mencken, for often then he could not go out into the yard to work, and the days dragged heavily. The chief consolation was the fireplace. He told me how he and his brother especially enjoyed it on very cold or snowy evenings. They would stand at the window watching the snow for a while and then return to sit before the fire. . . .
>
> What remained to him of his old joys was music; many mornings he told me how he had listened for a couple of hours the night before and how superb it had been. Yet in truth he had left to him something the average man never acquires—the capacity to enjoy the commonplace activities of life. Though these, of course, could not make up for his inability to work, they helped. One lovely autumn morning, with the sky clear, the breeze cool, and the sun warm, Mr. Mencken sat over in the sun so that it fell on his back. "Well, this is very nice. This is fine. This ought to make us feel good. . . . You know, I always enjoyed life in all its forms. I've always taken a great pleasure in getting up in the morning, having breakfast, and settling down to work. I had a good time while it lasted."

Then, toward the end of 1955, came a final blessing. Rummaging through boxes of old manuscripts in the basement of Hollins Street, Mrs. Lohrfinck stumbled on a folder full of typescripts of the note-book entries he had been polishing in the mid-forties, together with the preface he had written for the volume he planned to call *Minority Report*. The project had been forgotten at the time of his stroke, but it turned out to be almost complete save for the final cull, so Mencken and his stenographer settled in for one last job of work: She would read him each entry, after which he decided whether or not to include it. His involvement, if limited, was no sham. "He alone did the deciding," August said, "and he was not only fully competent to do it but he allowed no one to influence him." All told he cut eighty-six notes, among them these three:

> The whole policy of Roosevelt II, whether in domestic or foreign affairs, was founded upon the fanning of hatreds—the first and last resort of unconscionable demagogues, at all times and everywhere. This fanning, of course, was done to the tune of loud demands for tolerance.

> The poetasters who imitate Thomas S. Eliot spend a large part of their time writing manifestoes [sic] explaining their poetical theories. The poetry magazines and the "literary" quarterlies published at cow-town "universities" are full of such rumble-bumble. All it teaches a rational reader is that its writers are jackasses. It stands precisely on all fours with the highfalutin glosses prepared by third-rate "modern" artists for their exhibitions of daubery.

> It is always well to distinguish between the character of a nation and the character of its people. The Germans, taken together, practise hardness and have a considerable talent for it, but individually they are mainly sentimentalists. Something of the same sort is true of most other nations, including the Jews. It reveals itself in the fact that every one, as the saying goes, has a pet Jew. The explanation here is that the average educated Jew tends to be an endurable enough fellow, despite his obvious failings, whereas Jewry as an organized body is almost unqualifiedly unpleasant.

What remained was controversial enough, and also vintage Mencken. He had never published an extensive selection of notebook entries, and *Minority Report* proves that while his gift for aphorism was only just adequate, he had a remarkable knack for the writing of *pensées*. These five-finger exercises, most no longer than a paragraph or two, reduce his thinking to its essence, paring away the frills and furbelows of his twenties essays without losing the pungent flavor of his style (just as he had slimmed down some of his most celebrated pieces for inclusion in the *Chrestomathy*, in every case strengthening them). The best of them are as memorable—and revealing—as anything he wrote:

> It is a folly to try to beat death. One second after my heart stops thumping I shall not know or care what becomes of all my books and articles. Why, then, do I try to keep my records in order, and make plans for their preservation once I am an angel in Heaven? I suppose the real reason is that a man so generally diligent and energetic as I have been finds it psychologically impossible to resign the game altogether. I know very well that oblivion will engulf me soon or late, and probably very soon, but I simply can't resist trying to push it back by a few inches. Civilized man, indeed, is essentially indomitable. He refuses to yield to the natural laws that have him in their grip. His life is always a struggle against the inevitable—in Christian terms, a rebellion against God. I'll know nothing of it when it happens, but it caresses my ego today to think of men reading me half a century after I am gone. This seems, superficially, to be mere vanity, but it is probably something more. That something is a sound impulse—the moving force behind all cultural progress— to take an active hand in the unfolding of human life on this sorry ball. Every man above the level of a clod is impelled to that participation, and every such man desires his contribution to last as long as possible.

Alfred Knopf was thrilled to hear of the rediscovery of *Minority Report*, and he set the book in type as fast as August could mail him installments of the finished manuscript. In January he announced its coming publication: "Henry is as eager to have the book in his hands as the greenest author of a first novel. He wants to see it, he says, before the angels see him." The *Sun* sent a reporter to talk to Mencken, who lowered himself into his easy chair, lit a cigar, and said, "It will be nice being denounced again." The interview, which ran in Baltimore on January 26, was picked up by the wire services the next day, helping to spread the word about a book that would become a bestseller when it was published in May.

Mencken spent the afternoon of January 28 listening to the Metropolitan Opera's Saturday matinee broadcast of *Meistersinger* on the radio. He had never been particularly fond of opera, or the music of Richard Wagner—he once had the cheek to suggest that Wagner should have turned the second act of *Tristan und Isolde* into a double concerto for violin and cello, by no means a bad idea—but the Met was on its best behavior that afternoon, and he enjoyed himself thoroughly. Louis Cheslock stopped by in the evening, and was dismayed to learn that his friend was suffering from chest pains. "Louis, this is the last you'll see me," he announced when he came downstairs, then threw up in a basin and felt better. Though he skipped his evening cigar, he downed three martinis "rather quickly" as he chatted about *Meistersinger* and *Minority Report*. He said good-bye at nine—an hour earlier than usual—and went upstairs to his third-floor bedroom to listen to more music on the radio, his substitute for late-night reading. Then he put out the light and went to sleep, perhaps sensing with relief that his too-long life had reached its merciful end.

"QUACKS AND SWINDLERS, FOOLS AND KNAVES"

The Mencken file in the *Sunpapers* morgue contained this memorandum:

Baltimore, June 1, 1943
Save in the event that the circumstances of my death make necessary a news story it is my earnest request to my old colleagues of the Sunpapers that they print only a brief announcement of it, with no attempt at a biographical sketch, no portrait, and no editorial.

H. L. Mencken

Some imperatives, however categorical, are meant to be ignored, and the editors of the *Sun* brushed this one aside without a second thought. The main story in Monday's paper broke the news simply: "Henry Louis Mencken, newspaper man, critic, wit and Baltimore's best-known writer, died early yesterday in his sleep at his house on Hollins street." But it was accompanied by a two-column portrait (in which Mencken, as was his wont when posing for formal photographs, contrived to look puzzled) and a page-one obituary by Hamilton Owens that ended on a personal note: "This is not the place or the time for any final appraisal of H. L. Mencken's contribu-

tion to the culture of the United States. But whatever the conclusions of the students of such things, his colleagues on *The Sun* had
the benefit, over many years, of association with one of the most
stimulating personalities of their time." Owens's tribute was illustrated by a reproduction of the Nikol Schattenstein painting that
now hangs in the Mencken Room and a portfolio of other images
from the not-so-distant past: Mencken being arrested in Boston for
selling a copy of the issue of the *American Mercury* containing
"Hatrack"; Mencken at the Scopes trial; Mencken ecstatically downing a glass of beer at the Hotel Rennert the night Prohibition was
repealed; Mencken at his typewriter, chewing a cigar; Mencken and
Sara Haardt on their wedding day. The editorial page, for whose
normal contents he had had nothing but scorn, carried a tribute of its
own: "To many outsiders who had been scourged by his pen, he was
the embodiment of evil, as their anguished efforts at reprisal showed,
but to those of his colleagues who knew that his battle was waged
against sham and hypocrisy and not against individuals as such, he
was simple, compassionate and even humble. In short; a warm, loyal
and understanding friend. We shall not soon see his like again."

Most of America's major newspapers ran similar salutes. "As
champion of the unfettered mind," said the *Boston Globe*, "he has few
equals." The *New York Times* described him as "an arch foe of every
sort of sham and hypocrisy. . . . We are impoverished by his death
just as we were enriched by his life"; the *Herald Tribune* called him
"the joyous voice of ridicule assailing the puritanical, blasting mealy-
mouthed politicians, taunting the do-gooders, fighting the censors. . . . We like to think that the Mencken zest for slaughtering
flummery hardened the American character against demagogy and
all forms of faking." One suspects that the author of "In Memoriam:
W.J.B." would have taken more pleasure in the vengeful last words
that trickled in from the editorial writers of Upper Sandusky, Big
Spring, and Miami Beach: "By his own admission, his life was aim-
less, so he spent it attempting to destroy what he disbelieved. The
Republican and Democratic parties which he ridiculed will perhaps

survive him many years. Religion will be a force for an eternity after he has been forgotten. . . . Henry L. Mencken is gone the way of all grass, but the Bible Belt is flourishing like the green bay tree, thank you. . . . In his final years he was a shriveled old man, and if unbeknownst to himself he possessed a soul, it was shriveled too." And he might well have laughed out loud at the text of the joint resolution passed three weeks after his death "expressing the sorrow of the General Assembly of Maryland over the passing of Henry Louis Mencken": "He was a citizen of Baltimore who always positively associated himself with that town. . . . He was involved in many controversies from which the surviving memories are his incorruptibility and courage, which were in the great tradition, and his humor, which was homeric."

One obituary that would have left him with mixed feelings appeared in the *Times* of London, house organ of the nation he so cordially detested:

Mencken's importance in American letters was for long obscured by his form and style of expression. Superficially his writing exhibited some of the characteristics of popular journalism in the United States. He relied, for instance, on the use of provocative titles, the freest American vernacular, wild and wilful exaggeration, and a slap-bang verbal gusto. But the style, with Mencken, was very much the man. If, in the end, it was the sheer energy of his diction, rather than the substance of his thought, that gave him his popular standing, the thought itself still mattered, and it had a deep and pervading influence on American literary standards. It was as a pioneer of the modern American literary idea that Mencken deserved—and secured—recognition.

The next day a small band of family, friends, and colleagues joined August, Gertrude, and Charlie in the chapel of a Hollins Street funeral parlor to say good-bye. Hamilton Owens led a delegation from the *Sunpapers;* Alfred Knopf and James Cain represented the world of literature; a sprinkling of Saturday Night Club members was on hand, including Ed Moffett, the oldest surviving member, and Louis Cheslock, who had seen more of Mencken in the past eight years than anyone other than August and his servants. In 1927 he had issued a "clarion call to poets" for an agnostic funeral service that was "free from the pious but unsupported asseverations that revolt so many of our best minds, and yet remains happily graceful and consoling," but none having obliged, his final instructions to August were passed on to Hamilton Owens. "August asked me to stand up for a few minutes," Owens told the mourners, "and repeat what most of you already know. His brother Henry, orally and in writing, said he wanted no funeral service of any kind. All he wanted was that a few of his old friends gather together and see him off on his last journey. That we are doing."

It did little to ease their sorrow. "Somehow, we were made to sub-serve a gag," Cain recalled, "and the effect wasn't so much bleak as blank." Cain, Owens, Knopf, and Frank Kent went straight from the funeral parlor to Marconi's, there to drown their sorrows in loud reminiscence, while August and Charlie accompanied the coffin to Loudon Park Cemetery, where it was cremated and the ashes placed in the family plot next to those of Sara. Mencken had gone there in 1945, ten years after her death, and gazed at the lone white marble stone that bore the family coat of arms and the names and dates of the dead. "There is room left for all the rest of us," he wrote in his diary that day. "My own name will be there soon enough."

Mencken's will was probated two weeks after his death, and the contents announced in the *Sunpapers* the following day. The estate was initially valued at three hundred thousand dollars. His library, "family papers," and correspondence with Marylanders were left to the Pratt, his remaining letters to the New York Public Library. Alfred Knopf received half of his Knopf stock, August the other half. Cash bequests were made to Charles and his daughter, Virginia, and to Mrs. Lohrfinck; August received a four-fifths interest for life in the residuary trust estate, Gertrude one-fifth. Emma Ball received five hundred dollars in cash and a fifty-dollars-a-month income. The eventual residue was divided between the Pratt and Johns Hopkins Hospital, which Mencken said had been "of more use and benefit to me during my lifetime than any other public institutions of Baltimore." The residual bequests were absolute, save for one condition: "I request that neither bequest shall be designated by my name, or by any other name embodying 'Mencken.' "

The Pratt's Mencken Room, in which Mencken's manuscripts, scrapbooks, private papers, and personal library were deposited for the use of scholars, was opened in April by Alistair Cooke, who dedicated it to "the comfort of sinners and the astonishment of the virtuous," adding that it would be "supreme blasphemy if anyone came to the Mencken Room solemnly, timidly, pedantically or reverently." *Minority Report* was widely excerpted in American newspapers, received mixed but generally favorable reviews, and briefly appeared on national bestseller lists, ultimately selling 22,500 copies. And in June, Charles Angoff published *H. L. Mencken: A Portrait from Memory*. Mencken had long assumed that Angoff, who had been writing critically about his old boss at increasingly frequent intervals since 1938, was waiting for him to die before bringing out a full-length memoir, and had no illusions about its likely contents. "You know, Angoff's going to write a book about me, and he's going to denounce me bitterly," Mencken said to August a few months before his death. "I know he is. It'll be a sour book. He'll denounce me bitterly." And so he did. *H. L. Mencken: A Portrait from Memory* is a full-length portrait of the artist as an anti-Semitic boor.

The resentment of an obscure man of letters who envied the fame of his illustrious patron is evident throughout Angoff's book, in which no opportunity is missed to enumerate Mencken's alleged literary shortcomings: "Many of the writers—at least the first-rate writers—whom Mencken praised were not always happy about his praise or very comfortable in his presence. They apparently did not feel that he really understood the process of writing, its spiritual and

psychological propulsions, its moral overtones, its esthetic essence."
Also present are errors and inconsistencies that call into question the
authenticity of Angoff's renderings of his conversations with
Mencken, which he claimed were based on "my notes made at the
time and upon my memory." Whatever the actual sources of his
Boswellian accounts of Mencken's table talk, the results are uncon-
vincing. Some passages ring true, but the overall effect is overdrawn,
more like a secondhand caricature than a portrait from life: "Say,
Angoff, you have a mighty fine tenor voice. You're very gifted in that
line. I'll see Gatti-Casazza about getting you into the Metropolitan.
He and I go to the same whorehouse."

Some of the excesses of Angoff's "portrait" were Mencken's own
fault. Witness, for instance, the book's climactic scene, in which the
two men visit the grave of Edgar Allan Poe:

> He began to laugh. "Every now and then," he said, "some
> members of the Saturday Night Club and I taxi over here and
> we pay our respects to Poe. Way back, we used to pour a bottle
> of whiskey on top of his grave. But when liquor became expen-
> sive and hard to get, at that, except from the best bootleggers, I
> suggested another mark of respect." He laughed again. "I sug-
> gested that we all piss on his grave!"
>
> I was so shocked by his vulgarity that I felt my stomach sink,
> and my face must have changed color, for Mencken said:
> "You're white, Angoff. That wine affecting you?"
>
> "No. I was just thinking about what you said. Really,
> that's . . ."
>
> "Cheap?" asked Mencken, in a strange tone of voice.
>
> "Well . . . it is a little rough to take."
>
> "Hell, you're sentimental, Angoff."
>
> "I suppose so."

H. L. Mencken: A Portrait from Memory received mostly negative
reviews, several written by old friends and colleagues of its subject.

The most venomous reaction came from Alfred Knopf: "The laws of libel forbid my saying what I think of [the book], but you can get the idea from Churchill's references to Mussolini in his wartime speeches." But even after allowing for malice and exaggeration, much of Angoff's book is in its broad outlines plausible enough, in particular the chapter on "The Jews," in which he specifically charges Mencken with anti-Semitism, offering in evidence his versions of their private conversations: "Angoff, Hitler is a jackass. But he isn't altogether crazy in what he says about the Jews. I understand the Jews make up about 10 per cent of the population of Berlin, yet 90 per cent of the lawyers are Jews. Also more than 90 per cent of the doctors are Jews. And so on. Do you think that's right? . . . The Jews in Berlin and in all of Germany are crowding the professions." In 1956 reviewers dismissed this passage and others like it as incredible, but numerous parallel passages from Mencken's sealed papers are now available to show that Angoff, whatever his other inaccuracies and distortions, was a basically reliable source when it came to the subject of Mencken and the Jews.

One final piece of news remained to be broken. The Pratt announced on July 13, 1957, that Mencken had left it a four-volume typescript, *My Life as Author and Editor*; a three-volume typescript, *Thirty-five Years of Newspaper Work*; and a five-volume diary. Nothing more was known about the nature of this material save that it had been dictated to Mrs. Lohrfinck "at intervals between the early 1930's and 1948," and Mencken had placed all of it, as well as the letters he left to the New York Public Library, under time seal. The letters were to remain sealed until 1971, the diary until 1981, the typescripts until 1991. According to the terms of the bequest, the diary and memoirs would ultimately be available "only to graduate students or those of a higher grade engaged in serious critical or historical investigation, approved after proper inquiry by the chief librarian." The manuscripts, nailed inside wooden crates sealed with metal strips, were duly deposited in a fireproof vault at the Pratt, there to await the inspection of literary historians of another time.*

With the publication of *Minority Report* and the dedication of the Mencken Room, Henry Louis Mencken officially became part of the ages. Between his death in 1956 and the appearance in 1989 of his diaries, two full-scale biographies, two book-length memoirs, nine posthumous collections, and four major volumes of letters were published. Two paperback anthologies continued to sell in large numbers: Alistair Cooke's *The Vintage Mencken*, which had sold 128,436 copies as of 1988, and James T. Farrell's selection of *Prejudices*, which had sold 70,622 copies. But the brief uproar over *H. L. Mencken: A Portrait from Memory* merely served to underline the extent to which its subject had receded into the shadows of American memory in the seven years since his enforced retirement. *Minority Report*, the book he had hoped would cause him to be "denounced again," brought only posthumous praise; Angoff's charges of anti-Semitism were dismissed as the sour fruit of jealousy. And while some of Mencken's work was still read, it was for the most part neither taught nor discussed in print. Rarely did any volume of the *Reader's Guide to Periodical Literature* published between 1956 and 1990 contain more than one or two Mencken entries, and whole years passed without a single listing. He was, it seemed, little more than a nostalgic relic, fondly remembered but no more to be taken seriously than Calvin Coolidge.

What made people start paying attention to Mencken once again, appropriately enough, was not a sober critical revaluation but a hot-tempered journalistic fistfight. When his diary was opened in 1981, each volume bore a label restating the terms of the original bequest:

---

*He had also left duplicate copies to the New York Public Library and Dartmouth College in order to circumvent the possibility that the Pratt's collection of his papers "may be destroyed in some future war or in a revolution. The United States has got through two World Wars without suffering a scratch at home, but this is not likely to happen the next time. If there is any raid on American libraries by radicals my papers will be among the first destroyed."

The material therein was only to be made available to serious scholars. But the board of trustees of the Pratt concluded that Mencken had not intended this language to forestall publication, and the Maryland attorney general agreed. In 1985 Charles Fecher, editor of *Menckeniana* and author of an important study of Mencken's life and work, began preparing the diaries for publication by Knopf. Rumors soon began to circulate that they were filled with anti-Semitic and racist statements, and when *The Diary of H. L. Mencken* came out at the end of 1989, it was prefaced by an introduction in which Fecher suggested as much: "Let it be said at once, clearly and unequivocally: Mencken was an anti-Semite."

This statement was immediately and precisely qualified ("To be sure, he was not one in the sense that Hitler was, or such men as Richard Wagner or Houston Stewart Chamberlain. He would have defended the civil rights of Jews as promptly and vigorously as those of any other group. . . . he knew large numbers of Jewish people, and appears to have been on the most cordial terms with nearly all of them"), and there was in fact comparatively little in the diaries likely to startle readers familiar with Mencken's published writings. But a new generation of reporters and columnists, ignorant of his work and its context and predisposed to regard those accused of racism as guilty until proven innocent, seized on Fecher's "clear and unequivocal" charge and ignored his qualifications. The headlines left little room for doubt, or even discussion: MENCKEN'S DARK SIDE/LEGEND'S RACIST DIARIES STUN BALTIMORE (*Washington Post*); H. L. MENCKEN: PORTRAIT OF A BIGOT/LITERARY ICON WAS RACIST, DIARY SHOWS (*New York Newsday*); MENCKEN THE HORRIBLE: BITTER HUMOR CONCEALED SOUL OF A BIGOT (*Chicago Tribune*); A BIGOT'S NOTES TO HIMSELF (*Kansas City Star*). One commentator stated flatly in an article published on the op-ed page of the *Washington Post* that "those who defend a writer such as H. L. Mencken must be said to possess an antisemitic sensibility themselves." And the Baltimore *Sun* was quick to condemn its most distinguished staffer, publishing a story headed "Dimming of a Legend: Mencken Diary Reveals Racism, Anti-Semitism, Spite for Friends."

More carefully considered responses appeared in due course, including a letter to the *New York Review of Books* signed by Louis Auchincloss, Ralph Ellison, John Kenneth Galbraith, John Hersey, Norman Mailer, Arthur Miller, Arthur Schlesinger, Jr., William Styron, and Kurt Vonnegut: "We wish to express our dismay at the overreaction to the *Diary* of H. L. Mencken. The *Diary* does indeed contain discourteous remarks about Jews and blacks. It also contains discourteous remarks about most races, nations and professions. . . . Whatever Mencken's 'prejudices'—the word he himself used to describe his essays—he was a tremendous liberating force in American culture, and should be so celebrated and remembered." But the controversy raged on. A year after the publication of the diaries, the typescripts of *My Life as Author and Editor* and *Thirty-five Years of Newspaper Work* were unsealed, and subsequently appeared in abridged trade editions. Both memoirs contained numerous passages that reinforced the general impression of Mencken's bigotry, so much so that Joseph Epstein, a longtime admirer of Mencken who had defended him in a widely noted review of *The Diary of H. L. Mencken* published in 1990, felt obliged to reverse himself five years later: "It is, alas, impossible any longer to let Mencken off the hook on the charge of anti-Semitism. . . . For all his high spirits, his immense talent, his considerable intelligence, Mencken, when he felt himself cornered by events, decided to blame the Jews." No less troubling was the additional evidence of the sometimes coldly opportunistic view of friendship seen in both the memoirs and the diary. It is sad to think what many of his closest friends would have said about him had they lived to read what he said about them.

By now it would be possible to bring out a new *Schimpflexicon* consisting solely of post-1989 articles about Mencken. Yet for all the anger, interest in his work has not abated. If anything, the publication of his sealed papers served to reestablish him as a key figure in American cultural life. Overnight it mattered once again what a half-forgotten Baltimore newspaperman thought about blacks and Jews, Dreiser and Fitzgerald, Bryan and Roosevelt. Today the "Mencken industry," as it is known among scholars, is bigger than ever. New collections of his writings continue to appear; new biographers flock to the Pratt to conduct their inquiries. When Richard Posner surfed the World Wide Web to determine which "public intellectuals" had been most frequently cited there between 1995 and 2000, Mencken ranked thirty-seventh with 2,462 mentions, the sixth most cited dead intellectual—less often than George Orwell, George Bernard Shaw, H. G. Wells, Timothy Leary, or C. S. Lewis, but more than W. H. Auden, Bertolt Brecht, Ayn Rand, Aldous Huxley, Thomas Mann, E. M. Forster, Ezra Pound, Jean-Paul Sartre, John Maynard Keynes, Lillian Hellman, Albert Camus, W. E. B. DuBois, Allen Ginsberg, and Marshall McLuhan.

BUT WHO NOW *ADMIRES* H. L. MENCKEN? AND WHAT IS THE NATURE OF his continuing influence on American life and letters?

Mencken himself thought his ultimate significance would be political: "At the moment, with the Roosevelt crusade to save humanity in full blast, my ideas are so unpopular that it is impossible, as it was from 1915 to 1920, for me to print them. But when the New Deal imposture blows up at last, as it is sure to do soon or late, they may have a kind of revival." Such a revival seemed possible when, two months before Mencken's death, *National Review*, the magazine that served as a rallying point for the postwar conservative movement in America, was launched by William F. Buckley, Jr., who in 1952 had briefly served as associate editor of a later incarnation of the *American Mercury*. Yet Mencken was never fully accepted by American conservatives, mainly because of his hostility to religion. (Had they known of the extent to which his work in the twenties helped lay the intellectual groundwork for the America-hating adversary culture of the sixties, they might have repudiated him altogether.) Nor did the right-wing populists of the post-Reagan era find his antidemocratic philosophy palatable. Newt Gingrich described him as "this miserable, mean person, who wrote with a deep cynicism . . . and despised people." Not until libertarianism emerged as a full-fledged wing of the conservative movement did he become a true influence on the right.

Mencken's conservatism, like so many other aspects of his thought, can only be fully understood in the context of his own time. Unlike the first postwar generation of American conservatives, his principal target was not socialism but democracy—or, more precisely, the unrestrained populism to which direct democracy can lead. His critique of democracy is uncomplicatedly elitist: By placing power in the hands of the unlettered and envious majority, it threatens "the two greatest intellectual possessions of modern man . . . the idea of personal freedom and the idea of the limitation of government." Few writers have put the case against the common man so memorably, whether in his mordant accounts of democracy in action or his theoretical discussions of the inherent flaws of democratic rule: "There is only one honest impulse at the bottom of Puritanism, and that is the impulse to punish the man with a superior capacity for happiness—to bring him down to the miserable level of 'good' men, *i.e.*, of stupid, cowardly and chronically unhappy men. And there is only one sound argument for democracy, and that is the argument that it is a crime for any man to hold himself out as better than other men, and, above all, a most heinous offense for him to prove it."

Mencken also viewed World War II from an isolationist vantage point common enough to conservatives of his generation, though it has few open adherents today, now that our knowledge of Hitler's intention to wipe out European Jewry has negated any possible reservations about the moral necessity of America's entry into the war. His postwar silence about the Holocaust is harder to excuse, though not to understand. As a battle-scarred survivor of the anti-German frenzy that gripped the United States between 1915 and 1920, he must have found it all but impossible to accept that the people whose culture he regarded as supremely enlightened were capable of installing and supporting a regime that slaughtered six million Jews; as a staunch believer in the gospel of science, he must have been deeply shaken, perhaps even to the point of denial, to have learned of the depraved ends to which the German scientific establishment was so readily turned by Hitler and his henchmen. Not that he should have been surprised by any manifestation of human depravity. "I do not aspire to set up any doctrine of my own," he told Burton Rascoe in 1920. "My notion is that all the larger human problems are insoluble, and that life is quite meaningless—a spectacle without purpose or moral." Yet it is difficult to reconcile his blank nihilism with the deep-dyed conservatism of the Baltimore burgher who reveled in tradition, preferably German. Be that as it may, he found no difficulty in posing as a nihilistic conservative, and his failure to recognize the absurdity of that position stands in the way of his being taken seriously as a political philosopher. Here as elsewhere his wide-ranging but unsystematic self-education, which emphasized *belles lettres* at the expense of history and philosophy, may have led him astray, or at least narrowed his perspective.

By the same token one need not accept René Wellek's exaggerated claim that Mencken "cannot survive as a literary critic" to see the truth in his final judgment on Mencken: "We have [him] in a nutshell when he says: 'A horse-laugh is worth ten thousand syllogisms. It is not only more effective; it is also vastly more intelligent.' It flattered and flatters an audience to feel superior to the 'swinish multitude,' the 'herd,' the 'booboisie,' the professors, politicians, clergymen, and sundry that Mencken professed to despise without the need of theorizing or even too much strenuous thinking." Mencken's criticism, when directed toward the authors and departments of literature that he found most sympathetic, remains both readable and illuminating. Still, there was, as Wellek implies, something of the bully in him. He was marvelous with the easy targets— and it should never be forgotten that many of the targets that now seem to us "easy" were in his day immensely powerful and dangerous, as his skirmishes with the censors make clear. But in literature and politics alike, he characteristically steered clear of those who would not yield so readily to his *ad hominem* brawling, and his rejection of "constructive criticism" ("I do not object to being denounced, but I can't abide being school-mastered, especially by men I regard as imbeciles") was simply an evasion of the critic's responsibility of working out an affirmative vision against which the failings of the objects of his criticism are to be measured. One thinks of what may be his most quoted remark:

> Q. If you find so much that is unworthy of reverence in the United States, then why do you live here?
> A. Why do men go to zoos?

It was both funny and memorable—but not, in the end, quite good enough.

All this points to a fundamental inadequacy in Mencken's thought: a skepticism so extreme as to issue in philosophical incoherence. It is useful in this connection to compare him to Samuel Johnson, his eighteenth-century counterpart. Like Johnson, Mencken was resolutely unsentimental, ebulliently grim, full of the sanity that comes from an unswerving commitment to common sense. But for Johnson "the mind can only repose on the stability of truth," while Mencken found nothing to be "wholly good, wholly desirable, wholly true." This unequivocal rejection of the possibility of ultimate truth, a position irreconcilable with his scientific rationalism, left him with nothing but a concept of "honor" as shallow as the Victorian idea of progress in which he believed so firmly (and so paradoxically). Though he was for the most part a genuinely honorable man, honor for Mencken would seem to have been little more than a higher species of etiquette. In 1917 he wrote of himself: "His moral code . . . has but one item: keep your engagements." No more revealing thing has ever been said about H. L. Mencken.

Mencken's lack of consistency has led many intellectuals to conclude that he need not be taken seriously, and that his work will not last. Edgar Kemler claimed as early as 1950 that "except for *The American Language*, the *Days* books, and a few selections" from his other books, he "produced no works likely to endure," and judging by many of the diary entries he made in the forties, Mencken himself might have agreed: "In the end every man of my limited capacities must be forgotten utterly." In his most clear-eyed moments, he saw himself as a journalist pure and simple. He turned his prose out by the ream, was always happy to submit to the attentions of line editors and fact checkers, and exhibited none of the tics of the compulsively self-important "artist."* Yet his journalism, written in haste and without pretensions, continues to give pleasure to countless readers untroubled by its internal contradictions. To read a volume of his letters makes one gluttonous for another, heftier collection, and at least half of the occasional essays included in the six series of *Prejudices*, which enshrine his reflections on everything from the literary offenses of Theodore Dreiser to the genius of Beethoven, are as readable today as they ever were. Four decades after his death, he still passes the Holden Caulfield test with flying colors: Anyone untroubled by the endless scruples of "sensitivity" would most definitely want to call him up after reading "A Girl from Red Lion, P.A.," the wonderfully silly chapter of *Newspaper Days* about a milkmaid from lower Pennsylvania who, having surrendered her virtue to her boyfriend, concluded that she had no choice but to board the next milk train to Baltimore and present herself at the nearest house of ill fame, there to live the "life of shame" to which she had sentenced herself. (Miss Nellie, the madam of the house, tipped off Mencken, who came straight over to help talk her unexpected guest into going home on the next train.)

---

*"During all my days of writing for the *Sunpapers*, in fact, I never made any objection to the cutting of my copy," he wrote in *Thirty-five Years of Newspaper Work*. "It seemed to me then, and it seems to me now, that what goes into a newspaper ought to be determined by the editors thereof, and that they should not be burdened with contributors beyond their control."

Mencken's commentaries on politics and culture also continue to be read and relished, despite innumerable attempts to dismiss them as period pieces, many of them by otherwise friendly critics: "What of the American Presidents whom he castigated as a matter of standing policy—Harding, Coolidge, Hoover? They survive, if at all, only as minor names in history texts . . . Does anyone have any really strong feelings about Socialism in a world where one-third of the earth's surface and one-half of its people are ruled by Communism? Even in the remoter reaches of the Bible Belt, are there very many Fundamentalists left?" If this paragraph, written by Charles Fecher in 1978, sounds oddly dated today—much more so than Mencken's own writings—it is a tribute not so much to Mencken's prescience as to the fact that he was primarily interested not in individual politicians, or even in politics as such, but in the American national character. For him the United States was a land in which puritanism is constantly at war with hedonism (and, analogously, collectivism with individualism), and his great achievement as a social critic was to isolate and dramatize this struggle in a permanently memorable way.

His war against the puritans is what his liberal admirers have in mind when they argue that he should be "celebrated and remembered" as "a tremendous liberating force in American culture." Nowhere has this belief been explained more dramatically (or, in light of the charges of racism so often leveled at him, more relevantly) than in Richard Wright's *Black Boy*, a copy of which can be found on the shelves of the Mencken Room. Mencken himself marked this passage from the final chapter:

> A block away from the library I opened one of [Mencken's] books and read a title: *A Book of Prefaces*. I was nearing my nineteenth birthday and I did not know how to pronounce the word "preface." I thumbed the pages and saw strange words and strange names. I shook my head, disappointed. I looked at the other book; it was called *Prejudices*. I knew what that word meant; I had heard it all my life. And right off I was on

guard. . . . That night in my rented room, whle letting the hot water run over my can of pork and beans in the sink, I opened *A Book of Prefaces* and began to read. I was jarred and shocked by the style, the clear, clean, sweeping sentences. Why did he write like that? And how did one write like that? . . . He was using words as a weapon, using them as one would use a club. Could words be weapons? Well, yes, for here they were. Then, maybe, perhaps, I could use them as a weapon? No. It frightened me. I read on and what amazed me was not what he said, but how on earth anybody had the courage to say it.

Yet Mencken's "liberating force," as this passage suggests, is less a function of his particular convictions than of the firmly balanced prose rhythms and vigorous diction in which they are couched. It is, in short, a triumph of style. The fact that this triumph was the work of a common newspaperman has long served to obscure its singularity, especially among academic critics. "The smell of the city room," Charles Angoff wrote in 1938, "was in everything he put between book covers." But what Angoff meant as a deadly criticism is in fact central to Mencken's appeal. It was the discipline of daily journalism that freed him from the clutches of the genteel tradition. The city room was for Mencken what Europe was for Henry James: the great good place where he became himself.

He was, of course, something more than a memorable stylist, if something less than a truly wise man. Daniel Aaron speaks of "the great comic writer who as time passes will be remembered less for what he said than how he said it," but his charm is inseparable from his habits of thought. However perverse or excessive his underlying ideas may be, they retain much of their impelling force. One cannot help admiring the stubborn way in which Mencken the self-made philosopher grapples, in his unpretentious, take-no-prisoners way, with the permanent things: the limits of art, the rule of law, the meaning of life. The simplicity, one comes to realize, is inseparable from the message. In Mencken, style and content are one, and the resulting alloy is more than merely individual. It is a matchlessly exact expression of the American temper. He was to the first part of the twentieth century what Mark Twain was to the last part of the nineteenth—the quintessential voice of American letters. Perhaps even a sage of sorts, too, though an altogether American one, not calm and reflective but as noisy as a tornado: witty and abrasive, self-confident and self-contradictory, sometimes maddening, often engaging, always inimitable.

Such praise would have been repulsive to a man who fancied himself a member of the international aristocracy of superior beings, but the passing of time has made it easier to see how very American he was. Invited in 1930 to contribute his "inmost credo" to a symposium whose other participants included Albert Einstein, John Dewey, H. G. Wells, Theodore Dreiser, Irving Babbitt, Bertrand Russell, Hilaire Belloc, Beatrice Webb, and Lewis Mumford, he put the whole of it into three crisp sentences: "I believe that it is better to tell the truth than to lie. I believe that it is better to be free than to be a slave. And I believe that it is better to know than to be ignorant." It is a credo without shadows or second thoughts. No one but an American—and no other American—could have written it.

MENCKEN SPENT MUCH OF HIS LIFE THINKING ABOUT HIS DEATH. He passed endless hours putting his literary remains in order; reduced by illness to uselessness and despair, he spent seven hard years longing for the arrival of "the inescapable last act." And he wrote his own epitaph, the one composed in 1921 that appears at the end of the *Chrestomathy*, the last book he completed before his stroke: "If, after I depart this vale, you ever remember me and have thought to please my ghost, forgive some sinner and wink your eye at some homely girl." When he died, nearly every obituary writer in America remembered it, and quoted it.

Yet he left behind another "epitaph" that, if less saucy, is more revealing, reminding us that beneath the raucous exterior of his public persona was a very different man indeed. He insisted that he wrote only for his own pleasure and never to make the world a better place. "My writings, such as they are, have had only one purpose: to attain for H. L. Mencken that feeling of tension relieved and function achieved which a cow enjoys on giving milk," he said in remarks prepared for the Associated Press for use in his obituary. But his half-century of dogged labor at the typewriter betrayed a different motive. Perhaps because it is so inconsistent with his public image, even down to the choice of author whose work occasioned it, he relegated this paragraph to his private papers, leaving it for the resurrection men to uncover and ponder:

Says Elizabeth in Jane Austen's *Pride and Prejudice*, XI:

I hope I never ridicule what is wise or good.

This was really Jane speaking for herself. I believe I might say the same thing, and put it in the past tense. So far as I can recall I have never thrown a dead cat at a single honest and intelligent man. My sneers and objurations have been reserved exclusively for braggarts and mountebanks, quacks and swindlers, fools and knaves.

SOURCE NOTES

References to books are to the specific editions cited in the bibliography. When no author is given, the book is by H. L. Mencken (HLM).

Except as otherwise indicated, all documents on deposit in the Enoch Pratt Free Library's H. L. Mencken Collection (EPFL) were written or compiled by Mencken. All of Mencken's newspaper stories and magazine articles are preserved in the Mencken Collection's scrapbooks and are cited by name and date of publication only. (Where the date is unavailable, the scrapbook is cited.)

ABBREVIATIONS

AN1925	*Autobiographical Notes, 1925* (EPFL)
AN1941	*Autobiographical Notes, 1941–* (EPFL, unpaginated)
Diary	*The Diary of H. L. Mencken* (published version; citations are by date of entry only)
Diary/TS	*The Diary of H. L. Mencken* (typescript, EPFL; citations are by date of entry only)
HD	*Happy Days*
HD/AC	*Happy Days: Additions, Corrections and Explanatory Notes* (EPFL, unpaginated; citations are to *HD* page numbers discussed in entry)
HLMC	HLM's clipping books (EPFL, unpaginated; citations are by name and date of original publication)
Letters	*Letters of H. L. Mencken*
Mercury	*American Mercury*
MC	*A Mencken Chrestomathy*
ML	*My Life as Author and Editor* (published version)
ML/TS	*My Life as Author and Editor* (typescript, EPFL)
MS	Mencken and Sara

ND	*Newspaper Days*
ND/AC	*Newspaper Days: Additions, Corrections and Explanatory Notes* (EPFL, unpaginated; citations are to *ND* page numbers discussed in entry)
New Letters	*New Mencken Letters*
P/1	*Prejudices* (number indicates series)
RAM	*Reminiscences of August Mencken* (transcript of interviews conducted with August Mencken in 1958 for the Columbia University Oral History Project; copy in EPFL)
SS	*Smart Set*
SSC	*H. L. Mencken's* Smart Set *Criticism*
Sun/A.M.	Baltimore *Sun* (morning edition)
Sun/P.M.	Baltimore *Evening Sun*
35	*Thirty-Five Years of Newspaper Work* (published version)
35/TS	*Thirty-Five Years of Newspaper Work* (typescript, EPFL)

PROLOGUE
"A PERMANENT OPPOSITION"

DOCUMENTS

The account of the Gridiron Club dinner is based in large part on HLM's own account in *35*, 266–69. All unattributed quotes are from this source.

HLM's voice can be heard on *H. L. Mencken Speaking* (Caedmon SWC 1082), an interview recorded in 1948.

BOOKS

Allen, *Only Yesterday;* Asbell, *When F.D.R. Died;* Bode, *Mencken;* Brayman, *The President Speaks Off-the-Record;* Caro, *Means of Ascent;* Conrad, *Portable Conrad;* Cooke, *Six Men;* Ferguson, *The Otis Ferguson Reader;* Ickes, *Secret Diary;* Kemler, *The Irreverent Mr. Mencken;* Lehman, *Signs of the Times;* Manchester, *Disturber of the Peace;* Mayfield, *The Constant Circle; Menckeniana: A Schimpflexikon;* Tully, *F.D.R., My Boss;* White, *Poems and Sketches of E. B. White;* Wilson, *The Bit Between My Teeth;* Winfield, *FDR and the News Media;* Wolfskill and Hudson, *All But the People.*

NOTES

"I am going": HLM to Harry Leon Wilson, Dec. 7, 1934; *Letters*, 381. **"All save a small minority"**: "Journalism in America," *P/6*, 16–17.

A term he coined: HLM to Charles Green Shaw, Dec. 2, 1927; *Letters*, 305. **"All government"**: "Government: Its Inner Nature," *MC*, 145. (This is the final version of an essay originally published in *SS* in 1919 and revised for *P/3* as "Matters of State: *Le Contrat Social*.")

"In all my life": *35/TS*, 405. **He voted for Roosevelt**: HLM publicly confessed to having voted for FDR in "Vive le Roi!" *Sun/P.M.*, May 1, 1933. **"The**

imbecility of the New Deal": "Right, Left, Right, Left," *Sun/P.M.*, Oct. 22, 1934. **"We have had so many"**: "Roosevelt," *Sun/P.M.*, Jan. 2, 1934.

"A man who has lied": *Notes*, 104. **"Finding virtues"**: *35/TS*, 405.

"The chiropractors want her": Mayfield, 209. **Dr. Benjamin Baker**: Manchester, 272.

On their arrival: Mayfield, 209.

"I did not approach him": *Diary*, Dec. 9, 1934. Unlike the account in *35*, this entry breaks off immediately after HLM's description of his backstage encounter with FDR. (It is not known why the Dec. 9 entry is incomplete.) **Mencken had no way of knowing**: Winfield, 60. FDR's confrontational tactics eventually proved effective. By 1938 the skits had been toned down noticeably. Ibid., 73–74.

Well over four hundred guests: The banquet hall held 490. For a complete guest list, see "Gridiron Club Roasts New Deal And G.O.P. as 'Widows' Hold First Masque at White House," *Washington Post*, Dec. 9, 1934 (HLMC).

Mencken's 1942 draft card: Described in James H. Bready, "The Menckens' Era Ends at 1524 Hollins Street," *Sun Magazine*, Aug. 6, 1967 (HLMC). **"What I saw"**: Cooke, 90–91.

Among the papers: See *Clippings, Cartons and Souvenirs Including Music, 1893–1938* (EPFL). **"The whole Puritan scheme"**: *ML*, 177. **"The wretched old crab"**: Obituary, *Miami Beach Sun*, Feb. 1, 1956 (HLMC).

"Mr. President, Mr. Wright": The three surviving drafts of HLM's speech are in *Miscellaneous Speeches 1913–38* (EPFL). The salutation is from the draft marked "final form." (All sources agree that HLM began the speech in this manner.) Other quotations follow the text preserved in the Gridiron Club files and cited by Brayman, 255–56. **The raucous, gravelly voice**: *H. L. Mencken Speaking*. **"Cleverly cynical"**: Ickes, 242.

Roosevelt began: Brayman, 262. The official "text" of FDR's speech, as preserved in the Franklin D. Roosevelt Library at Hyde Park, consists of a complete speech draft containing no HLM quotations, a page of notes in FDR's hand, an alternative draft of the speech's opening and three pages of quotations from HLM's writings. Brayman follows the text preserved in the Gridiron Club files. (The passage from HLM's "Journalism in America" quoted here is from *P/6*, 15.) Audience reaction to FDR's speech, as well as HLM's own reaction, is described by Bode, 310–11, who interviewed the journalists Arthur Krock and Marquis Childs, both of whom were present. According to Ickes, "I looked over at Mencken two or three times while the President was speaking and it was clear to see that he didn't like it at all. He seemed to me to be distinctly put out." Ickes, 242. **"His writings give me a chuckle"**: In fact Roosevelt read the morning edition of the *Sun*, not the evening edition, which carried HLM's Monday Articles. See Tully, 76–77.

"I'll get the son of a bitch": Kemler, 271–72.

Several reporters came up to Mencken: "After [FDR] sat down," HLM recalled in *35*, "Wright asked him why he had replied so ill-humoredly to my good-humored speech, and he replied that he couldn't help it, for the text was before him and he couldn't think of anything else to say." **"Simply smeared"**: Ickes, 242. **"I did not really intend"**: Quoted in *MS*, 511.

"We'll even it up": This was John Owens, editor of the *Sun*. See Bode, 311.

"**I got in a bout**": Mayfield, 210. "**Dr. Roosevelt is no longer the demigod**": "1936," *Sun/P.M.*, Aug. 26, 1935; *A Carnival of Buncombe*, 295. "**The greatest President since Hoover**": "Three Years of Dr. Roosevelt," *Mercury*, Mar. 1936, 264. (The MS. of this article is dated "Dec. 26–30, 1935.")

"**I find it difficult**": Undated memorandum, *Minority Report/HLM's Notes/Unpublished Material* (EPFL). "**He had every quality**": *Diary*, Apr. 15, 1945.

"**Did you really dislike Franklin Roosevelt**": Richard L. Dunlap, "The Sage at Dusk: An Account of H. L. Mencken's Last Interview," *Menckeniana* 35.

"**The most powerful personal influence**": Walter Lippmann, "H. L. Mencken," *Saturday Review of Literature*, Dec. 11, 1926 (HLMC). "**All Government**": "The Constitution," *Sun/P.M.*, Aug. 19, 1935.

"**Without question, since Poe**": "Mencken Through the Wrong End of the Telescope," *New Yorker*, May 6, 1950; Wilson, 31.

The hundreds of signed presentation copies: These also include books by distinguished critics. Morton Dauwen Zabel's *Portable Conrad*, for instance, is inscribed to "Mr. H. L. Mencken, pioneer American interpreter of Conrad; with gratitude and best regards," and Edmund Wilson's *Memoirs of Hecate County* is inscribed to "H. L. Mencken, who encouraged but finally rejected my first sex story."

In 1926 some five hundred editorials: Publisher's note, *Menckeniana: A Schimpflexikon* (n.p.). "**If Mencken only ran about**": E. F. Keene, *Concord* (N.H.) *Monitor*; ibid., 16. "**He seems to love**": Harold N. Coriell, *New York Herald Tribune*; ibid., 50. "**Mencken, with his filthy verbal hemorrhages**": Editorial, *Jackson* (Miss.) *News*; ibid., 24. "**Mencken is an outstanding, disgusting example**": Editorial, *Richmond* (Va.) *Times-Dispatch*; ibid. 49.

"**Poetry, religion, and Franklin D.**": White, 43. "Hopkins" and "Tugwell" were Harry Hopkins and Rexford Guy Tugwell, two of FDR's most prominent and controversial advisers.

"**He wrote like a bat out of hell**": "In the American Language," *New Republic*, Feb. 19, 1940; Ferguson, 196. "**Mencken's was the best prose**": Joseph Wood Krutch, "This Was Mencken: An Appreciation," *Nation*, Feb. 11, 1956 (HLMC). "**Mencken, Critic of All**": Obituary, *New Orleans Item*, Jan. 30, 1956 (HLMC). "**He calls you a swine**": Walter Lippmann, "H. L., Mencken," *Saturday Review of Literature*, Dec. 11, 1926 (HLMC).

The "comic mask" of which Edmund Wilson spoke: In "H. L. Mencken," *New Republic*, June 1, 1921 (HLMC).

"**Esthetic problems interest me**": HLM to Ivan J. Kramoris, Jan. 8, 1940; *New Letters*, 455. "**Large parts**": *Diary*, May 26, 1945.

"**When I think of anything**": *HD*, 196. "**Nothing moves me so profoundly**": *AN1925*, 155. **A complete edition of Oscar Wilde**: James H. Bready, "The Menckens' Era Ends at 1524 Hollins Street," *Sun Magazine*, Aug. 6, 1967 (HLMC). It is still there, together with HLM's sets of Nietzsche, Kipling, Conrad, Ibsen, Goethe, Schiller, Shakespeare, Herbert Spencer, James Branch Cabell, and Jane Austen. "**The motive of the critic**": "Footnote on Criticism," *P/3*, 84–85. "**Here,**" he writes: *Notes*, 208. "**He was the only man**": Lehman, 155–56.

"Constitutionally unable to believe": *AN1941*. **"I have little belief":** HLM to Charles Green Shaw, Dec. 2, 1927; *Letters*, 307. **"One of the greatest":** "Types of Men: The Scientist," *P/3*, 270. **"If a man, being ill":** "Dives into Quackery: Chiropractic," *P/6*, 224. **"I am thus inclined":** HLM to A. G. Keller, Jan. 22, 1943; *New Letters*, 514.

"His readers": Manchester, 326. **"He believes and stoutly maintains":** This was George Jean Nathan, quoted in Mayfield, 262. **Twelve-hour days at the typewriter:** Mayfield, 216, 222. **"I get little pleasure":** *AN1941*. **"Nine-tenths of the people":** *Diary*, Nov. 19, 1946. **"Never was a man":** Hamilton Owens, "A Personal Note," *Letters*, x. **"A time-server":** *Diary*, Oct. 24, 1945.

"The existence of most human beings": *Minority Report*, 39. **"Like Nietzsche, I console myself":** *AN1941*.

"The masses are animals": Manchester, xx. **"Inherently incapable of civilization":** "Advice to Young Men," Mar. 29, 1939; *Minority Report/HLM's Notes/Unpublished Material* (EPFL). **"It is amazing":** *Diary*, May 31, 1940.

"Mencken's vigour is astonishing": Conrad to George T. Keating, Dec. 14, 1922; *Portable Conrad*, 751.

"The Dayton Trial": Allen, 12. **"The only modern inventions":** This comes from HLM's written answers to questions supplied by Roger Butterfield for a 1946 "interview" published in *Life* (*Miscellaneous Typescripts, Carbons and Clippings, 1946–1948*, EPFL). **Those who knew its editor best:** This includes, among many others, his wife. See Mayfield, 173–74. **"If I had my life":** *HD*, ix.

CHAPTER ONE
BIRTH OF A BOURGEOIS, 1880–1899

BOOKS

Beirne, *The Amiable Baltimoreans*; Garraty, *The New Commonwealth*; Goldberg, *The Man Mencken*; Johnson, *The Sunpapers of Baltimore*; *Maryland: A Guide to the Old Line State*; Lippman, *The Sugar House*; McPherson, *Battle Cry of Freedom*; Owens, *Baltimore on the Chesapeake*; Schlereth, *Victorian America*; Simon, *Homicide*; Sutherland, *The Expansion of Everyday Life*; Twain, *Collected Tales, Sketches, Speeches & Essays 1852–1890*; Wilson, *Patriotic Gore*.

NOTES

"The charm of getting home": "On Living in Baltimore," *P/5*, 241.

Long after his death: For a description of HLM's neighborhood today, see Lippman, esp. 71–72. For a description of modern-day Baltimore from a policeman's point of view, see Simon, esp. 139, 160–61, 330–31 and 455–56.

Mencken's hometown has not changed: I lived in Baltimore from 1991 to 1995, not far from 704 Cathedral Street, where HLM lived from 1930 to 1936. **The family home survives:** In the spring of 2002 the city of Baltimore, which owned 1524 Hollins Street and had maintained it as a museum until financial problems forced its closing, was in discussions with the management of the *Sun*, which sought to preserve the house and make it available as a center for literary and academic activities. Negotiations were still under way as this book went to press.

"Fat, saucy and contented": *HD*, vii. **"The two-story houses":** "Aesthetic Diatribe," *Sun/P.M.* Feb. 7, 1927.

"Such undertakings as these": Owens, 294–95. **"A slow, plodding, dull town":** Johnson, 129. (This section of *The Sunpapers of Baltimore* is by Gerald Johnson.) **The term was in use:** *HD*, 120. **A major center of Know-Nothing activity:** See Beirne, 150–51; *Maryland: A Guide to the Old Line State*, 212; and Owens, 261–64. (For a general discussion of the Know-Nothing movement, see McPherson, 135–44.)

What followed was even more divisive: See Johnson, 128–29; Owens, 2–3; and McPherson, 284–90. **The irredentist sentiments:** For a brief discussion of "Maryland! My Maryland!," see Wilson, 400–401.

HLM's maternal grandfather was stoned: *HD/AC*, 92. (HLM alludes to the 1861 riot in *HD*, 119–20, but makes no direct mention of the Know-Nothing riots.)

"His father was August Mencken": *Miscellaneous Statements and Interviews, 1924–1936* (EPFL), 173. **"A violent and highly effective lampoon":** *AN1925*, 9.

"In Germany": *AN1941*. **"In his later life":** *HD*, 92. (HLM indicates that the specific cause of Burkhardt's emigration was the Dresden Revolution in *AN1925*, 28.) **Rated by Dun & Bradstreet:** *AN1925*, 36.

"In his later years": *HD/AC*, 248. **"A project of foreign nihilists":** *HD*, 32. **August's code of ethics:** Ibid., 251.

"I never saw him": *AN1941*. **"He instilled into me":** *ND/AC*, 4.

Born in 1858: *HD*, 7. **"My mother was a great worrier":** *AN1941*.

"I was on the fattish side": *HD*, 6–7. **"It was rare for a solvent person":** *HD/AC*, 7.

This last detail is telling: For descriptions of everyday American life around the time of HLM's birth, see Garraty, Schlereth, and Sutherland, all passim.

"We admired Buffalo Bill": *HD*, 49.

"I can't recall": Ibid., 63. **"A billion polecats":** Ibid., 70. **"The Baltimoreans of those days":** Ibid., 63. **"This town is anything but perfection":** Undated HLM column (c. 1912–14) in *Sun/P.M.*, quoted in John Dorsey (ed.), "Mencken's Baltimore," supplement to *Sun/A.M.*, Sept. 8, 1974 (HLMC).

"Placid, secure, uneventful and happy": *HD*, vii.

"A strange, wild land": Ibid., 8. **"There was room in it":** *AN1925*, 47–48. **"Even in the dead of Winter":** *HD*, 8. **"The Hollins street neighborhood":** Ibid., 8–9. **"My grandfather's house":** *AN1925*, 50, 48–49.

"We were paddled freely": *AN1941*. **"Dreadfully round-shouldered":** *AN1925*, 58. **His first-term report card:** Ibid., 60.

"The professor and his goons": *HD*, 27. **Enough German to quote Goethe in the original:** HLM quotes from *Faust*, without attribution and with slight inaccuracies, in his diary entry for May 31, 1940. (See p. 382.) **"A sort of star pupil":** *AN1925*, 60.

He never learned to dance: *HD*, 194. **"Preposterous lady Leschetizkys":** Ibid., 195: Theodor Leschetizky was one of the most celebrated European piano teachers of the nineteenth century. His students included Ignace Jan Paderewski and Artur Schnabel. **"Fifty or sixty pounds":** Ibid., 197. One of Mencken's pieces is reproduced in Goldberg, 88. **"My early impulse to compose":** Ibid., 196, 198.

"I had not gone further": Ibid., 167.

"Probably the most stupendous event": Ibid., 163. **"How he stood above and apart":** "Our One Authentic Giant," *SS*, February, 1913; *SSC*, 180.

"Well, I'll be durned!": *HD*, 168. **"An almost daily harrying":** Ibid., 174. **"I began to inhabit":** Ibid., 175.

That summer the Menckens rented a country house: In 1925, HLM told Isaac Goldberg that his first visit to the *Ellicott City Times* took place in the summer of 1888 (not 1889, as *HD* has it), and that it was this experience that caused him to ask for a printing press for Christmas. See *AN1925*, 52–53. **"I was captivated":** *HD*, 214. **"It was the printing-press that left its marks":** Ibid., 217.

"There was just as much delight": *ML/TS*, 12–13. **"I had, in those days":** *AN1941.* **"I was then at the brink":** *HD*, viii.

"Thackeray, Addison and Shakespeare": *ML/TS*, 25. **"I had read almost the whole canon":** *AN1925*, 72. **"The chief decoration":** *ML*, 5.

The family's daily routine: For a description of life at Hollins Street during this period, see *RAM*, 32–38. **"The calorie content probably ran to 3000":** *HD*, 62.

"The teens are at once too grotesque": *HD/AC*, v.

"The gossips of the little village": *Earliest Attempts at Verse and Prose, 1895–1901* (EPFL), 38. **"The greatest Englishman":** *Sun/P.M.*, May 4, 1925.

"All I learned at the Polytechnic": Ibid., 66–67. **"It seemed to me nonsensical":** Ibid., 67–68.

"An amazingly unliterary environment": *AN1941.* **"Jones will be here":** "Journalism in Tennessee," Twain, 312.

"He was naturally pretty well dashed": *ND/AC*, 4. **"As if to console me":** *AN1925*, 40.

"O sea-sick flag!": *ML*, 3. **"The big guns roar":** *ML/TS*, 12. **"You see the interesting side of things":** *Earliest Attempts at Verse and Prose, 1895–1901* (EPFL), 64, 71.

His application: *Souvenirs of Childhood and Schooldays, 1880–1896* (EPFL), 81.

On the evening of December 31, 1898: HLM's fullest account of his father's death is in *ND/AC*, 1.

"On the Monday evening immediately following": *ND*, 3. **One of Harry's aunts:** *HD/AC*, 13.

"I ceased to be a child: *HD/AC*, v. **"There was a time":** *AN1941.*

"A cough that persisted for years": *ND/AC*, 1.

CHAPTER TWO
REPORTER AND EDITOR, 1899–1906

BOOKS

Goldberg, *The Man Mencken*; Weisberger, *The American Newspaperman*.

NOTES

"I was eighteen years": *ND*, 3.

"In the present book": Ibid., ix, x.

Arthur Hawks testified: *AN1925*, 81. "I was a success from the start": *AN1941*.

One historian of American newspapers: Weisberger, 139. The same historian: Ibid., 157.

"Had I any newspaper experience": *ND*, 4.

"An exhibition of war scenes": *Morning Herald*, Feb. 24, 1899. "I was up with the milkman": *ND*, 8.

His first face-to-face meeting: AN1941.

Sixty cents a night: Ibid., 10. "He seemed to lead": *Morning Herald*, Apr. 21, 1899.

"City assignments of the kind": *ND*, 10–11. "The fugitive four-footed hoboes": *Morning Herald*, n.d.; *Early Newspaper and Magazine Work* (EPFL), 64–66.

"He didn't bawl out": *AN1925*, 85. Eight dollars a week: Ibid. HLM later recalled being paid "$7 a week, with the hope of an early lift to $8 if I made good"; *ND*, 13.

He would keep one on the floor: James H. Bready, "The Menckens' Era Ends at 1524 Hollins Street," *Sun Magazine*, Aug. 6, 1967 (HLMC). "To see and savor": "The New York Sun," *A Second Mencken Chrestomathy*, 361–62.

No time to read anything: *ML/TS*, 42. "The heavy reading of my teens": *ND*, ix. "No less than four": Ibid., 25. "My first hanging": *ND/AC*, 25.

"I made up my mind": Ibid., 37–38.

"The old-time journalist's concept": Ibid., xi.

"Forget politics": Ibid., 65.

More than an accomplished hack: "The Cook's Victory" is reprinted in Goldberg, 313–24. "Those stories that I wrote": *AN1925*, 109.

"Sweethearts often quarrel": *Morning Herald*, Nov. 4, 1900.

"It is a wise candidate": Ibid., Oct. 28, 1900. "Some day the critics": Ibid., Nov. 4, 1900. "Not bad criticism": *AN1925*, 86.

The most promising of Mencken's early columns: A selection of "Untold Tales" is reprinted in Goldberg, 343–370. "The word reporter": *Sunday Herald*, July 14, 1900.

"The fashion at the time": *Typescripts of Early Fiction* (EPFL), 4. "Now though I was full joyous": Ibid., 19.

"The first job of a reviewer": Ibid., 111–12.

He sold the first of several stories: *ML*, 45–47.

"It gradually lost its humorous character": *AN1925*, 87. "A proposed victim is known": *Morning Herald*, Feb. 23, 1902.

"A young reporter": *ND*, 261.

"My dislike": *AN1941*.

"The edition was limited": *ML/TS*, 35–37.

"I am convinced": Ibid., 27.

"These verses are salable": *Chicago Record-Herald*, Aug. 17, 1903 (HLMC).

"Mr. Mencken is a dramatic critic": *Cleveland Leader*, July 19, 1903 (HLMC).

"Sterile Philistinism": *ML/TS*, 58.

"I began business" Ibid., 87–88. "To this day": *ND*, 136.

"Since 1910": Ibid., 313. "All the ideas that are thought of": *AN1925*, 178–79.

"The Bend in the Tube": The story is reprinted in Goldberg, 325–39.

"Settled and middle-aged": *ND*, 278. "Swell pickings": Ibid., 276–77.

"Twisted as if it had died": Ibid., 302.

His work was "brilliant": *Editor & Publisher*, Feb. 3, 1906 (HLMC).

"He proposed to pay me": *AN1925*, 108–9.

"I owed to her": 35, 155. "A woman who, though she lacked the polish": *ND*, 230. "Whenever I got home": *AN1941*.

"With perspiration standing out": *Sunday Herald*, June 24, 1904; *The Impossible H. L. Mencken*, 225, 229.

His name was on the letterhead: *ND*, 305. "Grafting as an art": *Evening Herald*, Aug. 24, 1905. "There are many things": Ibid., Mar. 14, 1905. "A gentleman of bulging brow": Ibid., July 8, 1905. "The best way": Ibid., Sept. 26, 1905. "Beside Mark Twain": Ibid., Nov. 29, 1905.

"The average newspaper reporter": Ibid., Sept. 9, 1905.

"I was so enchanted": *ND*, 307.

"A little handbook": *George Bernard Shaw: His Plays*, vii. Only fifteen had been published: *ML*, 75. "In all the history": *George Bernard Shaw: His Plays*, xxix.

"His private opinions": Ibid., xxiii, xxiv.

Mencken's book received serious attention: *ML*, 82–83.

His first and last check: Ibid., 89. "The book, on re-reading": *AN1925*, 125–26.

"Tomorrow the property": *ND*, 311.

He saw nine men hanged: Introduction, *Early News Stories, Baltimore Morning Herald, 1899–1901* (EPFL).

An opinion from which he never wavered: "Thus his painful psychologizings, translated into plain English, turn out to be chiefly mere kittenishness—an arch tickling of the ribs of elderly virgins—the daring of a grandma smoking cigarettes" ("Our Literary Centers," *SS*, Nov. 1920; *SSC*, 14).

CHAPTER THREE
COLUMNIST AND CRITIC, 1906–1914

BOOKS

Dolmetsch, *The Smart Set;* Epstein, *Partial Payments;* Johnson, *The Sunpapers of Baltimore;* Nathan, *A George Jean Nathan Reader;* Nathan, *The World of George Jean Nathan;* Staggs, *All About "All About Eve";* Swanberg, *Dreiser;* Wilson, *The Devils and Canon Barham.*

NOTES

"THE SUN, in those days": Johnson, 242. (This section of *The Sunpapers of Baltimore* was written by HLM.) **"Somewhat stodgy"**: Ibid., 262. **"A good club"**: Ibid., 265.

"The attempt to make": *35*, 4. **He was doubling his paycheck**: Ibid., 11.

"Any literate boy": *AN1925*, 125. **"Poor old Nietzsche"**: " 'Hypocrites' Is Here," *Sun/A.M.* Mar. 5, 1907.

"The audience was not large": "Frank Keenan at Ford's," *Sun/A.M.*, 1910 (no further dating), *Editorials and Dramatic Reviews, Baltimore Sun, 1906–1910* (EPFL), 157. **"Grossly offended"**: *AN1925*, 93.

"Next to nothing": *AN1925*, 125. **"Unfortunately we do not know"**: *Educational Review*, May 1908 (HLMC). The review was unsigned, but HLM "got word" that Butler had written it (*ML/TS*, 101).

Belief in the gospels: *The Philosophy of Friedrich Nietzsche*, 128. **"Beginning allegro"**: Ibid., 133. **Even in popular treatments**: Among those books that make no mention of HLM are Walter Kaufmann's *Portable Nietzsche* (1954), Ronald Hayman's *Nietzsche: A Critical Life* (1980), and *What Nietzsche Really Said*, by Robert C. Solomon and Kathleen M. Higgins (2000). **"He applied the acid"**: *The Philosophy of Friedrich Nietzsche*, 39.

His debt to Nietzsche: HLM to Edward Stone, Mar. 1, 1937; *Letters*, 414. **"He believed that there was need"**: *The Philosophy of Friedrich Nietzsche*, 73. **"Such a man"**: Ibid., 97–98.

"Something to wield his sword upon": Ibid., 320–21.

"I'm H.L. Mencken from Baltimore": Nathan, *World*, 174. This is Nathan's version, which contains factual inconsistencies (among other things, he did not begin publishing in *SS* until Oct. 1909) and may be apocryphal. HLM's various accounts are less specific but equally inconsistent, and there is no surviving contemporary correspondence describing the meeting.

"In American colleges": Dolmetsch, xx.

The model for Addison DeWitt: Staggs, 242. **He was born**: Nathan, *Reader*, 15–16.

Nathan's influence: Nathan, *World*, xvi. **"Inferior"**: Ibid., xxvii. **"A pansy paraphrase"**: Ibid., xx. **"About as close to the obscene and ridiculous"**: Ibid., 449. **"Bosh sprinkled with mystic cologne"**: Ibid., xxi. **"The story, for all its basic triviality"**: "Scott Fitzgerald and His Work," *Chicago Tribune*, May 3,

1925; *H. L. Mencken on American Literature*, 113. **"Men go to the theatre"**: Ibid., 480.

At the high-water mark: *Pistols for Two*, 5, 6, 8, 9, 10, 18.

"A pernickety and snubbing little man": Wilson, 102. **"The surface of life"**: Nathan, *World*, xxvi.

"Nothing in this world": Ibid., 44.

"The great problems of the world": Ibid. **"His world was limited by New York"**: 35, 134. **"A slight fellow"**: *ML*, 206. **"He has denied, in recent years, that he is a Jew himself"**: In 1938 Mencken would record in his diary a conversation with the silent-film actress Lillian Gish, one of Nathan's old girlfriends, in which she claimed that he not only "gave her his word . . . that he had no Jewish blood" but asked her to vouch for the fact so that he could rent an apartment in a restricted building on Park Avenue. The landlord refused to take her word for it, and he remained in his suite at the Royalton Hotel (*Diary/TS*, Oct. 26, 1938). **"What's your attitude toward the world?"**: Nathan, *Reader*, 174.

"The Good, the Bad and the Best Sellers": *SS*, Nov. 1908; *The Young Mencken*, 100–09. **"A certain amount of stage-fright"**: *ML*, 11. **A novel a day**: "A Novel a Day," *SS*, Sept. 1912; *A Second Mencken Chrestomathy*, 310–11.

"Now and then, however": "A Road Map of the New Books," *SS*, Jan. 1909; *SSC*, 35.

"No more than a name": *ML*, 25. **Mencken's Conrad**: "The Monthly Feuilleton—IV," *SS*, Dec. 1922; "Joseph Conrad," *MC*, 519. **"Always being swallowed up"**: "The Motive of the Critic," *New Republic*, Oct. 26, 1921; "Footnote on Criticism," *P/3*, 87, 91.

Few contemporary scholars seem to know about it: HLM is mentioned only in passing in such otherwise authoritative studies as Norman Sherry's *Conrad* (1972), Zdzislaw Najder's *Joseph Conrad: A Chronicle* (1983) and *The Oxford Reader's Companion to Conrad* (2000).

"The frauds and dolts": Ibid., 130.

A search of the World Wide Web: www.google.com, Sept. 12, 2000. **"Incurable antipathy"**: Swanberg, 527. This comes from a statement that HLM wrote to be read at Dreiser's funeral (it was not used).

"Insatiable appetite": Ibid. **A more recent critic**: Epstein, 260, 278.

"We like sentiment": Dreiser to Charles G. Ross, Aug. 16, 1909; Swanberg, 127.

"Bristled with gay phraseology": Dreiser, "Henry L. Mencken and Myself"; *Dreiser-Mencken Letters*, 738. **"Instantaneous friendship"**: Ibid., 739.

"There appeared in my office": Ibid.

"A somewhat solemn fellow": *ML*, 28.

"The American, try as he will": *A Book of Prefaces*, 200.

"You see, Mencken, unlike yourself": Dreiser to HLM, Mar. 27, 1943; *Dreiser-Mencken Letters*, 689–90.

"A rich magazine editor": HLM to Dreiser, Apr. 1, 1943; *New Letters*, 521. **He could have no respect**: HLM never demanded or accepted advances for his books (*ML*, 331).

"Considerable self-interest": *ML*, 130. **"If only for the sake"**: Ibid. **"What I needed"**: Ibid., 132.

Dreiser would later claim: "I suggested that as intriguing as anything would be a Book Department with a really brilliant and illuminating reviewer. Instantly the one name that appealed to me as ideal for this work was that of Mencken. I insisted that he could not do better than get this man and that he should engage him at once" (Dreiser, "Henry L. Mencken and Myself"; *Dreiser-Mencken Letters*, 740). For HLM's account, see *ML*, 10. **"In memory of furious disputations"**: *Dreiser-Mencken Letters*, 4. **"Puerile philosophizing"**: *ML*, 133.

"Immutable": *Men Versus the Man*, 69. **"The first-caste man"**: Ibid., 152. **"The book seems somewhat archaic"**: *AN1925*, 127–28.

"He addressed them familiarly": *HD*, 239.

"Let us confess it": *Sun/P.M.*, Apr. 18, 1910. **"A cardinal socialistic doctrine"**: Ibid., Apr. 19, 1910. **"There is no space here"**: Ibid., Apr. 21, 1910. **Next to none have survived:** *The Young Mencken* contains a small selection of "H. L. M." columns.

"The father and mother of joy": "On Alcohol," *Sun/P.M.*, May 6, 1911; *The Young Mencken*, 203.

"A private editorial column": *AN1925*, 94. **"Sixty cents reward"**: "The Free Lance," *Sun/P.M.*, June 10, 1911. **"The truth that survives"**: Ibid., Oct. 23, 1915. **"I went fer to git it"**: Ibid., May 24, 1911.

"I never mentioned any contemporary public character": *AN1941*. **"All the worst attacks"**: *AN1925*, 190.

"Before it had gone on a year": *35*, 33.

"Two men with sharp and naughty pens": *New Orleans Times-Democrat*, Mar. 7, 1911 (HLMC).

"The truth will out": *Dreiser-Mencken Letters*, 63.

"You know I am one of those": Ibid., 64.

"The story comes upon me": Ibid., 68.

"If the Harpers want": Ibid., 77.

"I confess that my cheeks colored some": Ibid., 78.

"If you miss reading 'Jennie Gerhardt' ": *SS*, Nov. 1911; *SSC*, 244–48.

Nowhere does he say anything to the contrary: This is not true of Dreiser's later novels. For example, HLM's praise of *The Financier*, both in print and in his letters to Dreiser, is inconsistent with what he said to others in private (see HLM to Harry Leon Wilson, Dec. 10, 1912; *Letters*, 27).

"Who, indeed, was I": *ML*, 139–40. The Fox sisters were a pair of fraudulent spiritualist mediums who appeared throughout the United States in the second half of the nineteenth century.

"I never drink alone": *AN1925*, 164. **"I hope you have not been leading"**: This speech from *The Importance of Being Earnest* is included under "Hypocrisy" in HLM's *New Dictionary of Quotations*, 565.

"No mean revenue": *ML*, 41–42.

"Violently against virtually everything": Ibid., 41. **He gave Wright the boot:** After a long stretch of ill-paid obscurity, Wright became one of the most popular mystery novelists of the twenties, chronicling the adventures of Nietzschean supersleuth Philo Vance under the pen name of "S. S. Van Dine."

"R. L. Menecken": *Los Angeles Examiner,* Nov. 8, 1913 (HLMC). **"A member of the staff":** "Newspaper Morals," *Atlantic Monthly,* Mar. 1914 (HLMC). **"What distinguished writers have said":** HLMC. **He later claimed:** *ML,* 44.

A counterproposal: See Dolmetsch, 46–47, and *ML,* 46–48, for accounts of the negotiations.

"Our policy, I needn't say": HLM to Ellery Sedgwick, Aug. 25, 1914; *Letters,* 49. **"Rubbish":** *ML,* 49. **"The constitution of the United States":** *SS,* Nov. 1914, 18. **"So this was the end!":** Ibid., 31; *The Young Mencken,* 395.

"Our first issue": *ML,* 54. **"If the thing goes through":** HLM to Theodore Dreiser, Aug. 11, 1914; *Letters,* 46.

CHAPTER FOUR
AT THE *SMART SET,* 1915–1918

BOOKS

Chesterton, *The Illustrated London News 1929–1931;* Dreiser, *American Diaries;* Dreiser, *A Gallery of Women;* Hobson, *Mencken;* Martin, *In Defense of Marion;* Mitgang, *Dangerous Dossiers;* Wellek, *A History of Modern Criticism;* Wilson, *The Bit Between My Teeth;* Wilson, *The Shock of Recognition.*

NOTES

"I note what you say": HLM to William Saroyan, Jan. 25, 1936; *New Letters,* 373.

"I read all the manuscripts": "A Personal Word by H. L. Mencken" (n.d.), *Pamphlets and Leaflets, 1915–36* (EPFL); *ML,* 419–20.

"If you have ever given": HLM to Theodore Dreiser, Apr. 6, 1915; *Letters,* 63. **"Under you and Nathan":** Dreiser to HLM, Apr. 20, 1915; ibid., 65–66.

"I am sorry": HLM to Theodore Dreiser, Apr. 22, 1915; ibid., 67. **"Civilized in point of view":** *ML,* 59. **"Fostering and wet-nursing the national letters":** Ibid., 113. **"Deliberately mystifying":** Ibid., 62.

"They were not much interested": Malcolm Cowley, "The Smart Set Legend," *New Republic,* Jan. 16, 1935 (HLMC). **"We want to devote":** HLM to Nathan, Feb. 2, 1915; Cornell University Library.

"An author should be judged": *ML,* 131. **"In one of his columns:** "From the Diary of a Reviewer," *SS,* Sept. 1921, 141. **"Not bad in the sense that Dreiser's writing is often bad":** "The Anatomy of Ochlocracy," *SS,* Feb. 1921, 143. **"Beautiful balderdash":** "The Poet and His Art," *P/3,* 154. **"The harsh facts of life":** "The Novel," ibid., 208. **"Sharp and more or less truculent dissent":** *ML,* 174. **"He seized every chance":** Ibid., 25. **"You will escape":** HLM to Louis Untermeyer, Nov. 25, 1919; *New Letters,* 117.

"Even today he has kept": Wellek, 10. **"Mencken fulfilled an important function"**: Ibid., 9.

"The infinitely picturesque and brilliant life": "The Curious Republic of Gondour," *SS*, Oct. 1919; *SSC*, 185, 184. **"It took an ex-newspaperman"**: Wilson, *The Bit Between My Teeth*, 31. **"Its aim is to fill the breast"**: "Mush for the Multitude," *SS*, Dec. 1914; *SSC*, 167.

"A man's women folk": *In Defense of Women*, 3. **"It is my experience"**: *AN1925*, 188. **"A particularly strong sex drive"**: Hobson, 196.

"They were young adventurers": Dreiser, *A Gallery of Women*, 240.

"Certainly I am coming over": Martin, 5.

"My dear Miss Bloom: Beware!": Ibid., 6.

"Dear Miss: The idea is": Ibid.

"Once I get loose": Ibid., 7.

"It is no crime": Ibid., 8.

"Dear M: If you are not at home": Ibid., 9.

Marion scribbled a note: Ibid. **He knew his way around the whorehouses of Baltimore**: HLM describes his friendship with "Nellie d'Alembert," one of the city's leading madams, in "A Girl from Red Lion, P.A.," the fifteenth chapter of *Newspaper Days*. In his unpublished explanatory note to this chapter, he identifies her by name and describes in fuller detail the workings of prostitution in Baltimore at the turn of the century. See *ND/AC*, 227–38.

"Jennie Gerhardt, like Carrie Meeber": "A Novel of the First Rank," *SS*, Nov. 1911; *SSC*, 245. The Shudras are the lowest of the four Hindu castes.

"You are 80 times as clever": Martin, 11–12. **"I suppose he wanted a writer"**: Ibid., 388. **"Pending proofs that you have read"**: Ibid., 16.

"The morons": *ML*, 75. **"The louse magazines"**: HLM to Ernest Boyd, Oct. 18, 1916; *New Letters*, 62. **"Homicidal fiction"**: *ML*, 355. **"It sufficed to relieve Nathan and me"**: *AN1925*, 114.

"Tension in the right side of the tongue": *AN1925*, 161.

"On my own motion entirely": 35, 59. **"I had an uneasy fear"**: Ibid. **"If I ever get out of my present morass"**: HLM to Theodore Dreiser, c. Jan. 1915; *Letters*, 58. **"Horn into the war"**: Ibid.

"Ignoramuses of the petty trading class": Ibid., 54. **He heaped praise on the "new Germany"**: All remaining quotes in this paragraph are from "The Mailed Fist and Its Prophet," *Atlantic Monthly*, Nov. 1914; *The Young Mencken*, 425–41.

"Race consciousness": *ML*, 176. **"It suddenly dawned on me"**: Ibid., 175. **"The haunting fear"**: *MC*, 624.

"I have so warm an admiration": G. K. Chesterton, "Mr. Mencken and the New Physics," *Illustrated London News*, June 14, 1930; Chesterton, 325. **"I have always lived in the wrong country"**: *AN1941*. **"German notions of what is good and bad"**: "The Free Lance," *Sun/P.M.*, Oct. 16, 1915.

"Liberty up to the extreme limits": "H. L. Mencken, by Himself," *Nation*, Dec. 3, 1923; *A Second Mencken Chrestomathy*, 471. **"Liberty is not for slaves"**: Ibid., 472. **"Strongly in favor"**: HLM to Cal Tinney, Feb. 5, 1941; *New Letters*,

475. **"Unable to make the grade"**: Ibid. **"One who is willing"**: *A Second Mencken Chrestomathy*, 51.

"The Germans offer the world": "Are The Germans Immoral? Of Course!," *Sun/P.M.*, Nov. 11, 1915. **"It always amuses me"**: *AN1941*.

"At the bottom of Puritanism": "Notes For Proposed Treatise Upon The Origin And Nature Of Puritanism," *Sun/P.M.*, Oct. 25, 1915. **"England is the mother-country"**: "More Notes For A Work Upon The Origin And Nature Of Puritanism," *Sun/P.M.*, Nov. 2, 1915.

"D. got drunk upon his own story": HLM to Harry Leon Wilson, Dec. 10, 1912; *Letters*, 27. **"An exact statement of the passages"**: HLM to Theodore Dreiser, July 28, 1916; *Letters*, 87.

"A fight is the only thing": Dreiser to HLM, Aug. 4, 1916; *ML*, 161. **"Some of us may differ"**: *ML*, 163.

"May I suggest that your signature": HLM to William Allen White, Nov. 22, 1916; *New Letters*, 64–65.

"MENCKEN IS NOT NEUTRAL": *Sun/P.M.* Jan. 27, 1917; *Germany, 1917* (EPFL). **"I remember a moment'**: *AN1941*.

"During the last war": Mitgang, 172.

"The war actually treated me magnificently": HLM to Fielding Hudson Garrison, Nov. 17, 1919; *Letters*, 161. **"For the first and last time in my life"**: *ML*, 125.

"There has never been an occasion": *AN1941*. **"I can't understand the martyr"**: HLM to Burton Rascoe, c. summer 1920; *Letters*, 188. **"Unless the Germans shoot all their statesmen"**: Ibid., 188–89.

A good critic is a catalyst: "Critics and Their Ways," *New York Evening Sun*, July 1, 1918; *P/1* as ("Criticism of Criticism of Criticism"), 20–21.

"The most important book": *ML*, 173. **"American literature is set off sharply"**: *A Book of Prefaces*, 199, 225, 280.

"The most headlong and uncompromising attack": *ML*, 180. **"We remained on good terms"**: HLM to Burton Rascoe, c. summer 1920; *Letters*, 184. **"There are passages"**: *A Book of Prefaces*, 106.

"Less interesting and provocative": *New Republic*, Nov. 24, 1917 (HLMC). **"His continuous laudation"**: "Beautifying American Letters," *Nation*, Nov. 14, 1917 (HLMC). **"Hitting below the belt"**: *ML*, 187. **"It gave me a kind of authority"**: Ibid., 189–90.

"I had little if any prejudice against Jews": Ibid., 181–82. In *ML/TS*, the word "pushing" has been cancelled and replaced with "tactlessness." **"He realizes himself"**: *Diary*, Sept. 19, 1946. **"I don't want any author"**: *ML*, 225.

"The book is anything but academic criticism": Knopf advertisement for *A Book of Prefaces*, 1917 (HLMC).

"Mencken is there with her sister": Dreiser, *American Diaries*, 169.

"You infuriate me": Marion Bloom to HLM, Nov. 1, 1916; Martin, 44. **"Let me know how many days"**: HLM to Marion Bloom, c. Nov. 1916; ibid., 44–45.

"He sat in her studio": Marion L. Bloom, "Reflection," *SS*, Mar. 1917, 31; Martin, 56.

"**Can he so easily close a door**": Marion Bloom to Estelle Bloom Kubitz, c. 1917–18; Martin, 70.

"**Perhaps one of the chief charms of women**": *In Defense of Women*, 41, 8–9.

"**Now that women have the political power**": Ibid., 163.

"**Wild joy**": Martin, 78. "**I shall not forget, my dear**": HLM to Marion Bloom, Sept. 20, 1918; ibid., 77–78.

CHAPTER FIVE
BECOMING A LEGEND, 1918–1923

BOOKS

Cooke, *Six Men*; Hoopes, *Cain*; Kazin, *On Native Grounds*; Lynd, *Middletown*; Martin, *In Defense of Marion*; Nathan, *A George Jean Nathan Reader*.

NOTES

"**There is something delightful**": *MC*, vii.

"**The best wines and liquors I could find**": *ML*, 289–90. "**We live in a land**": *MC*, vii. "**Between Wilson and his brigades**": "A Carnival of Buncombe," *Sun/P.M.* Feb. 9, 1920; *A Carnival of Buncombe*, 9.

"**The new espionage act**": HLM to George Sterling, May 6, 1918; *New Letters*, 87–88.

"**I have a hand for a compromise dialect**": *The American Language* (1919), vii. "**American is now so rich**": "The American: His Language," *SS*, Aug. 1913; *The Young Mencken*, 318, 332, 330.

"**Certain salient differences**": *The American Language* (1919), v. "**A general impatience**": *The American Language* (1921), 173.

"**When things get so balled up**": "The Declaration of Independence in American," ibid., 388; *MC*, 583.

"**This study shows a certain utility**": *The American Language* (1919), vii. "**According to a recent historian**": *The American Language* (1921), 58.

"**Next after the use**": *The American Language* (1963), 441. "**Virtually every book and pamphlet**": *ML*, 299. "**Linguistic science in North America**": *The American Language* (1963), v–vi.

"**Every Puritan is not necessarily a *wowser***": Ibid., 317.

"**In England all branches of human endeavor**": Ibid., 17–18, 31.

"**The American is not, of course, lacking**": Ibid., 99. This passage was carried over essentially intact from the first edition.

"**The business of introducing the American language**": Ibid., 84, 85.

"**They gave me some uneasiness**": *ML*, 298. "**Tart letters from Anglomaniacs**": HLM to Carl Van Vechten, May 19, 1919; *Letters*, 147.

"**Ten years ago Adolf was driving a wagon**": HLM to Philip Goodman, Aug. 1918; *Letters*, 125. **It sold only a thousand copies**: Knopf took over *In Defense of Women* after Goodman went out of business, and made it a success.

"H. L. Mencken will surprise you": *New York Evening Sun*, Dec. 16, 1918 (HLMC).

"150,000 civilized Americans": HLM to Harry Leon Wilson, Jan. 26, 1915; *Letters*, 59. **"There were some things in it"**: *ML*, 254. **Willa Cather:** See HLM's "Sunrise on the Prairie," *SS*, Feb. 1919; *SSC*, 266. **"The Smart Set is addressed"**: *Pamphlets and Leaflets, 1915–36* (EPFL). "This one was sent to Henrik Ibsen," HLM scrawled on his file copy of this form rejection slip.

"The gala days": Nathan, "H. L. Mencken (1880–1956)"; Nathan, 175. **"1. The editorial chambers are open daily"**: *ML*, 427–29.

"Boys dying like flies": "Marion Bloom's Journal," Oct. 14, 1918; Martin, 94. **"It was no uncommon thing"**: "Marion Bloom's Journal," Oct. 20, 1918; ibid., 97. **"The horrors and privations"**: HLM to Ernest Boyd, Mar. 13, 1919; ibid., 114. **"I begin to see I do not want to marry you"**: Marion Bloom to HLM, Mar. 1919 (draft); Ibid., 115–16.

"You have made a piece of literature": HLM to Marion Bloom, Mar. 24, 1919; ibid., 114.

"She lashes herself into a fury": Ibid. 118–19.

"I can't understand your attitude": Ibid., 123–24.

"What is this Christian Science stuff": Ibid., 126.

"I am surely not going to try": Ibid.

Mencken's attitude toward doctors: *RAM*, 47–48.

"Is it desirable to preserve the lives": "A Free State Measure," *Sun/P.M.*, Feb. 28, 1927; revised as "Christian Science," *MC*, 345.

"We have got along the past few years": HLM to Marion Bloom, Nov. 2, 1922; Martin, 182. **"I'd bet a lot that my lack of virginity"**: Marion Bloom to Estelle Bloom Kubitz, Mar. 20, 1921; ibid., 150.

"It is the close of a busy and vexatious day": "Répétition Générale," *SS*, Aug. 1921; *In Defense of Women*, 169–70.

"The model for my portrait": *ML*, 87. **"Marion Bloom was a woman I knew"**: HLM to Bradford F. Swan, Feb. 18, 1948; *New Letters*, 587–88.

"What am I to say?": HLM to Marion Bloom, c. Aug. 1920 (n.d.); Martin, 198–99. **"Little will that gal ever suspect"**: HLM to Estelle Bloom Kubitz, Aug. 10, 1923; ibid., 199.

"Butcheries of some of the elder demigods": *ML*, 304.

"For easy reference": Ibid., 302. **"Such books are mere stinkpots"**: HLM to Fielding H. Garrison, Sept. 28, 1919; ibid., 303–4.

"The Emerson cult": *P/1*, 192. **"Frost? A standard New England poet"**: Ibid., 89–90. **"The prophesying business"**: Ibid., 31. **"Always the ethical obsession"**: Ibid., 189.

"Most of the contents": *ML*, 304.

"A smartly phrased series": *Boston Evening Transcript*, Oct. 25, 1919 (HLMC). **"A book full of sound, hard sense"**: Ibid., Nov. 1, 1919 (HLMC). **"There is no man living"**: *Chicago Tribune*, n.d. (HLMC). **"I have discovered something"**: *ML*, 321.

"The first trombone blasts": Kazin, 205. **"The best American novel that I have seen of late"**: "Books More or Less Amusing—II," *SS*, Aug. 1920, 140.

"I have no definite duties": *AN1925*, 96–97. **"My own connection with the paper"**: HLM to Ernest Boyd, Jan. 26, 1920; *Letters*, 171.

"All of the great patriots": "A Carnival of Buncombe," *Sun/P.M.*, Feb. 9, 1920; *A Carnival of Buncombe*, 5. **"The genial dampness continues"**: "It's All in Wilson's Hands, Mencken Concludes After Looking Around at Frisco," *Sun/P.M.*, June 29, 1920; *The Impossible H. L. Mencken*, 246.

"Let us have our hangings": "A Great Moral Sport," ibid., Apr. 25, 1921.

"The Jews of Detroit": "On Censorships," ibid., June 27, 1921.

"We suffer most when the White House bursts with ideas": "The Coolidge Mystery," *Sun/P.M.*, Jan. 30, 1933; *The Impossible H. L. Mencken*, 418. **"We suffer most, not when the White House"**: "Coolidge," *MC*, 254.

"The total of my published writings": Ibid., vii. **"Everything else was supplementary"**: *RAM*, 16–17.

The daily details of that life: The most detailed account of HLM's regular routine is found in his typewritten responses to the questions submitted by Roger Butterfield for his 1946 *Life* "interview" (*Miscellaneous Typescripts, Carbons and Clippings, 1946–48*, EPFL). See also *AN1925*, 75–76. **All of which he stamped:** Philip Wagner, "Mencken Remembered," *American Scholar*, Spring 1963 (HLMC). **"Mencken carried it to the extreme"**: Cooke, 102. **One young staffer claimed:** Wagner. **"My manuscript done on the typewriter"**: *AN1925*, 202. **His handmade silk pajamas:** *Diary*, Apr. 15, 1943. **Reading himself to sleep by gaslight:** *ML*, 361.

Books were his closest companions: HLM describes his reading habits in *AN1925*, 76–77.

"Two hours to actual playing": *Heathen Days*, 89.

"The club has always had": Ibid., 91. **"Literati of the third, fourth and fifth rate"**: *ML*, 309. **"I am a one hundred percent American!"**: *New Dictionary of Quotations*, 37. The second stanza runs as follows: "I am an anti-Darwin intellectual. / The man who says that any nice young boy or gal / Is a descendant of the ape / Shall never from Hell's fire escape."

On a typical evening in 1934: Louis Cheslock, unpublished diary entry, June 23, 1934 (EPFL). **"Henry's humming as he plays"**: Ibid., Oct. 16, 1943. **"From the first notes it was obvious"**: Cheslock, "Postlude (The Saturday Night Club)"; *H. L. Mencken on Music*, 212–13. **"A lusty *shit!*"**: Cheslock, unpublished note, n.d. (EPFL).

"The world presents itself to me": *Minority Report*, 47. See also *Diary*, July 19, 1944. **"So far as I can make out"**: HLM to Isaac Goldberg, May 6, 1925; *H.L. Mencken on Music*, 201. **"There is no sneaking"**: "Beethoven," *Sun/P.M.*, Apr. 24, 1922; *MC*, 526.

"My gregariousness is satisfied": *AN1941*. **"People in general interest me"**: Unpublished *Minority Report* note, n.d. (EPFL). **"The late Albert Hildebrandt"**: "The End of a Happy Life," *Sun/P.M.*, Nov. 21, 1932; *H.L. Mencken on Music*, 216–17.

"I began to work during the war": *Middletown*, 28–29.

"Crude and childish": "Night Club," ibid., Sept. 3, 1934; *The Impossible H. L. Mencken*, 470. **"An old longing"**: "Movies," *Sun/P.M.*, May 30, 1921. **"My surroundings have helped to make me"**: *AN1941*.

"**Because he wrote**": *Vanity Fair*, Feb. 1921 (HLMC). "**A genuine artist**": Edmund Wilson, "H. L. Mencken," *New Republic*, June 1, 1921 (HLMC). "**I had suddenly become**": *ML*, 322.

"**That many soldiers' lives**": *The American Credo*, 145. "**The readiness of the inferior American**": Ibid., 95, 100.

"**No man of active, original and courageous mind**": "An Editorial Memorandum," 35, 364–65.

"**It is the prime function**": Ibid., 100.

"**He had lately finished a novel**": *ML*, 332–33.

"**Grab hold of the bar-rail**": Nathan, 181. "**The best thing of its sort**": HLM to Sinclair Lewis, Oct. 27, 1920; *Letters*, 206. "**A phenomenon of the first order**": "Peasant and Cockney," *Sun/P.M.*, Jan. 3, 1921. "**It presents characters that are genuinely human**": "Consolation—I—An American Novel," *SS*, Jan., 1921; *SSC*, 280, 282.

"**I know the Babbitt type**": "Portrait of an American Citizen," *SS*, Oct. 1921; *SSC*, 283, 285.

"**The year 1921**": *ML*, 359. "**We turned up relatively little stuff that lifted us**": Ibid., 371. "**Editorially, 1922 was an average year**": Ibid., 387.

"**I am sick of the Smart Set**": HLM to Sara Haardt, July 10, 1923; *MS*, 84.

"**I might want to change the Table of Contents page**": HLM to George Jean Nathan, May 3, 1923; *Letters*, 249.

"**Mr. Mencken, or Mr. Nathan**": *Boston American*, Dec. 23, 1920 (HLMC). "**And the two they talked of the aims of Art**": Berton Braley, "Three—Minus One," *New York Sun*, Dec. 6, 1920 (HLMC).

"**I aimed all my professional shots**": *ML*, 329. "**I am preparing to assassinate the Smart Set**": HLM to Max Brödel, July 30, 1923; *Letters*, 252. "**I shall try to cut a rather wide swathe**": HLM to Upton Sinclair, Aug. 24, 1923; ibid., 258–59.

"**What we need is something that looks highly respectable**": HLM to Theodore Dreiser, Sept. 10, 1923; ibid., 260. "**Very simple and beautiful**": HLM to Howard Parshley, Oct. 6, 1923; *New Letters*, 176. "**A bit Frenchy**": HLM to Gamaliel Bradford, Jan. 2, 1924; *Letters*, 263.

"**Glancing back over the decade and a half**": "Fifteen Years," *SS*, Dec. 1923; *SSC*, 326, 329.

"**I have believed all my life**": Statement "written for the Associated Press, for use in my obituary," Nov. 20, 1940; *A Second Mencken Chrestomathy*, 491. "**What ails the beautiful letters of the Republic**": "The National Letters," *P/2*, 87.

"**Here the general average of intelligence**": "On Being an American," *P/3*, 13–14.

This essay originated in a Monday Article: "On Being an American," *Sun/P.M.*, Oct. 11, 1920; *The Impossible H. L. Mencken*, 25–29.

"**The real event of the year**": F. Scott Fitzgerald to Bernard Vaughan, Dec. 1923; Nathan, 208.

HLM had lunch at Marconi's: Hoopes, 106–7.

CHAPTER SIX
AT THE *AMERICAN MERCURY,* 1924–1928

DOCUMENTS

The Mencken Collection contains a copy of an unpublished memoir by Alfred A. Knopf, cited as "Knopf" (EPFL).

William Jennings Bryan's voice can be heard on *In Their Own Voices: The U.S. Presidential Elections of 1908 and 1912* (Marston 52028-2), a two-CD set containing transfers of ten commercial recordings made by Bryan in 1908.

BOOKS

Allen, *Only Yesterday;* Angoff, *H. L. Mencken;* Bode, *Mencken;* De Casseres, *Mencken and Shaw;* Goldberg, *The Man Mencken;* Howe, *Holmes-Laski Letters;* Johnson, *A History of the American People;* Kunkel, *Genius in Disguise;* Larson, *Summer for the Gods;* Marquand, *Point of No Return;* Martin, *In Defense of Marion;* Mayfield, *The Constant Circle;* Nathan, *A George Jean Nathan Reader;* Singleton, *H. L. Mencken and the* American Mercury *Adventure;* Sobel, *Coolidge;* Spivak, *The American Mercury Reader;* Stearns, *Civilization in the United States;* Stenerson, *Critical Essays on H. L. Mencken;* Wilson, *The Shores of Light.*

NOTES

"You mean that one in the magazine with the green cover?": Sobel, 13. The article was Frank R. Kent's "Mr. Coolidge" (*Mercury,* Aug. 1924).

"Every flapper carries one": *Huntington* (W.Va.) *Advertiser,* May 8, 1924 (HLMC). **"Clyde was a dear, poky place":** Marquand, 264. (The ellipsis is in the original.)

"It is doubtful": "You Are Invited to Subscribe for the *American Mercury*" (n.d.); "Clippings, Carbons and Souvenirs, Including Music, 1893–1938" (EPFL). **"Emotional and aesthetic starvation":** Stearns, vii.

He told Dreiser: HLM to Theodore Dreiser, Oct. 4, 1923; *Dreiser-Mencken Letters,* 503. **Main Street had sold 390,000 copies:** Allen, 198. **An initial press run of 5,000 copies:** This and all otherwise unsourced details about the operations of the *Mercury* are from Singleton, passim. **"We have word this morning":** HLM to Philip Goodman, Dec. 28, 1923; *Letters,* 262.

"The older I grow": "Clinical Notes," *Mercury,* Jan. 1924, 75. **"Final triumph of Calvinism":** "Americana," ibid., 48.

"The Editors are committed": "Editorial," ibid., 27–30.

"He was fortunate": Isaac R. Pennypacker, "The Lincoln Legend," ibid., 7. **"The general tendency of the human race":** Harry E. Barnes, "The Drool Method in History," ibid., 31. (For Barnes's later career as a Holocaust denier, see p. 379.)

"When he went to Harvard": Ernest Boyd, "Aesthete: Model 1924," ibid., 51.

"Bernard Shaw, seeking to demonstrate": L. M. Hussey, "The Pother About Glands," ibid., 93.

"I saw my article appear in print": "Hazlitt for Mencken," *Time*, Oct. 16, 1933 (HLMC). **"We do not expect Mr. Mencken to understand"**: "The American Mercury," *New Republic*, Feb. 6, 1924 (HLMC).

"The undergraduates must have their gods": "Another God," *Daily Iowan*, Dec. 6, 1924 (HLMC).

"A native born American": "Henry L. Mencken," *Knoxville* (Tenn.) *Sentinel*, Mar. 2, 1925 (HLMC). **"The uplift has damn nigh ruined the country"**: HLM to Upton Sinclair, Dec. 10, 1924; *Letters*, 273.

"He is recruited from people": James M. Cain, "American Portraits: I. The Labor Leader," *Mercury*, Feb. 1924, 196. **"An *affaire flambée*"**: George Jean Nathan, "The Theatre," ibid., 241.

"The newspaperman in Mencken": George Jean Nathan, "The Happiest Days of H. L. Mencken," *Esquire*, Oct. 1957; Nathan, 177.

"When the *American Mercury* at last came out": 35, 125, 134.

"He turned out to be far too narrow": Ibid. **"The busiest of my whole life"**: Ibid., 133.

"A magazine is a despotism": *Diary*, Sept. 13, 1939.

"After a year's hard experience: HLM to George Jean Nathan, Oct. 15, 1924; Mencken Collection, New York Public Library. **"Our differences are too far apart"**: HLM to George Jean Nathan, Oct. 19, 1924; *Letters*, 269–70.

"The unreasoning and often insulting attitude": George Jean Nathan to HLM, n.d. (c. Oct. 1924); Mencken Collection, New York Public Library. **"Violent and long-continued resistance"**: 35, 134. **"What will you take, in cash"**: HLM to George Jean Nathan, Dec. 25, 1924; *New Letters*, 184.

"In friendship there must be no thought of money": *AN1925*, 197.

"His name was removed from the directory": Knopf, 33–34. **"It seems perfectly clear to me"**: Ibid., 35.

HLM'S dislike of homosexuals: In *ML*, for instance, he refers to Lord Alfred Douglas, Oscar Wilde's lover, as a "filthy homo" (111) and makes the following remark about Somerset Maugham: "He was reputed in New York to be a homosexual of the school of Hugh Walpole, and I was thus somewhat shy of his society" (373). **"One of my convenient habits"**: "Advice to Young Men," May 17, 1939, *Minority Report/HLM's Notes/Unpublished Material* (EPFL).

"I always felt": Theodore Dreiser to George Jean Nathan, Oct. 7, 1933; Singleton, 133.

"I get a sense": Rollin Kirby, "Cézanne and His Crowd," *Mercury*, May 1928, 76–77. **"Isadora simply loved to prance"**: "Two Enterprising Ladies," *Mercury*, Apr. 1928, 507.

In 1924 Sinclair Lewis wrote a piece: Sinclair Lewis, "Main Street's Been Paved," *Nation*, Sept. 10, 1924; Johnson, 723.

"Somehow, I have a feeling": HLM to Ernest Boyd, Jan. 28, 1926; Mencken Collection, New York Public Library.

HLM never spent more than $2,500 a month: *Diary*, Oct. 23, 1942. **"A new kind of editor"**: De Casseres, 66–67. **"Remarkable for the definiteness"**: Edmund Wilson, "The Literary Worker's Polonius: A Brief Guide for Authors and

Editors"; Wilson, 595. **"I append my usual hopes"**: HLM to Burton Kline, May 21, 1931; *New Letters*, 252.

"The old lady in Dubuque": "Announcing a New Weekly Magazine, *The New Yorker*"; Kunkel, 440. **"Observe closely the common middle-class American"**: Harvey Fergusson, "American Portraits: II. The Washington Job-Holder," *Mercury*, Mar. 1924, 345. **"I presume that country grand juries"**: H. Rob. Keeble, "The Grand Inquest of the Country," ibid., June 1925, 142. **"Censorship is the Cuckoo Klux Klan"**: James R. Quirk, "The Wowsers Tackle the Movies," ibid., July 1927, 349.

"Love is a wild thing": Sara Haardt, "Fragment," *Smart Set*, Dec. 1919, 128.

"Sara Haardt's as suave as silk": Mayfield, 22. **"The greatest living writers in America"**: Ibid., 31. **"A soulful highbrow"**: Ibid., 33.

"Before I get stuck in Alabama": Sara Haardt to HLM, May 20, 1923; *MS*, 77.

"I shall be at Domenique's": Sara Haardt to HLM, May 26, 1923; ibid., 78.

"Let us have another palaver": HLM to Sara Haardt, June 5, 1923; ibid., 81.

"Let me see the novel": HLM to Sara Haardt, July 26, 1923; ibid., 86.

"I am working day and night": HLM to Sara Haardt, Aug. 17, 1923; ibid., 88.

"The review I am simply wild about": Sara Haardt to HLM, Aug. 22, 1923; ibid., 89.

"Since I am looking forward so very much": Sara Haardt to HLM, Sept. 13, 1923; ibid., 93.

"Wohledle, Hochehr- und Tugenbelobte": HLM to Sara Haardt, Nov. 15, 1923; ibid., 104.

"I refuse to believe that you are ill": HLM to Sara Haardt, Feb. 23, 1924; ibid., 118.

"For a woman she had a good sense of humor": *RAM*, 75.

"The Mercury boil is about to bust": HLM to Sara Haardt, Oct. 19, 1924; *MS*, 171. **"You will live to be the oldest woman in Alabama!"**: HLM to Sara Haardt, Feb. 25, 1925; ibid., 196. **"Friend Sary, I miss you like hell"**: HLM to Sara Haardt, Mar. 4, 1925; ibid., 199.

"First and last, many papers subscribed": *35*, 135. **" 'Tough Guy' of American Criticism"**: *New York Times*, Nov. 2, 1924 (HLMC).

"Diligent and reasonably competent": *35*, 134. **"My slave"**: *RAM*, 20. See also HLM to Sara Haardt, Jan. 16, 1925; *MS*, 187. **"Mencken got up, opened the door of the bathroom"**: Angoff, 22.

Mencken told Angoff dirty jokes: According to August Mencken, "When he would tell Angoff some horrible, foul, dirty story, he did it really to shock Angoff, because he was not in the habit of telling those stories and detested them. I have never heard him actually, in my whole life, tell what are known as 'stories.' If somebody told one that had microscopic humor in it—most of them don't have that much—he might laugh, but he didn't like them. . . . So if he told one to Angoff, it was really to kid Angoff, and Angoff, I gather, must be a simple-minded fellow, and he didn't know he was being kidded and took the whole thing in" (*RAM*, 24). **"I can't understand how anybody can be an anti-Semite"**: Angoff, 161. **"He sometimes wrote about Jews"**: Ibid., 162.

"**A Jew of the better sort**": *ML*, 411. "**G. plans an elaborate work**": HLM to Sara Haardt, Mar. 2, 1925; *MS*, 197. "**Inasmuch as he knows no more about me**": HLM to Sara Haardt, Mar. 4, 1925; ibid., 199. **A two-hundred-page autobiographical manuscript**: *AN1925*. "**A piercing vision**": Goldberg, 290. "**God help you if you try to read it**": Mayfield, 91.

The Tennessee senate passed a law: For the Scopes trial and its intellectual history, see Larson, passim.

"**Evolution was coming to be recognized**": Ibid., 20–21.

Phonograph records: *In Their Own Voices*.

"**If such people were lower animals**": Larson, 27.

"**I repeat that it eases and soothes me**": "Dives into Quackery: 1. Chiropractic," *P/6*; *MC*, 349.

"**Evolution represents man as reaching his present perfection**": Larson, 39. "**As true or as a human fact**": Ibid., 47. "**A scientific soviet**": Ibid., 45.

"**No principle is at stake in Dayton**": "In Tennessee," *Nation*, July 1, 1925; *35*, 139–40.

"**State sovereignty is a question of fact**": Oliver Wendell Holmes, Jr., to Harold Laski, Sept. 15, 1916; Howe, 21. "**The theory that the common people know what they want**": "Sententiae," *MC*, 622.

Mencken took a sharply different tack: *35*, 137–38.

He also claimed to have persuaded Darrow: This was what he told Huntingdon Cairns. See Bode, 265. "**I have got myself involved**": HLM to Sara Haardt, May 27, 1925; *MS*, 211.

"**Far off in a dark, romantic glade**": "The Hills of Zion," *P/5*; *MC*, 394, 396.

"**A comic scene? Somehow, no**": "Yearning Mountaineers' Souls Need Reconversion Nightly, Mencken Finds," *Sun/P.M.* July 13, 1925; *The Impossible H. L. Mencken*, 580. (This passage also appears in the *P/5* version of "The Hills of Zion.")

"**It was hard to believe**": "In Memoriam: W. J. B.," *Mercury*, Oct. 1925; *MC*, 245.

"**Eighteenth-century qualities**": "The All-Star Literary Vaudeville," *New Republic*, June 30, 1926; Wilson, 236. "**I have never consciously tried**": "On Happiness," *Sun/P.M.*, May 9, 1927; "The Tight-Rope," *A Second Mencken Chrestomathy*, 483. "**I had never been on close terms**": HLM to Gamaliel Bradford, July 28, 1925; *New Letters*, 189. "**The job before democracy**": "Bryan," *Sun/P.M.*, July. 27, 1925; *The Impossible H. L. Mencken*, 608. "**We killed the son-of-a-bitch**": Mayfield, 90. "**Vague, unpleasant manginess**": *MC*, 246. "**A typical Puritan**": "The Archangel Woodrow," Ibid., 248.

"**Getting Scopes acquitted**": *35*, 137–38.

The Web site of Bryan College: www.bryan.edu.

"**A new insight into the nature of democracy**": Mayfield, 92. "**The phrase appears word for word**": "Democracy, as I see it, is simply the organized hatred of the lower orders. . . . I have exposed this idea at various times and at great length, but never, so far as I can recall, in a formal essay. It will be at the bottom of my projected book on democracy" (HLM to Percy Marks, Feb. 3, 1925; *Letters*, 278). "**As I worked in my third-floor office**": *35*, 154.

"A stupendous debt": Ibid., 154–55.

The fireplace was usually lit at 1524 Hollins Street: "When I think of the enormous pleasure that little fireplace has given us, and then figure out how little it has cost, I am impressed all over again with the fact that the most durable delights of life are cheap . . . [at] a total cost of not more than $150 we have had a cheery fire burning for nearly 28 years. It burns, on an average, about 100 nights a year, which works out to 5¹/₂ cents a night" (*AN1941*). **"I told Dreiser that I feared she might die":** *Diary*, Apr. 18, 1931.

"His customary attitude to the world": *ML*, 399. **"A man of active and resilient mind":** "Répétition Générale," *SS*, July 1919; "The Friend," *MC*, 16. **"What a woman is gone!":** HLM to Marion Bloom Maritzer, Dec. 14, 1925; Martin, 208.

"Rescued me from myself": *35*, 155.

He installed his cluttered mahogany desk: August Mencken kept HLM's second-floor office intact after his death in 1956. Immediately following August's death in 1967, the whole house was photographed and its contents described in detail. See James H. Bready, "The Menckens' Era Ends at 1524 Hollins Street," *Sun Magazine*, Aug. 6, 1967 (HLMC). **Five different encyclopedias:** They were the fourteenth edition of the *Encyclopaedia Britannica*, the *Chambers* and *New International Encyclopedias*, the *Catholic Encyclopedia* and the *Encyclopedia of Religion and Ethics*. See Philip Wagner, "Mencken Remembered," *American Scholar*, Spring 1963 (HLMC).

"From the Christians and their God": Herbert Asbury, "Hatrack," *Mercury*, Apr. 1926; Spivak, 142–43. HLM chronicled the "Hatrack" case exhaustively in a 1937 manuscript published in 1988 as *The Editor, the Bluenose, and the Prostitute* (cited as *Bluenose*). Asbury wrote about it in "The Day Mencken Broke the Law," *Mercury*, Oct. 1951 (HLMC). It is also discussed in Angoff, 40–53. This narrative draws on all three accounts.

"A Methodist vice-hunter": James D. Bernard, "The Methodists," *Mercury*, Apr. 1926, 431. **"Filthy and degrading descriptions":** *Bluenose*, 50.

"If Chase were permitted": Ibid., 52. **"Few doctrines seem to me":** HLM to Burton Rascoe, c. summer 1920; *Letters*, 188. **"I was now free to take chances":** *Bluenose*, 53.

"On the ground that we had missed": Ibid., 79.

"A sensational and pornographic magazine": Ibid., 126. **"Unredeemed dirt":** *Bluenose*, 112. **"The general smugness and lack of intellectual enterprise":** "Journalism in America," *P/6*, 28, 29.

"It remains, to me at least": *35/TS*, 471.

"If he retains his rectitude": *Notes on Democracy*, 110.

"Until he got to Washington": Ibid., 125–26.

"Sordid and degrading motives": Ibid., 131.

"When I protested": Ibid., 134.

"I am not engaged": *Notes on Democracy*, 195. **"I enjoy democracy immensely":** Ibid., 211.

"Far more an artist than a metaphysician": HLM to Sara Haardt, Aug. 31, 1925; *MS*, 227. **"The man is bigger than his ideas":** Walter Lippmann, "H.L.

Mencken," *Saturday Review of Literature*, Dec. 11, 1926 (HLMC). **"An ideal monster"**: Edmund Wilson, "Mencken's Democratic Man," *New Republic*, Dec. 15, 1926; Wilson, 297. **"There is no evidence"**: *Notes on Democracy*, 112. **"The people cannot look to legislation"**: Sobel, 84.

"A grand tour of the South": 35, 156, 160.

"I am innocent, mon colonel": HLM to Ernest Boyd, July 26, 1926; *New Letters*, 200. **"I love you loving me"**: Aileen Pringle to HLM, n.d. (c. Sept. 1926); Martin, 244. **"In spite of the mosquitoes"**: Aileen Pringle to HLM, n.d. (c. Sept. 1926); ibid., 245.

"I sat under Aimee": HLM to Raymond Pearl, Nov. 1, 1926; *New Letters*, 205. **"Sooner or later the movies will have to split"**: "The Low-down on Hollywood," *Photoplay*, Apr. 1927; "Interlude in the Socratic Manner," *A Second Mencken Chrestomathy*, 201–2.

"A man of relatively civilized feelings": "Valentino," *Sun/P.M.*, Aug. 30, 1926; *MC*, 283–84.

"A performer who pleases the generality of morons": "Interlude in the Socratic Manner," *A Second Mencken Chrestomathy*, 203. **"I AM MASHED ON YOU"**: Telegram from HLM to Aileen Pringle, Dec. 31, 1926; Martin, 260. **"Two eminent female stars"**: HLM to Marion Bloom Maritzer, Sept. 20, 1926; ibid., 252. (The other one was Lillian Gish, who was in fact in love with George Jean Nathan, though she and HLM would become good friends.) **"The public prints have probably informed you"**: HLM to Marion Bloom Maritzer, Jan. 13, 1927; ibid., 266. **Aileen made the mistake of mentioning marriage:** She mentioned it to other people, too, including HLM's friend Carl Van Vechten, who recorded in his daybook entry for Mar. 4, 1928, that "Aileen tells me she is in love with Mencken and wants to marry him, but she won't live with his sister in Baltimore." Years later, she told Bruce Kellner, Van Vechten's biographer, that "he couldn't stand the climate in Hollywood and she couldn't stand the climate in Baltimore and so they split. Aileen made clear to me that by 'climate' they were not talking only about the weather. She hated Hollywood glitz as much as he did, but found it preferable to the prospect of life in Baltimore" (Bruce Kellner, private communication). **"You are still young, and beautiful"**: HLM to Aileen Pringle, n.d. (c. Feb. 28, 1927); ibid., 278–79.

"I see Sara Haardt here very often": HLM to Joseph Hergesheimer, July 27, 1927; *New Letters*, 211. **"Insufferable"**: Sara Haardt to HLM, Oct. 18, 1927; *MS*, 301. **"A man in full possession"**: Hannah Stein, "H. L. Mencken (Bachelor) Says: 'Most Men Marry When Women Make Up Their Minds That They Shall Marry—and Then There Is No Escape!,'" *Philadelphia Public Ledger*, Dec. 18, 1927 (HLMC).

"He I suspect would prove": Justice Holmes to Harold Laski, Jan. 28 and Feb. 10, 1920; Howe, 236, 240.

"Old Hank's portrait": Marion Bloom Maritzer to Estelle Bloom Kubitz, July 26, 1927; Martin, 298. (Schattenstein's portrait of HLM now hangs in the Mencken Room of the Enoch Pratt Free Library.)

"I have done a great deal less": Memo, Jan. 1, 1927 (EPFL). The complete text is reprinted in *Menckeniana* 33.

"**I am going on furlough**": "The Choice of a Career," *Chicago Tribune,* Jan. 29, 1928.

"**Up to this time**": Allen, 201–2.

"**Addressed to the intellectual left wing**": Ibid., 199. "**I gather from both of them**": "The Library: Fiction," *Mercury,* May 1928, 127.

"**Nearer to intellectual vaudeville**": Irving Babbitt, "The Critic and American Life," *Forum,* Feb. 1928; Stenerson, 79, 82.

CHAPTER SEVEN
MARRIAGE AND THE CRASH, 1929–1935

BOOKS

Angoff, *H. L. Mencken;* Haardt, *Southern Album;* Haardt, *Southern Souvenirs;* Hoopes, *Cain:* Hughes, *American Visions;* Lipstadt, *Denying the Holocaust;* Martin, *In Defense of Marion;* Mayfield, *The Constant Circle;* Posner, *Law and Literature;* Stenerson, *Critical Essays on H. L. Mencken.*

NOTES

Sometimes as many as fifty: *Diary/TS,* July 24, 1944. "**By the way, what is the best current book of Catholic doctrine?**": HLM to Msgr. J. B. Dudek, Jan. 2, 1929; *Letters,* 311.

"**She is bearing the thing very bravely**": HLM to Sara Mayfield, June 25, 1929; *New Letters,* 235.

"**No longer in his intellectual class**": Marion Bloom to Betty Adler, Nov. 3, 1968 (EPFL); Martin, 352. "**I have a notion**": HLM to Marion Bloom Maritzer, Nov. 25, 1928; ibid.

He said he had been told: Mayfield, 37. "**Wrecked me for a week**": *35,* 188.

"**If I ever marry**": HLM to Marion Bloom Maritzer, Feb. 4, 1927; Martin, 274. "**The fact that marriage between older persons**": "Répétition Générale," *SS,* Aug. 1922, 52; ibid., 178.

Whatever the extent of their physical relationship: Neither HLM nor Sara Haardt made any statement on this subject that has survived. According to Sara Mayfield, they were "affectionate, and quietly demonstrative with each other; but during their long courtship, they had become mentally and emotionally so closely knit that their marriage merely set a physical seal on a romantic attraction" (Mayfield, 181). "**Anything but photogenic**": *35/TS,* 606. "**I can recall no single moment**": *Diary,* May 31, 1940. "**Her indefinable pleasant thoughtfulness**": HLM, "Preface"; Haardt, *Southern Album,* xxii–xxiii. "**I could have latched on**": Sara Haardt, "Dear Life," *Story,* Sept. 1934; Haardt, *Southern Souvenirs,* 307–8.

One of her stories: "Little White Girl," *Scribner's,* Apr. 1934; ibid., 70–78. "**An important contributor**": Ann Henley, "Introduction"; ibid., 2. "**The ether poured out of the cone**": "Dear Life," ibid., 307, 310.

Bragging to Philip Goodman: HLM to Philip Goodman, May 1, 1933; *Letters,* 365.

"THE BOOK WAS FINISHED": Telegram, HLM to Sara Haardt, Nov. 29, 1929; *MS*, 415.

"It is seldom discussed": *AN1941*. "The best popular account we have": Granville Hicks, "Primer to Religion," *Forum*, Apr. 30, 1930 (HLMC). "The gleam of fanaticism": Reinhold Niebuhr, *Atlantic Monthly*, June 1930; Stenerson, 95.

"Quite devoid of the religious impulse": *Treatise on the Gods* (1930 edition), vii.

"Chapter by chapter, as it was written": Angoff, 148, 149. He also claimed that Philip Goodman dismissed the book as "the work of a charlatan. He's cribbed and misunderstood his betters" ibid., 150).

"I have been torn all my writing life": AN1941. "The God of the Episcopalians": *Treatise on the Gods* (1930 edition), 64. "Men may live decently without it": Ibid., 352.

"Lush and lovely poetry": Ibid, 345–46. Faint echoes of Nietzsche: For a concise summary of Nietzsche's complex attitudes toward Judaism and the Jews, see Posner, 146–49.

"A chance reference": Mayfield, 152. "Don't forget that my book is about religion": Joseph Brainin, "Is H. L. Mencken an Anti-Semite?," *Pittsburgh Jewish Chronicle*, Apr. 11, 1930 (HLMC).

A defense his admirers would echo: "While some of Mencken's remarks on the Jews were no more flattering than those on the Methodists and Baptists, he certainly had no such intention. It is undeniably true that he frequently used the term 'the Jews' as a pejorative for movie moguls, theatrical producers, and promoters of all kinds; but he intended to reflect upon their profession, not their religion" (Mayfield, 152).

"Ever since the rise of the Coolidge prosperity": 35, 189. "The virtue of the story": "Fiction by Adept Hands," *Mercury*, Jan. 1930, 127. "How lovely you are": HLM to Sara Haardt, Dec. 27, 1929; *MS*, 416. "Your letter was lovely": HLM to Sara Haardt, Jan. 1, 1930; ibid., 421. "I miss you dreadfully": HLM to Sara Haardt, Jan. 5, 1930; ibid., 424.

"I am so happy I am dizzy": Sara Haardt to HLM, Apr. 26, 1930; ibid., 443. An abstract painting by Thomas Hart Benton: Benton had illustrated *Europe after 8:15*, and HLM bought "Bubbles" from him for $50 around 1915 as a favor to their then-mutual friend Willard Huntington Wright (*ML*, 211–12). One of Benton's few surviving abstract canvases, it was a pastiche of the "synchromistic" style of Stanton MacDonald-Wright, Willard's brother. HLM presented it to the Baltimore Museum of Art in 1947. For a reproduction, see Hughes, 443. "Barring acts of God": HLM to Edith Lustgarten, July 28, 1930; *New Letters*, 244.

"My fear was that the newspapers": 35, 202. "I formerly was not as wise": *Sun/A.M.*, Aug. 3, 1930 (HLMC). "Any of the traditional fluster": United Press interview with HLM and Sara Haardt, Aug. 28, 1930 (HLMC).

"The marriage of a first-rate man": *In Defense of Women*, 83. "Bellicose Mencken Will Trip to Altar": *Meridian* (Miss.) *Star*, Aug. 11, 1930 (HLMC). "Busting into the church": HLM to Max Brödel, Aug. 20, 1930; *New Letters*, 246.

"No Hearst photographer": *35/TS*, 606–7. **That morning he had bundled up Aileen Pringle's letters:** She saved the wrapping papers, which are postmarked from Baltimore at 8:30 P.M. on August 27 (Martin, 359).

He was doing all his serious writing at night: HLM to A. O. Bowden, Apr. 12, 1932; *New Letters*, 262. **"I don't think that we ever bored each other":** *Diary*, May 31, 1940.

"I expected to make rather heavy weather": HLM to Philip Goodman, Aug. 27, 1931; Mayfield, 189. **"She never uses coercion on me":** HLM to Philip Goodman, Aug. 29, 1931; *Letters*, 333. **"Marriage hasn't changed Henry a bit":** Sara Haardt Mencken, 1932 interview with Julia Blanshard (HLMC).

"My friend Mencken": "Nathan and Mencken Part," *New York Times*, Mar. 1, 1930 (HLMC). **"They were certainly on the best of terms":** Mayfield, 172. **"George tells me that he wants":** HLM to Blanche Knopf, Mar. 13, 1930; *Letters*, 315.

"Wielded an influence in this country": Ernest Boyd, "H. L. Mencken, Homo Americanus," *Vanity Fair*, Mar. 1932 (HLMC).

"Inner doubts about his own grasp": Angoff, 208.

"Strongly opinionated articles": Clipping from *The Writer's Market*, 1930 edition (HLMC). **"The Word of God came to the United States Military Academy":** Lloyd Lewis, "The Holy Spirit at West Point," *Mercury*, Nov. 1930, 353.

"This article is so good": HLM to Jim Tully, Nov. 21, 1932; *New Letters*, 274. **"You couldn't argue with him":** Hoopes, 191. **"Can't Henry see that this silly gag":** ibid.

Mencken was unapologetic: *Minority Report*, 240.

"Utterly unable, once the times grew darker": Hoopes, 193. **"The factitious, drug-storish 'superiority'":** "Uprising in the Confederacy," *Mercury*, Mar. 1931, 380–81.

"Plans for clearing out": *35/TS*, 644.

"I must consult authorities at every step": *Diary*, Sept. 6, 1931.

Willa Cather was at the same performance: Ibid., Nov. 27, 1931.

"My inclination is to take over the house": Ibid., Jan. 18, 1931. **"The depression was nonexistent":** Angoff, 211.

"Having the same emotions": Edmund Wilson, "Brokers and Pioneers," *New Republic*, Mar. 23, 1932 (HLMC). **"Life grows sweeter to him now":** "Mellow Mencken," *Omaha World Herald*, Jan. 20, 1932 (HLMC).

"It is only the story": "What Is Going On in the World," *Mercury*, Mar. 1930, 259–61.

"My total income for 1932": HLM to Friedrich Schoenemann, Jan. 25, 1933; *New Letters*, 280. **"Greatly exaggerated, mainly by interested parties":** HLM, interview by H. Allen Smith, May 25, 1932 (HLMC).

"Pale and somewhat pathetic": "An American Troubadour," *Sun/P.M.*, Dec. 27, 1920. **"He is one":** "Where Are We At?", *Sun/P.M.*, July 5, 1934. **"The pathetic mud-turtle":** "A Time to Be Wary," *Sun/P.M.*, Mar. 13, 1933. **"I believe that, taking everything into consideration":** HLM, United Press interview, Nov. 8, 1932 (HLMC).

"He was built for lyrical work": "Mencken Tells How Magic Word 'Beer' Brought the Cheers," *Sun/P.M.*, Oct. 26, 1932; *The Impossible H. L. Mencken*, 327.

The two men had been bickering: Angoff, 225, 223. **"He had decided"**: *RAM*, 20.

"I had a low view of Roosevelt": 35, 228. **"What I here suggest"**: "A Time to Be Wary," *Sun/P.M.*, Mar. 13, 1933; *A Carnival of Buncombe*, 274–75.

"The German news": HLM to Julia Collier Harris, Mar. 24, 1933; *New Letters*, 284. **"I am keenly conscious"**: HLM to Edwin Emerson, May 12, 1933; ibid., 287.

"Professional kikes": 35, 227. **"It was plain from the start"**: Ibid., 226.

"Not many observant Marylanders": "The Eastern Shore Kultur," *Sun/P.M.*, Dec. 7, 1931. **"I emerge cherishing a hope"**: "The Spanish Idea of a Good Time," *Mercury*, Dec. 1932, 507.

"The white man is actually superior": "Advice to Young Men," Mar. 22, 1939; *Minority Report/HLM's Notes/Unpublished Material* (EPFL). **"There was very little anti-Semitism"**: Untitled note (n.d.), Ibid. This note is one of several in a folder marked "Jews" on which someone other than HLM has written, "The nature of most of this material may make it unsuitable for publication."

"The troubles of the Jews": 35/TS, 678n. According to August Mencken, "The thing that broke them up—I know it, my brother told me all about it at the time—according to Angoff, they fell out over Hitler, which is pure nonsense. Hitler had nothing whatever to do with it. Goodman was a gambler. He was a fellow who backed plays, and he made a lot of money and lost a lot of money. At that time, Goodman was broke, but he had a great idea whereby he could make a great deal of money. And he wrote to my brother and asked him to loan him a very considerable sum of money. It wasn't small, it was enough to make you think; and my brother thought. I remember him struggling with that problem, trying to decide what to do. He didn't want to offend Goodman, but on the other hand he knew well that if he loaned him the money, he'd never see it again. And he finally decided that as long as he was going to lose Goodman's friendship, he might as well keep his money. He wrote to him and told him he couldn't afford to loan it to him. That's when Goodman started denouncing him as a Nazi" (*RAM*, 22–23). **"Germany is the appeasing refuge"**: *Mercury*, May 1933, i.

"He let a good deal of wind": "Notes and Comment," *New Yorker*, Oct. 22, 1933 (HLMC). **"THE AMERICAN MERCURY will be seriously concerned"**: HLMC.

"Hard study and continuous application": "Ten Years," *Mercury*, Dec. 1933, 386. **"The disadvantage of the Jew"**: "Hitlerismus," ibid., 509–10. **When Knopf saw the galleys**: Alfred Knopf to HLM, Oct. 24, 1933; Mencken Collection, New York Public Library. **All he was trying to do**: HLM to Alfred Knopf, Oct. 26, 1933; Mencken Collection, New York Public Library.

It survived until 1980: At the end of 1952 the *Mercury* was bought by the first of several publishers who used it to print racist and anti-Semitic propaganda. In 1966 it was acquired by Willis Carto, who turned it into an organ of Holocaust denial. (One of his contributors was Harry Elmer Barnes, the "revisionist" historian of World War I whose article "The Drool Method in History" had appeared in the first issue of the *Mercury*.) Carto also operated the anti-Semitic

Noontide Press, whose publications included a reprint of HLM's *The Philosophy of Friedrich Nietzsche,* a circulating copy of which could be found in the stacks of the Enoch Pratt Free Library in 1995. Similarly some unwitting librarians subscribed to the *Mercury* well into the seventies without realizing the nature of its contents. (I saw a copy at the Naval Academy library in Annapolis, Md., in 1974.) Next to nothing has been written about HLM's posthumous influence on American political extremists, but it is known that Timothy McVeigh "quote[d], from memory, H. L. Mencken," according to Gore Vidal, who corresponded with the Oklahoma City bomber prior to his execution in 2001. See Harry Elmer Barnes, "Zionist Fraud," *Mercury,* Fall 1968; Lipstadt, 144–49; Vincent Fitzpatrick, "The American Mercury," *Menckeniana* 123; and Linton Weeks, "Coming Soon, Epilogue by Gore Vidal," *Washington Post,* May 7, 2001 (HLMC). **"Its angry crusadings":** "Hazlitt for Mencken," *Time,* Oct. 16, 1933 (HLMC).

"Our happiest time together": *Diary,* May 31, 1940. **"On the one hand, it is being planted":** "Erez Israel," *Sun/P.M.,* Apr. 9, 1934; *The Impossible H. L. Mencken,* 650–51. **He was inundated with offers:** *35/TS,* 771–73. **HLM told her he was about to go abroad:** HLM to Katharine White, Jan. 26, 1934; *New Letters,* 304.

"Competent in all qualities": Louis Cheslock, unpublished diary entry, Apr. 7, 1934 (EPFL). **"One is puzzled to note":** "Mr. Mencken and Other Warriors," *New Yorker,* Apr. 7, 1934 (HLMC). **"Moral science":** *Treatise on Right and Wrong,* vii. **"Mr. Mencken gives the impression":** Irwin Edman, "Treatise on Mencken," *Saturday Review of Literature,* Apr. 7, 1934 (HLMC).

"Sara has had a little setback": HLM to Blanche Knopf, Mar. 4, 1935 (carbon in HLM medical file, EPFL). **"She is still showing some temperature":** HLM to Blanche Knopf, Mar. 16, 1935 (carbon in HLM medical file, EPFL). **"Sara's recovery is proceeding with depressing slowness":** HLM to Blanche Knopf, Mar. 30, 1935 (carbon in HLM medical file, EPFL). **"I begin to suspect":** Ibid. **"Alarming symptoms":** *35,* 270. **"Sara has meningitis":** HLM to Max Brödel, May 29, 1935; *New Letters,* 354. **"I did not look upon her in death":** *Diary,* May 30, 1945. **"If only to get away":** *35,* 270.

"Now I feel completely dashed": HLM to Ellery Sedgwick, June 7, 1935; *Letters,* 393. **"The thing that obsessed her":** HLM, "Preface"; Haardt, *Southern Album,* xxii–xxiii.

"I was fifty-five years old": *AN1941.*

CHAPTER EIGHT
THE AGE OF ROOSEVELT II, 1935–1941

BOOKS

Cooke, *Six Men;* Decter, *An Old Wife's Tale;* Hoopes, *Cain;* Kemler, *The Irreverent Mr. Mencken;* Ketchum, *The Borrowed Years;* Lipstadt, *Denying the Holocaust;* Manchester, *Disturber of the Peace.*

NOTES

"The ghosts stalk that dreadful apartment": Hoopes, 273. **"In the weeks after Sara's death"**: HLM to Maud DeWitt Pearl, Nov. 20, 1940; *New Letters*, 469. Maud DeWitt Pearl was the widow of Raymond Pearl, a Johns Hopkins professor, Saturday Night Club member, and occasional *Mercury* contributor who had been among Mencken's very closest personal friends. A geneticist and biostatistician, Pearl published prolifically in professional journals and was also widely known in the thirties for his popular writings on science, but is now forgotten save by specialists in his fields of study.

"His infallibility has dropped from him": "1936," *Sun/p.m.*, Aug. 26, 1935; *A Carnival of Buncombe*, 295. **"He is directly and solely responsible"**: "Three Years of Dr. Roosevelt," *Mercury*, Mar. 1936, 264.

"Roosevelt would try to take the United States into it": 35, 272. **"The uproar being made by the Jews"**: HLM to Benjamin De Casseres, July 25, 1935; *Letters*, 392.

"Even more than a hero": Decter, 8. **The Trans-Lux:** Arno's socialites planned to hiss FDR in one of a chain of now-extinct movie houses that showed only newsreels. **A Roper poll:** Ketchum, 224. **Mencken's half-admiring Anglophobia:** HLM to Benjamin De Casseres, Sept. 7, 1935; *Letters*, 394.

"I am against the violation of civil rights": HLM to Upton Sinclair, May 2, 1936; *Letters*, 402.

"Getting rid of Hitler": "H.L. Mencken Wishes—'Jews Would Stop Alarming Goyim'," *American Hebrew and Jewish Times*, Sept. 20, 1935 (HLMC).

"This is the piece I threatened you with": HLM to Katharine White, Jan. 9, 1936; Bro. James Atwell, F.S.C., "Eclipse and Emergence," *Menckeniana* 24, **"That learning and virtue do not always run together"**: "Ordeal of a Philosopher," *New Yorker*, Apr. 11, 1936; "The Career of a Philosopher," *HD*, 277–78.

"Pretty well exhausted": *Diary*, May 23, 1936. **"The articles razzing the New Deal"**: Edmund Wilson, "Talking United States," *New Republic*, July 15, 1936 (HLMC).

August, his favorite sibling: "He has always been closer to me than either my brother Charlie or my sister. We not only think alike; we also look alike" (*HD/AC*, 95). **He quickly settled into a routine:** *RAM*, 14–17. **"My own marriage was too short"**: *Diary*, Feb. 5, 1942.

"A blank check from and upon the American people": "Coroner's Inquest," *Sun/p.m.*, Nov. 9, 1936; *A Carnival of Buncombe*, 332. **"It will come, inevitably"**: "Democratic Twilight," *Sun/p.m.*, Nov. 16, 1936. **"Young, goatish and full of an innocent delight"**: *ND*, v.

"Complaint: tickling in back of throat": HLM medical file (EPFL). **He claimed to have tried out his vasectomy:** Manchester, 322.

"At the moment, I feel pretty good": HLM to Harry Leon Wilson, Jr., Mar. 8, 1937; *New Letters*, 406. **"My conscience has me down"**: HLM to Katharine White, Sept. 1937; Atwell, "Eclipse and Emergence."

"I gave over six whole columns": 35, 303. **"The contents varied wildly from day to day"**: R.P. Harriss, "Life with Mencken," *Gardens, Houses and*

People, May 1949 (HLMC). **"Something fantastic and altogether delight-ful"**: Philip Wagner, "Mencken Remembered," *American Scholar*, Spring 1963 (HLMC).

"Ranged downward from third-rate": *35*, 305.

"A peculiar sensation": HLM medical file (EPFL). **"The German General Staff"**: *35*, 309. **"The roughing of Jews upset him"**: Ibid., 315–16. **"Germany looked like a church to me"**: "H. L. Mencken Returns from European Trip," *Sun/A.M.*, July 26, 1938.

"It is to be hoped that the poor Jews": "Help for the Jews," *Sun/A.M.*, Nov. 27, 1938; *The Impossible H. L. Mencken*, 635–38.

This little-known *Sun* column: Bode, for example, makes no mention of it.

"Floods of letters of protest and abuse": *35/TS*, 1198–99. **"The calamity of World War II"**: "Pilgrimage," *Heathen Days*, 276. **The Holocaust is conspicu-ously missing:** See HLM's typewritten responses to the questions submitted by Roger Butterfield for his 1946 *Life* "interview" (*Miscellaneous Typescripts, Carbons and Clippings, 1946–48*, EPFL).

"Rough jottings": HLM, covering note for *Minority Report/HLM's Notes/Unpublished Material* (EPFL). **"The movement toward realism and sense"**: This is one of four unedited notes apparently inserted by accident in the typescript of HLM's diary, immediately following the entry for Aug. 30, 1937 (EPFL). **"The Negro is held back"**: "Advice to Young Men," July 10, 1939, Ibid. **He now believed FDR was scheming:** "Warming Up the Bugles," *Sun/A.M.*, Dec. 18, 1938. **"The sharp, unyielding separateness"**: Untitled note (n.d.), *Minority Report/HLM's Notes/ Unpublished Material* (EPFL). **"The chief whooping for what is called racial tolerance"**: Untitled note (n.d.), ibid. **"I hate to have to ask a man"**: Untitled note (n.d.), ibid.

The language he used to criticize Jewish friends: "Goodman and I became friends almost immediately, and remained so until the shattering impact of Hitler made him turn Jewish on me" (*ML*, 208).

"Three years ago, even two years ago": Charles Angoff, "Mencken Twilight: Another Forgotten Man—That Enfant Terrible of Our Era of Nonsense," *North American Review*, Winter 1938–39 (HLMC).

"If the holy saints spare me so long": HLM to St. Clair McKelway, c. September 1940; Atwell, "Eclipse and Emergence." **"I was growing old at last"**: *35*, 319.

"A book to be read twice a year": *Atlantic Monthly*, Mar. 1940 (HLMC). **"In one chapter he confesses"**: "Mr. H. L. Mencken at School: Great Times with the Street Gang," (London) *Times Literary Supplement*, Oct. 12, 1940 (HLMC).

"At the instant I first became aware": "Introduction to the Universe," *HD*, 3–4.

"We have parted forever": *Diary*, May 31, 1940. *Deny yourself! You must deny yourself!*: HLM gives this as "Entbehren sollst; du sollst entbehren." (Goethe's original is *"Entbehren sollst Du! Sollst entbehren!"*)

"Almost incandescent": *35*, 337–38. **He made a joke of it:** Kemler, 282. **As he clumped heavily out of the hall:** "Mr. Mencken was a clumper, very heavy on his feet. Even from the other end of the restaurant, I knew him by the heavy

sound of his walk." Hazel Ashworth, "I Remember . . . Mencken and the Saturday Night Club Meetings," *Sun Magazine*, Mar. 12, 1972 (HLMC). (Mrs. Ashworth was a waitress and hostess at Schellhase's.) **"I was still able to cover a
story":** *35*, 339.

"Supporting Roosevelt II's effort": "The Baltimore Sunpaper," *A Second
Mencken Chrestomathy*, 364. **"I was resolved, I said":** *35*, 345.

"Win or lose, the United States": "Progress of the Great Crusade,"
Sun/A.M., Feb. 2, 1941.

"Yesterday was the first Sunday": HLM to Msgr. J. B. Dudek, Feb. 10, 1941;
New Letters, 477. **His opinions would become unpublishable:** HLM to Cal
Tinney, Feb. 5, 1941; Ibid., 475. **"In my brother's life, the *Sun* came first":**
RAM, 18.

"At large in a wicked seaport": *ND*, ix.

"I went out to Loudon Park Cemetery": *Diary*, Dec. 29, 1941.

CHAPTER NINE
SKEPTIC IN EXILE, 1942–1948

BOOKS

Dorsey, *On Mencken*; Kunkel, *Letters from the Editor*; Penn, *Passage*; Spivak, *The
American Mercury Reader*; Stannard, *Evelyn Waugh*; Waugh, *The Ordeal of Gilbert
Pinfold*; Waugh, *Sword of Honour*.

NOTES

"The war-time order against giving weather information": *Diary*, Mar. 31,
1942. **He kept up with war-related gossip:** *Diary/TS*, Nov. 1, 1943. **"I write to
every soldier":** "Mencken Will Go All Out Buying Beers for Army," Associated
Press dispatch, May 19, 1944 (HLMC). **"To me the war":** *Diary/TS*, Mar. 4, 1945.

"I was born here": *Diary*, Aug. 27, 1942. **He was pleasantly surprised by
how little effect the war had:** Ibid., July 12, 1943. **"Oakies, lintheads, hill-billies
and other anthropoids":** Ibid., Dec. 14, 1945. **No one he knew was killed or
wounded:** Ibid., Nov. 2, 1944. **He was content to stay home with August:** Ibid.,
Dec. 31, 1943.

"I think the war will never end": Bob Considine, "Nazis to Quit in '45, Japs
Soon After, Editors Say," *New York Mirror*, Apr. 26, 1944 (HLMC). **"The nation's
comical, warm-spirited, outstanding village atheist":** " 'Come In, Gents'," *Time*,
Mar. 1, 1943 (HLMC).

"The Germans had animated the myth": Ben Hecht, "Remember Us!," *Mercury*, Feb. 1943; Spivak, 51. **"The American people are now wholly at the
mercy":** *Diary*, Nov. 2, 1944.

He and August had promised each other: *AN1941*. **"Only by leading a
quiet life":** Ibid., June 1, 1942. **"In 1920 I was able to press that revival":** Ibid.

"The Congressman hunting for platitudes": *New Dictionary of Quotations*,
viii. **Well-chosen, well-organized, accurately attributed and dated:** I have long

found *A New Dictionary of Quotations* to be a highly satisfactory reference tool, though a few howlers did manage to slip past the editor's vigilant eye, including his attribution of William Blake's couplet "If the sun and moon should doubt/ They'd immediately go out" to "author unidentified." **"I thought it would be unseemly":** "Book to End Books," *Time*, May 11, 1942 (HLMC). **"The impression soon becomes inescapable":** Morton Dauwen Zabel, "Dictionnaire Philosophique," *Nation*, Oct. 17, 1942 (HLMC).

"The Resurrection men": *ML*, xvii. See also *The American Language* (1962), 341n.

"Time has released all confidences": *35*, 3–4. He did not quite accomplish this goal: Philip Wagner would live long enough to read what HLM wrote about him in *Thirty-Five Years of Newspaper Work*, and to be saddened by it.

"A sort of Mencken Encyclopedia": *Diary*, April 26, 1943. **Patterson tried to talk him into covering the conventions:** Ibid., Mar. 17, 1944.

"At 3 o'clock this morning": *Diary/TS*, May 3, 1943. **"Such things are disquieting":** Ibid., Mar. 20, 1945.

"When I dropped in yesterday": HLM to George Jean Nathan, Apr. 27, 1942; *New Letters*, 496. **A double portrait by Irving Penn:** For a reproduction, see Penn, 27. (This photograph, originally published in *Vogue*, can also be viewed at http://www.photocollect.com/archives/info/twomen.html.) **"The long-ago days when hope hopped high":** *Diary/TS*, Dec. 31, 1945. **"All the congratulations that the Western Union refused to send":** HLM to James M. Cain, July 24, 1944; *New Letters*, 545.

"Extreme neurotics, and given to hysterics": *Diary*, Jan. 21, 1942. **"Perfect and glorious":** "Bach at Bethlehem," *Sun/p.m.*, May 20, 1929; *H. L. Mencken on Music*, 24. **"I am going up to Boston":** Harold Ross to HLM, Apr. 8, 1943; Kunkel, 207.

"Without August I'd be lost indeed": *Diary*, Dec. 31, 1943.

"So long as Roosevelt is in office": Ibid., Apr. 1, 1945. **"His career will greatly engage historians":** Ibid., Apr. 15, 1945.

"Mencken, like Van Wyck Brooks": Edmund Wilson, "The Progress of the American Language," *The New Yorker*, Aug. 25, 1945 (HLMC). **He put aside *My Life*:** *Diary*, Oct. 16, 1945. **"Views on public questions":** Roger Butterfield, "Mr. Mencken Sounds Off," *Life*, Aug. 5, 1946 (HLMC). All quotes from HLM are from his original typewritten answers to questions supplied in advance by Butterfield (*Miscellaneous Typescripts, Carbons and Clippings, 1946–1948*, EPFL). **Chinese highbinders:** Yet again, one need only consult *The American Language* to learn that the word "highbinder" "was first used in 1806 to designate a variety of Irish gangster in New York; it was not applied to Chinese until the late 1870s." *The American Language* (1963), 264.

Life received 313 letters: Beulah Holland, "Report on the August 5 Issue," *Life* interoffice memo, Sept. 6, 1946 (EPFL).

"I awoke this morning": *Diary*, July 30, 1947. **"I fear I am in for it":** *Diary/TS*, July 31, 1947. **"Pile it up without plan":** *Diary*, Aug. 23, 1947.

In February he went to Florida: HLM to Bradford F. Swan, Feb. 18, 1948;

New Letters, 587–88. **"Strange errors in rhythm and time"**: Louis Cheslock, unpublished diary entry, Mar. 27, 1948 (EPFL).

"An appalling business": *Diary,* May 25, 1948.

"I was sitting quietly": "Television Lamps Stir Up 2-Way Use for Beer," *Sun/A.M.,* June 20, 1948; *The Impossible H. L. Mencken,* 382. **"Despite his braggadocio"**: *Diary,* July 16, 1948.

"Such types persist": "Mencken Finds Several Raisins in Paranoiac Confection," *Sun/A.M.,* July 25, 1948; *The Impossible H. L. Mencken,* 391.

"I quit in disgust": HLM to P. E. Cleator, Aug. 3, 1948; *New Letters,* 595.

"Simply to present a selection": *MC,* v–vi. **One carefully arranged volume:** Among the papers HLM presented to the Enoch Pratt Free Library are five boxes of unsorted manuscript material intended for use in a second *Chrestomathy.* In 1963 Betty Adler, then in charge of the Mencken Collection, proposed to Alfred Knopf that he publish a new anthology based partly on HLM's notebooks and partly on the *Second Chrestomathy* material. He rejected the proposal, and the material was shelved in the Mencken Room, where it went unexamined for twenty-nine years. I looked through it in the course of my research for this book and realized that it could be prepared for publication with comparatively little difficulty. (*A Second Mencken Chrestomathy* was published by Knopf in 1994.)

Mencken read the Kinsey Report: "All that humorless document proves is (a) that all men lie when they are asked about their adventures in amour, and (b) that pedagogues are singularly naïve and credulous creatures" (*MC,* 36n).

"The super- or ultra-explosion": "Mencken Says Country Jolly Well Deserves It," *Sun/A.M.,* Nov. 7, 1948; *The Impossible H. L. Mencken,* 392. **"Certainly it is astounding"**: "Mencken Calls Tennis Order Silly, Nefarious," *Sun/P.M.,* Nov. 9, 1948; ibid., 205.

The Loved One impressed him: HLM mentions the book in one of his last *New Yorker* pieces on linguistics, "Scented Words" (*New Yorker,* Apr. 2, 1949). **Baltimore would be Waugh's first stop:** Stannard, 224. **The arrangements were made:** Manchester, "Mencken in Person," Dorsey, 10–11. **"The enemy at last was plain in view"**: *Sword of Honour,* 15.

"His strongest tastes were negative": *The Ordeal of Gilbert Pinfold,* 11–12. **"All his fears and hatreds and prejudices"**: *Minority Report,* 20.

CHAPTER TEN
SCOURGE AT BAY, 1948–1956

DOCUMENTS

In addition to *RAM,* the principal documentary sources for this chapter are August Mencken's *Record of HLM's Health After the Stroke in November, 1948* (EPFL), an unpaginated collection of memos and other related material, cited as *Health*; Louis Cheslock's diary (EPFL), cited as "Cheslock" by date of entry only; and Alfred Knopf's unpublished memoir (EPFL), cited as "Knopf."

BOOKS

Cooke, *Six Men;* Dorsey, *On Mencken;* Goldberg, *The Man Mencken;* Hobson, *Mencken;* Manchester, *Disturber of the Peace.*

NOTES

"As I grow older": *Minority Report,* v. **"Moderation in all things":** Ibid., 239. **"I have never encountered":** *Diary/TS,* July 24, 1944.

HLM's stroke: *RAM,* 79–80, *Health,* and Cheslock, Dec. 6, 1948. Baker later denied having told August that HLM would not live through the night (Hobson, 500). **"Mr. Mencken has suffered a stroke":** Knopf, 38.

Waugh departed Baltimore: *On Mencken,* 11. Waugh made no mention of the cancelled lunch in his diary or contemporary correspondence. **They gathered to play the G Minor Symphony:** Cheslock, Nov. 27, 1948. **"Beyond critical analysis":** Goldberg, 179.

Johns Hopkins announced: Associated Press wire story, Nov. 30, 1948 (HLMC). **"Raising hell to get home":** *RAM,* 85–92. **"A high grade nursing home":** *Health.*

"I was amazed to find Henry": Cheslock, Jan. 7, 1949.

The last one, an expanded version: "The Life and Times of O.K.," *New Yorker,* Oct. 1, 1949. **"A severe stroke":** "Unregenerate Iconoclast," *Time,* July 11, 1949 (HLMC). **The wrong words came out:** See *Health* and Robert Allen Durr, "The Last Days of H. L. Mencken," *Yale Review,* Autumn 1958 (HLMC).

His doctors were briefly afraid: Louis Cheslock noted in his diary that according to one doctor, the only thing that could now be done for HLM was "simply to see that Henry's hope is kept alive in order to prevent his doing anything radical" (Cheslock, Oct. 1, 1949). **He tried to play a Strauss waltz:** Cheslock, June 4, 1949. **A full year went by:** Manchester, 317.

"If I could only read": Durr, "The Last Days." **"The trouble is, I didn't live long enough":** Robert McHugh, Associated Press wire story, June 10, 1955 (HLMC). **He started going to movies regularly:** See Durr, "The Last Days," and Bob Thomas, Associated Press wire story, May 12, 1949 (both HLMC).

His picture ran in the paper: "New Sunpapers Building Cornerstone Is Laid," *Sun/A.M.,* Oct. 26, 1949 (HLMC). **He took part in a discussion:** *Health.* **"If I could only read again!":** Cheslock, June 26, 1951.

"Mr. Mencken sends his best thanks": Mrs. John W. Lohrfinck to Fania Marinoff, Apr. 7, 1950; *Letters,* 505.

The whole thing was a mistake: *RAM,* 28–30, and Knopf, 38A–38B. **"He has been to his people":** Associated Press wire story, May 25, 1950 (HLMC).

"I certainly hope you don't stay your hand": HLM to Edgar Kemler, June 11, 1948; *Menckeniana* 4. **"A skeptic of the first rank":** Kemler, v. **"Since the advent of Hitler":** Ibid., 280–81.

"Mencken found it a very enlightening experience": Ibid., 294.

"An elder now himself": James H. Bready, "Mencken Has A Birthday," *Sun/A.M.,* Sept. 10, 1950 (HLMC). **"Well, I'm ready for the angels":** Manchester, 315.

"Hero-worship of such lush proportions": George Jean Nathan, "Mencken: A Portrait," *New York Times Book Review*, Jan. 14, 1951 (HLMC). "The writing is so Menckenian in style": Charles Angoff, "H. L. Mencken—Joyous Iconoclast," *Saturday Review*, Jan. 1951 (HLMC). "For several sentences he might talk": Manchester, 312, 315.

"Perhaps not since the days": Clifton Fadiman, *New Yorker*, Apr. 22, 1950 (HLMC).

"No chance": Associated Press wire story, Oct. 17, 1950 (HLMC). "His speech more coherent": Cheslock, Dec. 28, 1950. "My nurse called the cops in": Manfred Guttmacher, "Guttmacher/Mencken," *Menckeniana* 72.

"God, why didn't the quacks let me die?": Philip Wagner, "Mencken Remembered," *American Scholar*, Spring 1963 (HLMC). Manchester interviewed him: William Manchester, "Mencken, 73, Mellows; Does Not 'Dislike Ike'," *Sun/A.M.*, Jan. 11, 1954 (HLMC).

"Some time ago I put in a blue afternoon": "The Monthly Feuilleton—IV," *SS*, Dec. 1922; "Joseph Conrad," *MC*, 518. "Something obscene": Manchester, 322. "Children had become dear to him": Ibid., 319, and Cooke, 116–17.

"What's in the papers?": Durr, "The Last Days."

"He alone did the deciding": August Mencken, "Letters to the Editor: H. L. Mencken's 'Minority Report,'" *Gardens, Houses and People*, Sept. 1956 (HLMC).

Among them these three: Cancelled notes 176, 223, and 251, *Minority Report* MS. (EPFL).

"It is a folly to try to beat death": *Minority Report*, 238. (This is note 358 in its entirety.)

"Henry is as eager": Alfred A. Knopf, announcement of the publication of *Minority Report* (EPFL). "It will be nice being denounced again": Robert P. McHugh, "Mencken Eyes Book and Critics," *Sun/A.M.*, Jan. 26, 1956 (HLMC).

He once had the cheek: "The Bryan of Bayreuth," *A Second Mencken Chrestomathy*, 406. The Met was on its best behavior: It was, in fact, a near-perfect cast—Paul Schöffler as Hans Sachs and Lisa della Casa as Eva, with Rudolf Kempe conducting. "Louis, this is the last you'll see me": Cheslock, Jan. 28, 1956.

EPILOGUE
"QUACKS AND SWINDLERS, FOOLS AND KNAVES"

DOCUMENTS
August Mencken's annotated copy of Charles Angoff's *H. L. Mencken: A Portrait from Memory* is on deposit in EPFL.

BOOKS
Angoff, *H. L. Mencken*; Cooke, *Six Men*; Fecher, *Mencken*; Hoopes, *Cain*; Howe, *Holmes-Laski Letters*; Kemler, *The Irreverent Mr. Mencken*; Lipstadt, *Denying the Holocaust*; *Living Philosophies*; Nash, *The Conservative Intellectual Movement*; Pos-

ner, *Public Intellectuals;* Wellek, *A History of Modern Criticism;* Wilson, *The Shores of Light;* Wright, *Black Boy.*

NOTES

"Save in the event": *Diary,* June 2, 1943. HLM wrote and delivered this memo a year after his deteriorating health led him to start putting his private papers in order.

"Henry Louis Mencken, newspaper man": Anne W. Hutchison, "H.L. Mencken, Author, Dies at 75," *Sun/A.M.,* Jan. 30, 1956 (HLMC). **"This is not the place":** Hamilton Owens, "H.L.'s Pungent Pen a Challenge to Orthodox," ibid. **"To many outsiders":** "Henry Louis Mencken," ibid.

"As champion of the unfettered mind": *Boston Globe,* Jan. 30, 1956 (HLMC). **"An arch foe":** *New York Times,* Jan. 30, 1956 (HLMC). **"His was the joyous voice":** New York *Herald Tribune,* Jan. 30, 1956 (HLMC). **"By his own admission":** *Upper Sandusky* (Ohio) *Chief-Union,* Jan. 31, 1956 (HLMC). **"Henry L. Mencken is gone":** *Big Spring* (Tex.) *Herald,* Feb. 1, 1956 (HLMC). **"In his final years":** *Miami Beach Sun,* Feb. 1, 1956 (HLMC).

"Mencken's importance": *The Times* (London), Jan. 30, 1956 (HLMC).

HLM's funeral: This account is based on a story written by Hamilton Owens for the Associated Press, which appeared in client papers across the country on Feb. 1, 1956 (HLMC). **"Free from pious asseverations":** "Clarion Call to Poets," *P/6,* 105.

"Somehow, we were made": Hoopes, 468. **"There is room left":** *Diary,* May 30, 1945.

HLM's will: "Hopkins Hospital, Library Named in Mencken Will," *Sun/P.M.,* Feb. 14, 1956 (HLMC).

Opening of the Mencken Room: "Mencken Room Dedicated in Library Ceremony," *Sun/P.M.,* Apr. 18, 1956 (HLMC). Sales of *Minority Report:* "Royalty Statement," *Menckeniana* 107. **Angoff was waiting for him to die:** According to a typewritten note by August pasted into his copy of Angoff's book, HLM "believed the book would not be published in his life time. Although he had raised Angoff from obscurity and given him his big chance in life he knew the man." **"Angoff's going to write a book about me":** *RAM,* 35.

"Many of the writers": Angoff, 96. **Numerous errors and inconsistencies:** These are discussed in detail in Arthur M. Louis, "Mencken and Angoff," *Menckeniana* 110. **"My notes made at the time":** Angoff, 11. **"Say, Angoff":** Ibid., 95.

"He began to laugh": Ibid., 231–32. Here is August's response to this passage: "There is no truth in this Poe story and if told at all it was merely to shock Angoff. Poe's grave is in full view of a public street."

"The laws of libel": Alfred A. Knopf, "The Second Quarter: 1956," *The Borzoi Quarterly* (HLMC). August Mencken folded a copy of this pamphlet into his annotated copy of Angoff's book. **"Angoff, Hitler is a jackass":** Ibid. 164–65. Angoff died in 1979.

One final piece of news: "Mencken's Unpublished Works Secret to 1981, 1991," *Sun/A.M.,* July 14, 1957 (HLMC). **The manuscripts were duly deposited:** *Diary,* July 15, 1945.

Two paperback anthologies: "Royalty Statement," 13. **HLM's work was not taught:** No piece of writing by HLM was ever assigned in any course I took in high school or college. When I began work on this book in 1991, I was astonished by the large number of my college-educated friends and acquaintances who knew nothing whatsoever about him (as well as several who knew only that he had recently been charged with anti-Semitism).

"Let it be said": *Diary*, xix. **"To be sure, he was not one":** Ibid., xx. **"Mencken's Dark Side":** *Washington Post*, Dec. 9, 1989 (HLMC). **"Portrait of a Bigot":** *New York Newsday*, Dec. 9, 1989 (HLMC). **"Mencken the Horrible":** *Chicago Tribune*, Dec. 17, 1989 (HLMC). **"A Bigot's Notes":** *Kansas City Star*, Jan. 28, 1990 (HLMC). **"Those who defend a writer":** Doris Grumbach, "Mencken: Just Plain Antisemitism," *Washington Post*, Jan. 12, 1990 (HLMC). **"Dimming of a Legend":** *Sun/A.M.*, Dec. 7, 1989 (HLMC). (For an analysis of press reaction to the *Diary*, see Charles A. Fecher, "Firestorm: The Publication of HLM's Diary," *Menckeniana* 113.)

"We wish to express our dismay": *New York Review of Books*, Mar. 15, 1990 (HLMC). **"It is, alas, impossible":** Joseph Epstein, "Cornered by Events," (London) *Times Literary Supplement*, June 9, 1995, 14.

When Richard Posner surfed the Web: Posner, 209–11.

"At the moment": *Diary*, June 1, 1942. **Mencken was never fully accepted on the right:** HLM is mentioned only in passing in the standard history of post-war American conservatism to 1976 (Nash, passim). **"This miserable, mean person":** Susan Baer, "It's All Mencken's Fault," *Sun/A.M.*, Dec. 26, 1994 (HLMC).

"The two greatest intellectual possessions": HLM to Edgar R. Dawson, Dec. 3, 1937; *New Letters*, 417. **"There is only one honest impulse":** *SS*, Apr. 1920; "A Blind Spot," *MC*, 163.

Perhaps even to the point of denial: "I have to say that I don't remember Mencken ever admitting that the gas chambers and the concentration camps existed. That he half believed they did was obvious from the uneasiness with which he brushed aside any mention of Nazi brutality and in a weary grumble equated such rumors with the First World War legend that the Germans had made soap out of the bodies of Belgian and French civilians" (Cooke, 99). It should be noted that Cooke does not indicate when the encounter or encounters described here took place, and that many other Americans who remembered the false atrocity stories of World War I initially shared HLM's apparent skepticism about the extent of the Holocaust. (It should also be noted that HLM's stroke forced him to withdraw from public life in 1948, just three years after most Americans were first exposed to documentary evidence of the Holocaust.) See Lipstadt, 33–34.

"I do not aspire": HLM to Burton Rascoe, c. 1920; *Letters*, 188.

"We have Mencken in a nutshell": Wellek, 10. **"I do not object":** "Footnote on Criticism," *P/3*, 100. **"Q. If you find so much":** "Catechism," *P/5*, 304.

"Wholly good, wholly desirable, wholly true": *AN1941*. **"His moral code":** *Pistols for Two*, 23.

Kemler claimed: Kemler, v. **"In the end":** *Diary*, June 1, 1942. **He was always happy to submit:** *35/TS*, 371. **"The life of shame":** "A Girl from Red Lion, P.A.," *ND*, 233–34.

"What of the American Presidents": Fecher, 11.

"A block away from the library": Wright, 217–18.

"The smell of the city room": Charles Angoff, "Mencken Twilight: Another Forgotten Man—That Enfant Terrible of Our Era of Nonsense," *North American Review*, Winter 1938–39 (HLMC).

"The great comic writer": *Letters*, x.

"I believe": *Living Philosophies*, 193.

"The inescapable last act": "Clarion Call to Poets," *P/6*, 112. **"If, after I depart this vale":** "Epitaph," *MC*, 627. **Nearly every obituary writer remembered it:** See vols. 102 and 103 of his scrapbooks, passim (HLMC).

"My writings, such as they are": *A Second Mencken Chrestomathy*, 491. **"Says Elizabeth":** *AN1941*.

SELECT BIBLIOGRAPHY

BOOKS BY H. L. MENCKEN
(PUBLISHED BY ALFRED A. KNOPF UNLESS OTHERWISE INDICATED)

The American Credo: A Contribution Toward the Interpretation of the National Mind, "by George Jean Nathan and H. L. Mencken." New York, 1920.

The American Language: An Inquiry into the Development of English in the United States. New York, 1919. 2nd ed., "revised and enlarged," 1921. 3rd ed., "revised and enlarged," 1923. 4th ed., "corrected, enlarged and rewritten," 1936. *Supplement I* (to chapters 1–6 of 4th ed.), 1945. *Supplement II* (to chapters 7–13 of 4th ed.), 1948. "One-volume abridged edition" of 4th ed. and supplements, "abridged, with annotations and new material, by Raven I. McDavid, Jr., with the assistance of David W. Maurer," 1963.

A Book of Prefaces. New York, 1917.

A Carnival of Buncombe: Writings on Politics. Malcolm Moos, ed. Baltimore: Johns Hopkins University Press, 1956.

Damn! A Book of Calumny. New York: Philip Goodman, 1918.

The Diary of H. L. Mencken. Charles A. Fecher, ed. New York, 1989.

The Editor, the Bluenose, and the Prostitute: H. L. Mencken's History of the "Hatrack" Censorship Case. Carl Bode, ed. Boulder, Colo.: Roberts Rinehart, 1988.

Europe after 8:15 (written in collaboration with George Jean Nathan and Willard Huntington Wright). New York: Lane, 1914.

George Bernard Shaw: His Plays. Boston: John W. Luce, 1905.

Happy Days, 1880–1892. New York, 1940.

Heathen Days, 1890–1936. New York, 1943.

H. L. Mencken on American Literature. S. T. Joshi, ed. Athens: Ohio University Press, 2002.

H. L. Mencken on Music: A Selection of His Writings on Music Together with an Account of H. L. Mencken's Musical Life and a History of The Saturday Night Club. "Selected by Louis Cheslock." New York, 1961.

H. L. Mencken's Smart Set *Criticism* (2nd ed.). William H. Nolte, ed. Washington: Gateway Editions, 1987.

The Impossible H. L. Mencken: A Selection of His Best Newspaper Stories. Marion Elizabeth Rodgers, ed. New York: Anchor, 1991.

In Defense of Women (revised 1922 edition). Alexandria, Va.: Time-Life Books, 1963.

Making a President: A Footnote to the Saga of Democracy. New York, 1932.

Men Versus the Man: A Correspondence Between Rives La Monte, Socialist, and H. L. Mencken, Individualist. New York: Henry Holt, 1910.

A Mencken Chrestomathy. New York, 1949.

Mencken's Last Campaign: H. L. Mencken on the 1948 Election. Joseph C. Goulden, ed. Washington, D.C.: New Republic Books, 1976.

Minority Report: H. L. Mencken's Notebooks. New York, 1956.

My Life as Author and Editor. Jonathan Yardley, ed. New York, 1993.

A New Dictionary of Quotations on Historical Principles from Ancient and Modern Sources, Selected and Edited by H. L. Mencken. New York, 1942.

Newspaper Days, 1899–1906. New York, 1941.

Notes on Democracy. New York, 1926.

The Philosophy of Friedrich Nietzsche. Boston: John W. Luce, 1907.

Pistols for Two (signed by "Owen Hatteras"; written by Mencken in collaboration with George Jean Nathan). New York, 1917.

Prejudices (in six "series"). New York, 1919, 1920, 1922, 1924, 1926, 1927.

A Second Mencken Chrestomathy. Terry Teachout, ed. New York, 1995.

Thirty-five Years of Newspaper Work. Fred Hobson, Vincent Fitzpatrick, and

Bradford Jacobs, eds. Baltimore: Johns Hopkins University Press, 1994.

Treatise on the Gods. New York, 1930. Second edition, "corrected and rewritten," New York, 1946.

Treatise on Right and Wrong. New York, 1934.

Ventures into Verse. Baltimore: Marshall, Beek & Gordon, 1903.

The Young Mencken: The Best of His Work. Carl Bode, ed. New York: Dial, 1973.

PUBLISHED CORRESPONDENCE OF H. L. MENCKEN

Dreiser-Mencken Letters: The Correspondence of Theodore Dreiser and H. L. Mencken, 1907–1945. Thomas P. Riggio, ed. Two vol. Philadelphia: University of Pennsylvania Press, 1986.

Letters of H. L. Mencken (2nd ed.). Guy J. Forgue, ed. New foreword by Daniel Aaron. Boston: Northeastern University Press, 1981.

Mencken and Sara, a Life in Letters: The Private Correspondence of H. L. Mencken and Sara Haardt. Marion Elizabeth Rodgers, ed. New York: McGraw-Hill, 1987.

The New Mencken Letters. Carl Bode, ed. New York: Dial, 1977.

BOOKS BY OTHER AUTHORS

Allen, Frederick Lewis. *Only Yesterday: An Informal History of the Nineteen-Twenties*. New York: Perennial Classics, 2000.

Angoff, Charles. *H. L. Mencken: A Portrait from Memory*. New York: Thomas Yoseloff, 1956.

Asbell, Bernard. *When F.D.R. Died*. New York: Holt, Rinehart & Winston, 1961.

Beirne, Francis F. *The Amiable Baltimoreans*. New York: E.P. Dutton, 1951.

Bode, Carl. *Mencken*. Carbondale: Southern Illinois University, 1969.

Boyd, Ernest. *H. L. Mencken*. New York: McBride, 1925.

Brayman, Harold. *The President Speaks Off-the-Record: Historic Evenings with America's Leaders, the Press, and Other Men of Power, at Washington's Exclusive Gridiron Club*. Princeton, N.J.: Dow Jones Books, 1976.

Caro, Robert. *The Years of Lyndon Johnson: Means of Ascent.* New York: Alfred A. Knopf, 1990.

Chesterton, G. K. *The Illustrated London News 1926–1928* ("Collected Works, Vol. XXXIV"). Lawrence J. Clipper, ed. San Francisco: Ignatius Press, 1991.

———. *The Illustrated London News 1929–1931* ("Collected Works, Vol. XXXV"). Lawrence J. Clipper, ed. San Francisco: Ignatius Press, 1991.

Conrad, Joseph. *The Portable Conrad.* Revised ed. Morton Dauwen Zabel, ed. New York: Viking, 1969.

Cooke, Alistair. *Six Men.* New York: Alfred A. Knopf, 1977.

De Casseres, Benjamin. *Mencken and Shaw: The Anatomy of America's Voltaire and England's Other John Bull.* New York: Silas Newton, 1930.

Decter, Midge. *An Old Wife's Tale: My Seven Decades in Love and War.* New York: ReganBooks, 2001.

Dolmetsch, Carl R., ed. *The Smart Set: A History and Anthology.* "With an introductory reminiscence by S. N. Behrman." New York: Dial, 1966.

Dorsey, John, ed. *On Mencken.* New York: Alfred A. Knopf, 1980.

Dreiser, Theodore. *American Diaries 1902–1926.* Thomas P. Riggio, ed. Philadelphia: University of Pennsylvania Press, 1983.

———. *A Gallery of Women.* New York: Liveright, 1929.

Epstein, Joseph. *Partial Payments: Essays on Writers and Their Lives.* New York: W.W. Norton, 1989.

———. *Pertinent Players: Essays on the Literary Life.* New York: W.W. Norton, 1993.

Fecher, Charles A. *Mencken: A Study of His Thought.* New York: Alfred A. Knopf, 1978.

Ferguson, Otis. *The Otis Ferguson Reader.* Dorothy Chamberlain and Robert Wilson, eds. Highland Park, N.J.: December, 1982.

Fitzpatrick, Vincent. *H. L. Mencken.* New York: Continuum, 1989.

Garraty, John A. *The New Commonwealth, 1877–1890.* New York: Harper & Row, 1968.

Goldberg, Isaac. *The Man Mencken: A Biographical and Critical Study.* New York: Simon & Schuster, 1925.

Green, Harvey. *The Uncertainty of Everyday Life, 1915–1945*. (The Everyday Life in America Series, Richard Balkin, ed.) New York: HarperCollins, 1992.

Haardt, Sara. *Southern Album*. H. L. Mencken, ed. Garden City, N.Y.: Doubleday, Doran & Co., 1936.

————. *Southern Souvenirs: Selected Stories and Essays of Sara Haardt*. Ann Henley, ed. Tuscaloosa: University of Alabama Press, 1999.

Hobson, Fred. *Mencken: A Life*. New York: Random House, 1994.

Hoopes, Roy. *Cain: The Biography of James M. Cain*. New York: Holt, Rinehart & Winston, 1982.

Howe, Mark DeWolfe, ed. *Holmes-Laski Letters: The Correspondence of Mr. Justice Holmes and Harold J. Laski, 1916–1935*. 2 vol. Cambridge, Mass.: Harvard University, 1953.

Hughes, Robert. *American Visions: The Epic History of Art in America*. New York: Alfred A. Knopf, 1997.

Ickes, Harold L. *The Secret Diary of Harold L. Ickes: The First Thousand Days, 1933–1936*. New York: Simon & Schuster, 1953.

Johnson, Gerald W., Frank R. Kent, H. L. Mencken, and Hamilton Owens. *The Sunpapers of Baltimore*. New York: Alfred A. Knopf, 1937.

Johnson, Paul. *A History of the American People*. New York: HarperCollins, 1997.

Kazin, Alfred. *On Native Grounds: An Interpretation of Modern American Prose Literature*. New York: Reynal & Hitchcock, 1942.

Kemler, Edgar. *The Irreverent Mr. Mencken*. Boston: Little, Brown, 1950.

Ketchum, Richard M. *The Borrowed Years, 1938–1941: America on the Way to War*. New York: Random House, 1989.

Kunkel, Thomas. *Genius in Disguise: Harold Ross of* The New Yorker. New York: Random House, 1995.

Larson, Edward J. *Summer for the Gods: The Scopes Trial and America's Continuing Debate Over Science and Religion*. New York: Basic Books, 1997.

Lehman, David. *Signs of the Times: Deconstructionism and the Fall of Paul de Man*. New York: Poseidon, 1991.

Lippman, Laura. *The Sugar House*. New York: William Morrow, 2000.

Lipstadt, Deborah E. *Denying the Holocaust: The Growing Assault on Truth and Memory*. New York: Free Press, 1993.

Living Philosophies. "By Twenty-Two Representative Modern Thinkers." New York: Simon & Schuster, 1931.

Lynd, Robert, and Helen Lynd. *Middletown: A Study in Modern American Culture*. Rev. ed. New York: Harcourt, Brace & World, 1956.

Manchester, William. *Disturber of the Peace: The Life of H. L. Mencken*. 2nd ed. Amherst: University of Massachusetts Press, 1986.

Maryland: A Guide to the Old Line State. "Compiled by workers of the Writer's Program of the Works Projects Administration in the State of Maryland." New York: Oxford University Press, 1940.

Marquand, John P. *Point of No Return*. Boston: Little, Brown, 1949.

Martin, Edward A. *In Defense of Marion: The Love of Marian Bloom and H. L. Mencken*. Athens: University of Georgia Press, 1996.

Mayfield, Sara. *The Constant Circle: H. L. Mencken and His Friends*. New York: Delacorte Press, 1968.

McPherson, James M. *Battle Cry of Freedom: The Civil War Era*. New York: Oxford University Press, 1988.

Menckeniana: A Schimpflexicon. New York: Alfred A. Knopf, 1928.

Meyer, Karl E. *Pundits, Poets, and Wits: An Omnibus of American Newspaper Columns*. New York: Oxford University, 1990.

Mitgang, Herbert. *Dangerous Dossiers*. 2nd ed. New York: Primus/Donald I. Fine Books, 1996.

Nash, George H. *The Conservative Intellectual Movement in America: Since 1945*. New York: Basic Books, 1976.

Nathan, George Jean. *A George Jean Nathan Reader*. A. L. Lazarus, ed. Rutherford: Fairleigh Dickinson University, 1990.

———. *The World of George Jean Nathan: Essays, Reviews, and Commentary*. Charles S. Angoff, ed. New York: Alfred A. Knopf, 1952.

Nolte, William. *H. L. Mencken, Literary Critic*. Middletown Conn.: Wesleyan University, 1966.

Owens, Hamilton. *Baltimore on the Chesapeake*. New York: Doubleday, Doran & Co., 1941.

Penn, Irving. *Passage: A Work Record*. New York: Alfred A. Knopf, 1991.

Phillips, Cabell. *From the Crash to the Blitz, 1929–1939.* "The New York Times Chronicle of American Life." New York: Macmillan, 1969.

Posner, Richard A. *Law and Literature: A Misunderstood Relation*. Cambridge, Mass.: Harvard University, 1988.

——. *Public Intellectuals: A Study of Decline*. Cambridge, Mass.: Harvard University, 2001.

Rascoe, Burton, and Groff Conklin (editors). *The* Smart Set *Anthology*. New York: Reynal & Hitchcock, 1934.

Ross, Harold. *Letters from the Editor: The New Yorker's Harold Ross*. Thomas Kunkel, ed. New York: Modern Library, 2000.

Schlereth, Thomas J. *Victorian America: Transformations in Everyday Life, 1876–1915*. (Everyday Life in America Series, Richard Balkin, ed.) New York: HarperCollins, 1991.

Simon, David. *Homicide: A Year on the Killing Streets*. Boston: Houghton Mifflin, 1990.

Singleton, M. K. *H. L. Mencken and the* American Mercury *Adventure*. Durham, N.C.: Duke University, 1962.

Sobel, Robert. *Coolidge: An American Enigma*. Washington: D.C.: Regnery, 1998.

Spivak, Lawrence E., and Charles Angoff. *The American Mercury Reader*. Philadelphia: Blakiston Company, 1944.

Staggs, Sam. *All About "All About Eve."* New York: St. Martin's, 2000.

Stannard, Martin. *Evelyn Waugh: The Later Years 1939–1966*. New York: W. W. Norton, 1992.

Stearns, Harold E., ed. *Civilization in the United States: An Inquiry by Thirty Americans*. New York: Harcourt, Brace, 1922.

Stenerson, Douglas, ed. *Critical Essays on H. L. Mencken*. Boston: G. K. Hall & Co., 1987.

Sutherland, Daniel E. *The Expansion of Everyday Life 1860–1876*. (Everyday Life in America Series, Richard Balkin, ed.) New York: Harper & Row, 1989.

Tully, Grace. *F.D.R., My Boss*. New York: Charles Scribner's Sons, 1949.

Twain, Mark. *Collected Tales, Sketches, Speeches & Essays 1852–1890*. New York: Library of America, 1992.

Waugh, Evelyn. *The Ordeal of Gilbert Pinfold*. Boston: Little, Brown, 1957.

———. *Sword of Honour*. Boston: Little, Brown, 1961.

Weisberger, Bernard A. *The American Newspaperman*. (Chicago History of American Civilization, Daniel J. Boorstin, ed.) Chicago: University of Chicago Press, 1961.

Wellek, René. *A History of Modern Criticism, 1750–1950*. Volume Six: American Criticism, 1900–1950. New Haven: Yale University, 1986.

White, E. B. *Poems and Sketches of E. B. White*. New York: Harper & Row, 1981.

Wilson, Edmund. *The Bit Between My Teeth: A Literary Chronicle of 1950–1965*. New York: Farrar, Straus & Giroux, 1965.

———. *The Devils and Canon Barham: Ten Essays on Poets, Novelists and Monsters*. New York: Farrar, Straus & Giroux, 1973.

———. *Patriotic Gore: Studies in the Literature of the American Civil War*. New York: Farrar, Straus & Giroux, 1962.

———. ed. *The Shock of Recognition: The Development of Literature in the United States Recorded by the Men Who Made It*. Garden City, N.Y.: Doubleday, Doran, 1943.

———. *The Shores of Light: A Literary Chronicle of the 1920s and 1930s*. New York: Farrar, Straus Giroux, 1952.

———. *The Twenties: From Notebooks and Diaries of the Period*. Leon Edel, ed. New York: Farrar, Straus & Giroux, 1975.

Winfield, Betty Houchin. *FDR and the News Media*. Urbana: University of Illinois Press, 1990.

Wolfskill, George, and John A. Hudson. *All But the People: Franklin D. Roosevelt and His Critics, 1933–39*. New York: Macmillan, 1969.

Wright, Richard. *Black Boy: A Record of Childhood and Youth*. New York: Harper and Brothers, 1945.

INDEX

Aaron, Daniel, 347
Abhau, Anna Margaret, *see* Mencken, Anna
    Margaret Abhau
Abhau, Carl Heinrich, 23
"Absolution" (Fitzgerald), 204
Addison, Joseph, 168
Ade, George, 56, 117
Adventures of Huckleberry Finn (Twain), 29,
    35, 70, 89, 90, 99–100, 168, 326
Adventures of Tom Sawyer, The (Twain),
    35–36
Advice to Young Men (Mencken's planned
    book), 259
"Aesthete" (Boyd), 192
Age of Innocence, The (Wharton), 115–16
Algonquin Hotel, 141, 169–70, 210–11, 227
Allen, Frederick Lewis, 17, 221, 237–38
All God's Chillun Got Wings (O'Neill), 194,
    195, 197
"American, The: His Ideas of Beauty"
    (Mencken), 146
"American, The: His Language"
    (Mencken), 145–46
"American, The: His New Puritanism"
    (Mencken), 146
Americana (Mencken), 239
American Academy and National Institute
    of Arts and Letters, 321–22
American Civil Liberties Union (ACLU),
    217, 219, 221
American Credo, The (Mencken), 176
American Dialect Society, 148
American language, HLM's interest in, 10,
    35, 58, 101, 118, 145–50

American Language, The (Mencken), 10,
    146–50, 153, 159, 247, 295, 302, 303*n*,
    345
    revisions of, 30, 179, 259, 272, 276,
        281–82, 291, 304, 307
American Mercury, 5, 10, 16, 17, 102, 166,
    185–205, 209–12, 220, 224, 243, 250,
    253–62, 269–71, 274, 277, 290*n*, 341
    "Americana" in, 190, 194–95, 201, 231
    book reviews in, 191, 194, 196, 250, 269,
        270
    censorship and, 183, 225–28, 332
    circulation of, 189, 235, 237, 238, 259
    "Clinical Notes" in, 189–90, 198, 254,
        289
    editorial-prospectus for, 190–91
    failure of, 237–38
    first issue of, 186, 191–93, 196, 207
    Hazlitt's takeover of, 264, 270, 271–72
    HLM's desire to withdraw from, 259,
        264, 269–71
    Knopf's sale of, 270, 272
    Nathan forced out of, 195–200, 203, 209,
        254–55, 256
    origins of, 179–82
    overediting of, 204–5, 257–58
    payment policy of, 203
    Smart Set compared with, 189–91, 195,
        203, 257
American Mercury Reader, The (Angoff and
    Spivak, eds.), 300–301
American Spectator, 200
American Speech, 148, 303, 319
American Tragedy, An (Dreiser), 91, 93

Anderson, Sherwood, 10, 161–62, 204, 261
Angoff, Charles, xi, 85, 210–11, 225, 240,
  246, 249, 256, 260, 261, 272, 290–91,
  300–301, 302, 324, 335–36, 347
Antichrist, The (Nietzsche), 10, 176
Anti-Evolution Law, 212–13
anti-Semitism, 14, 137, 165, 211–12, 249,
  286–90, 296–97, 323, 335, 337–40
  Nazis and, 264, 266–70, 277–80, 286–88,
  290
Armstrong, Louis, 29, 238
Arno, Peter, 278
art, modern, 201, 238
Asbury, Herbert, 225–28
"As H.L.M. Sees It" column, 209–10
"Asses' Carnival, The" (Mencken), 201*n*
Associated Press, 24, 183, 255, 317, 325,
  349
Atlantic Monthly, 57, 107, 109, 124, 191,
  197, 243, 245, 291
atomic bomb, 307, 308
"At the Edge of the Spanish Main"
  (Mencken), 55
Aug. Mencken & Bro., 25, 38, 40, 41, 49,
  50, 223
Austen, Jane, 349
Austin, Chick, 238
Authors' League of America, 129, 130

Babbitt, Irving, 126, 238–39, 258, 266,
  348
Babbitt (Lewis), 178, 188, 255
"Babes in the Woods" (Fitzgerald), 114
Bach, Johann Sebastian, 131–32, 171
Badger, Richard G., 60
Baker, Benjamin, 3, 272, 317
Baker, Harry T., 205
Balanchine, George, 238
Balch, Jean Allen, 283–84
Ball, Emma, 310, 311, 335
Baltimore, Md., 19–54, 57–59, 63–70,
  73–80
  change in, 172–73
  City Hall in, 54, 56, 57
  crime in, 20, 48, 52
  great fire of 1904 in, 66–67, 78
  HLM's affection for, 67, 87, 99
  library in, 11, 20–21, 34, 36, 62, 73, 81,
    158, 251, 301, 303, 334, 335, 337–40
  as "Mobtown," 22, 23
  restaurants and saloons in, 20, 170,
    185–86, 207, 307, 334
  Union Square in, 19, 20, 21, 31–32, 171*n*,
    224, 299, 326
  violence in, 20, 22, 23, 52
Baltimore *American*, 42

"Baltimore and the Rest of the World"
  column, 55, 58–59
Baltimore *Evening Herald*, 68, 69, 74
Baltimore *Evening News*, 74, 78, 99
Baltimore *Evening Sun*, 3, 10, 106, 127,
  162–68, 217–18, 250, 293–94
  closing of, 77
  "The Free Lance" in, 100–101, 106, 108,
    121, 122–23, 125, 130, 131, 132
  "H.L.M." column in, 99–100, 102
  Monday Articles in, 1–2, 4, 8, 20, 163–67,
    172, 176, 178, 179, 196, 201*n*, 209,
    219–20, 233, 240, 263, 265, 267, 276,
    280, 284, 285, 312
Baltimore *Herald*, 291
  closing of, 74, 79
  see also Baltimore *Evening Herald*;
    Baltimore *Morning Herald*
Baltimore *Morning Herald*, 46–59, 61,
  73–74, 124
  HLM as city editor of, 63–68
  HLM as dramatic editor and critic of, 57,
    62–63, 68, 70
  HLM as managing editor of, 68–70
  HLM's City Hall beat at, 54, 56, 57
  HLM's editorial-page columns for,
    55–56, 58–59, 61
Baltimore *Morning Sun*, 287
Baltimore on the Chesapeake (Owens), 21
Baltimore Polytechnic School, 37–38, 40,
  47, 48
"Baltimore's New and Modern Dog
  Shelter" (Mencken), 50
Baltimore *Sun* (*Sunpapers*), xiii, 3, 29, 77–80,
  196, 206, 224, 282–88, 299, 303–4,
  310, 313, 326
  Bloom sisters' visit to, 119–20
  Grasty's reform of, 99
  Herald compared with, 57, 63
  HLM condemned by, 339–40
  HLM interviewed by, 330
  HLM's departures from, 131, 293–95,
    298, 311
  HLM's health problems and, 317, 323–24
  HLM's obituary in, 331–32
  HLM's start at, 74
  housecleaning of, 176–77
  as "paper of record," 78
  of today vs. HLM's day, 77
  see also Baltimore *Evening Sun*; Baltimore
    Morning Sun
Baltimore *Sunday Herald*, 63, 74
Baltimore *Weekly Herald*, 74
Bankhead, Tallulah, 206
"Barbarous Bradley, The" (Mencken), 110
Barnes, Harry Elmer, 192

Beerbohm, Max, 107, 109
Beethoven, Ludwig van, 14, 46, 132, 170, 171–72, 250, 312, 345
Behrman, S. N., 84
Bellow, Saul, 91
"Bend in the Tube, The" (Mencken), 64–65, 83
Benton, Thomas Hart, 251
Bernhardt, Sarah, 85
Bible, 168, 212, 213, 219, 222, 247
Bible Belt, 1, 195, 215, 333, 346
Bierce, Ambrose, 56, 117
Bismarck, Otto von, 27, 124
Black Boy (Wright), 346–47
Black Mask, 122, 175–76
blacks, 13, 16, 172–73, 203–4, 289, 313–14, 339, 340
Bloom, Adam, 119
Bloom, Estelle, *see* Kubitz, Estelle Bloom
Bloom, Marion, xi, 118–22, 138–43, 154–59, 205, 208, 224, 234, 236, 241, 242, 253, 254, 274
  marriage of, 158, 185, 206
  in World War I, 140–42, 143, 154
Blue Sphere, The (Dreiser), 113
"Boarding-House, A" (Joyce), 114
Bookman, 54
Book of Prefaces, A (Mencken), 94, 133–38, 146, 158*n*, 159, 160, 312, 346–47
Book-of-the-Month Club, 147, 291
Boston Evening Transcript, 161
Boston Globe, 332
Boston Transcript, 72
Boswell, James, 168
Bourne, Randolph, 136
Boyd, Ernest, 163, 192, 203, 212, 232, 255–56
Boyer, Norman, 103
Brahms, Johannes, 14, 28–29, 132, 250
Braley, Berton, 180–81
Branin, Joseph, 248
Brisbane, Arthur, 8
British English, 145
Brödel, Max, 169, 181, 273, 299
Brooks, Van Wyck, 189, 307
Brothers Karamazov, The (Dostoevsky), 29, 115
Broun, Heywood, 102
"Bryan" (Mencken), 220–21
Bryan, William Jennings (the Great Commoner), 58, 75, 80, 213–22
Buckley, William F., Jr., 341
Buck v. *Bell*, 214
Burgess, John, 191
Burley, Elsie, 310
Butler, Nicholas Murray, 81
Butterick Publications, 92, 96, 102

Cabell, James Branch, 84, 114, 117, 192
Cain, James M., 185–86, 194, 203, 232, 257, 258, 276, 305, 334
Callahan, Joe, 169
Capote, Truman, 313
"Carnival of Buncombe, A" (Mencken), 163
Carter, Robert I., 57
Cash, W. J., 204
Cather, Willa, 10, 114, 117, 152, 182, 194, 260
censorship, 127–33, 143, 145, 151, 165
  American Mercury and, 183, 225–28, 332
Chamberlain, Houston Stewart, 339
Chambers, Whittaker, 313
Chandler, Raymond, 122
Chase, Rev. J. Frank, 225–27
Cheslock, Louis, 170, 309, 318–19, 320, 325, 334
Chesterton, G. K., 125, 163–64, 193
Chicago Record-Herald, 62, 72
Chicago Tribune, 73, 161, 209, 237, 323, 339
Christianity, 82, 124, 194
Christian Scientists, 58, 155–56, 241, 263
Churchill, Winston, 9, 29, 314, 337
Circular Staircase, The (Rinehart), 89
Civic Biology, A (Hunter), 214
Civilization in the United States (Stearns, ed.), 188–89
civil rights, 16, 313–14
Civil War, U.S., 22–23, 27
Clark, Emily, 206
Clemens, Samuel, *see* Twain, Mark
Cleveland Leader, 61
Cody, Buffalo Bill, 28
Communists, communism, 261, 262, 266, 269, 346
Congress, U.S., 6, 98, 264–65, 279, 297
  see also House of Representatives, U.S.; Senate, U.S.
Conrad, Joseph, 10, 17, 90–91, 104, 107, 108, 109, 127, 134, 136, 183, 266, 326
conservatism, 258, 341–43
Considine, Bob, 300
Cooke, Alistair, 326*n*, 335, 338
"Cook's Victory, The" (Mencken), 54*n*
Coolidge, Calvin, 29, 143, 166–67, 187, 188, 228, 231, 249, 250, 257, 338, 346
Cosmopolitan, 42
Coughlin, Charles E., 268, 323
Coward, Noël, 86, 200
Cowell, Henry, 260
Cowley, Malcolm, 115, 312
Cozzens, James Gould, vii, 313
Crane, Stephen, 191
"Critics and Their Ways" (Mencken), 134
Crowe, Eugene F., 108, 122

Daily Iowan, 193–94
Damn! A Book of Calumny (Mencken), 150
Dana, Charles, 51
Darrow, Clarence, 213, 217, 219, 221, 222, 251
Darwin, Charles, 29, 39, 213, 214
Davis, Richard Harding, 40–41, 42
Day, Clarence, 281
Dearborn *Independent*, 165
"Dear Life" (Sara Haardt Mencken), 243–44
Death in the Afternoon (Hemingway), 267n
De Casseres, Benjamin, 277, 278n
de Man, Paul, 14
democracy, 124–27, 133, 149, 153, 194, 196, 216, 220–21, 222, 228–31, 341–42
Democrats, 68, 144, 163, 196, 262–63, 293, 310–11, 332–33
Denby, Hester, 310
Depression, Great, 2, 10, 182, 238, 249–50, 256, 260, 261–62, 264–65
De Voto, Bernard, 227
Dewey, Thomas, 311, 313
"Diamond as Big as the Ritz, The" (Fitzgerald), 179
Diary of H. L. Mencken, The, xii, 4, 9, 16–17, 118, 171n, 243, 253, 259, 273, 282n, 284, 288, 293, 297, 298, 306, 309, 310, 317, 334, 345
    prejudice in, 13, 137, 287, 337–40
Dickens, Charles, 36, 193, 202
Dietz, Howard, 151
Disturber of the Peace (Manchester), 324, 325
Doll's House, A (Ibsen), 97
Doubleday, 107–8, 251, 274
Dreiser, Jug (Mrs. Theodore), 105–6
Dreiser, Theodore, xi, 10, 62, 91–97, 102–6, 115, 123, 137, 139, 261, 348
    American Mercury and, 181, 182, 189, 192, 200, 204
    censorship and, 127–30, 133, 226
    death of, 305
    HLM's friendship with, 93, 95–96, 105–6, 200, 223–24, 274
    HLM's views on, 75, 92–97, 103–6, 134, 135–36, 345
    Smart Set and, 109, 110, 113–14, 117, 181, 183, 200
Dreiser Protest, 129–30
"Drool Method in History, The" (Barnes), 192
Dudek, Monsignor J. B., 240, 294–95
Duncan, Isadora, 201
Du Pont, Marcella, 284
Durr, Robert Allen, 326–27

Eastman, Max, 204
Eddy, Mary Baker, 58, 106, 156
Edison, Thomas, 28
Editor & Publisher, 67, 73–74
Edman, Irwin, 272
Eighteenth Amendment, 153, 214
Einstein, Albert, 76, 348
Eisenhower, Dwight, 298n, 325
elections, 22, 68, 75, 163, 165
    of 1932, 2, 7, 261, 262–64
    of 1936, 282–83
    of 1948, 10, 310–11, 313
Eliot, T. S., 86, 238, 260, 328
Ellicott City Times, 36
Elmer Gantry (Lewis), 235
Emerson, Ralph Waldo, 29, 160, 168
Episcopalianism, 247, 252, 268
Epstein, Joseph, 340
Espionage Act, 144, 145, 153
Esquire, 269
eugenics, 214–15
Europe after 8:15 (Mencken and Nathan), 107
Everybody's Magazine, 106
evolution, theory of, 212–19

Fables in Slang (Ade), 56
Fadiman, Clifton, 324
Farewell to Arms, A (Hemingway), 250
Farrell, James T., 338
Faulkner, William, 10
Faust (Goethe), 293
Fecher, Charles, 339, 346
Federal Bureau of Investigation (FBI), 132
Federal Reserve Board, 249–50
Ferguson, Otis, 12
Field, Eugene, 180
film, 85, 232–35, 320
Financier, The (Dreiser), 127–28, 135
Fitzgerald, F. Scott, 10, 11, 86, 114, 152, 161, 162, 173, 179, 182, 206, 260
    American Mercury and, 185, 204
Fitzgerald, Zelda Sayre, 206, 260
F. Knapp's Institute, 32–33, 36, 37, 73, 123
Flanner, Janet, 204
Flaubert, Gustave, 239
"Flight of the Victor, The" (Mencken), 57
Ford, Ford Madox, 107, 108
Ford, Henry, 76, 165, 174
Foster, William Z., 261
France, 141, 154, 173, 250, 286, 308
Frank Leslie's Popular Monthly, 57–58, 61, 67
Fundamentals, The, 213

Gallegher and Other Stories (Davis), 41
Gallery of Women, A (Dreiser), 119

Galsworthy, John, 108
Gastonia (N.C.) *Gazette*, 325
Gegner, Eva, *see* Mencken, Eva Gegner
General Electric Company, 173
"Genius," The (Dreiser), 128–30, 133, 135, 136
George Bernard Shaw (Mencken), 70–73, 81
German immigrants, 22–25, 33, 123
German Tourist Information Service, 269
Germany, 27, 123–27, 130–33, 135, 136, 138, 194, 328, 337
  HLM's family background in, 9, 24–25
  HLM's visits to, 123, 126, 130–31, 179, 286
Germany, Nazi, 9, 264, 266–70, 277–80, 286–88, 290, 293, 297, 301, 314, 322–23, 342–43
Gibbs, Wolcott, 204
Gingrich, Newt, 341–42
Ginsberg, Allen, 341
Gist of Nietzsche, The (Mencken, ed.), 97
Goethe, Johann Wolfgang von, 33, 293
Goldberg, Isaac, 24, 44, 67, 162, 192, 199, 212, 322
"Good, the Bad and the Best Sellers, The" (Mencken), 88–89
Goodman, Philip, 150–51, 189, 244, 253, 269, 290
Goucher College, 205, 206
government, HLM's views on, 2, 9–10, 97–98, 166, 259–60, 278
Grasty, Charles, 74, 99, 100
Great Britain, 131, 278*n*, 308, 314
  HLM's dislike of, 123, 127, 149, 277, 287, 295*n*
  HLM's visits to, 179, 250, 274
Great Gatsby, The (Fitzgerald), 86
Gridiron Club dinner (Dec. 8, 1934), 1–10, 273

Haardt, Sara Powell, *see* Mencken, Sara Powell Haardt
Hamburger, Samuel, 169
Hammett, Dashiell, 122
hangings, 52, 75, 164–65, 282
Happy Days (Mencken), 12, 21, 30–35, 37, 38, 46, 284, 291–92, 295, 298, 312, 345
Harding, Warren G., 174, 346
Harper's, 191
Harris, Frank, 107, 324
Hart, Moss, 169
"Hatrack" (Asbury), 225–29, 233, 332
Hawks, Arthur, 47
Hays, Arthur Garfield, 219, 226–27
Hazlitt, Henry, 264, 270, 271–72, 300
Heart of Darkness (Conrad), 90*n*

Heathen Days (Mencken), 170, 288, 303, 312, 345
Hecht, Ben, 301
Heliogabalus, a Buffoonery in Three Acts (Mencken and Nathan), 176
"Help for the Jews" (Mencken), 287–88
Hemingway, Ernest, 11, 91, 238, 250, 267*n*
Herald Publishing Company, 74
Hergesheimer, Joseph, xiii
Hicks, Granville, 245
Hildebrandt, Albert, 169
"Hills of Zion, The" (Mencken), 218–19
Hiss, Alger, 313
History of Modern Criticism, A (Wellek), 116
Hitler, Adolf, 9, 29, 211, 264, 266–70, 277–80, 286, 290, 293, 301, 314, 322–23, 337, 339
H. L. Mencken (Angoff), xi, 335–36
"H. L. Mencken Meets a Poet in the West Side Y.M.C.A." (White), 12
Hollywood, 232–35
Holmes, Oliver Wendell, 214, 215*n*, 216, 235
Holocaust, 267, 288, 308, 342–43
Holroyd, Michael, 72
Holy Rollers, 58, 218–19
Homo Sapiens (Mencken's planned treatise), 259
homosexuals, HLM's dislike of, 199
Hood, Gretchen, 241, 253
Hoover, Herbert, 8, 187, 188, 240, 249, 263, 265, 346
Hoover, J. Edgar, 132
Hopkins, Johns, 23
House of Representatives, U.S., 4, 201, 279
Howells, William Dean, 89, 117, 130, 160
Hughes, Langston, 204
Huneker, James, 117, 134, 192
Hunter, George William, 214
Hussey, L. M., 193
Huxley, Aldous, 114, 341
Huxley, Thomas Henry, 39

Ibsen, Henrik, 28, 97
Ickes, Harold L., 8
"Idyl" (Mencken), 39
I'll Take My Stand (manifesto), 258
Illustrated London News, 163–64
"I'm a Stranger Here Myself" (Lewis), 177
In Defense of Women (Mencken), 118, 133, 140–42, 150, 151, 157, 179, 242, 252, 259
individualism, 82, 97
Inherit the Wind (Lawrence and Lee), 221, 222

"In Memoriam: W.J.B." (Mencken), 220–21, 222, 230, 332
"Innocence in a Wicked World" (Mencken), 284
Interracial Conference of Baltimore, 173
Irish immigrants, 22
Irreverent Mr. Mencken, The (Kemler), 322–23, 324
isolationism, 9, 213, 322–23, 342
Italy, 297, 308, 323

Jamaica, 54–55
James, Henry, xiii, 29, 69–70, 71, 89, 188, 347
James, William, 281
Japan, 297, 308
jazz, 174, 201
Jazz Age, 174, 187
Jeffersonian democracy, 125, 126
Jennie Gerhardt (Dreiser), 102–6, 117, 127, 135
Jews, 13, 16, 87, 88, 92, 137, 151, 165, 211–12, 224, 247–49, 271*n*, 308, 328, 340
  Hollywood and, 233, 235
  Nazis and, 264, 266–70, 277–80, 286–88, 290, 301
  see also anti-Semitism
Johns Hopkins Hospital, 283, 284, 286, 291, 301, 316, 317, 318, 325, 335
Johnson, James Weldon, 16, 181
Johnson, Samuel, 149, 302, 344
Josey, Charles Conant, 191
"Journalism in America" (Mencken), 228
Joyce, James, 10, 114
Justice Department, U.S., 132

Kahn, Max, 208
Kahn, Otto, 268
Kansas City Star, 339
Kaufman, George S., 169
Kazin, Alfred, 161
"Keeping the Puritans Pure," 225
Kelly, Gene, 221
Kemler, Edgar, 322–23, 345
Kennicott, Carol, 11
Kent, Frank, 334
Kinsey, Alfred, 313
Kipling, Rudyard, 38, 42, 54, 55, 61, 62, 168
Kirby, Rollin, 201
Knapp, Friedrich, 32–33
Knopf, Alfred, 10, 24, 245, 260, 263–64, 269, 273, 284, 291, 295, 302, 309, 312, 317, 334, 339
  American Language and, 146, 148, 150, 159, 281, 307

American Mercury and, 179–80, 182, 189, 195, 197, 198, 199, 207, 228, 255, 259, 264, 269, 270, 272, 304
  Book of Prefaces and, 137–38
  In Defense of Women and, 157
  HLM's closeness to, 305–6
  as Jewish, 16, 137, 151, 211
  Menckeniana and, 11, 237
  Minority Report and, 312, 329–30
  Prejudices series and, 159, 161
Knopf, Blanche, 250, 255, 260, 273, 305
Knopf, Samuel, 259
Know-Nothing riots, 22, 23
Knoxville Sentinel, 194
Koppel, Holger A., 97
Koussevitzky, Serge, 260
Kraus, Karl, vii
Kreisler, Fritz, 250
Kristallnacht, 287
Krock, Arthur, 257
Krutch, Joseph Wood, 12, 204
Kubitz, Estelle Bloom, 119–22, 138, 139, 155–58, 236
Ku Klux Klan, 153, 228, 261, 266, 267

"Labor Leader, The" (Cain), 194
Ladies' Home Journal, 39
Landon, Alf, 283
Lardner, Ring, 189
Lawrence, D. H., 107
Lawrence, Jerome, 221
Lee, Robert E., 221
Leighton, Clare, 284
Lewis, Sinclair, 10, 11, 84, 117, 177–78, 202, 235, 255
Liberace, 326
Liberal Imagination, The (Trilling), 324
liberals, 125–26, 215*n*, 281, 286
libertarianism, 126, 256, 342
Library of Congress, 73
Life, 61, 288, 308–9, 314
Lincoln, Abraham, 22, 27, 191, 192
"Lincoln Legend, The" (Pennypacker), 192
Lindbergh, Charles A., 17, 174, 296–97, 323
Lippmann, Walter, 9, 12, 102, 231, 235, 238, 263
"Little Cloud, A" (Joyce), 114
Little Eyolf (Ibsen), 97
"Little White Girl" (Sara Haardt Mencken), 243
Lochner v. *New York,* 215*n*
Lohrfinck, Rosalind, 240, 260, 317, 321, 322, 327, 334, 337
London, Jack, 84, 117, 160

London Naval Conference, 250
Longfellow, Henry Wadsworth, 29, 202
Lord Jim (Conrad), 90, 91, 104, 168
Los Angeles Examiner, 107
Los Angeles Times, 106
Lost Generation, 161
Loudon Park Cemetery, 274, 297, 334
Loved One, The (Waugh), 314–15
Lustgarden, Edith, 251
lynching, 16, 172, 267n

MacArthur, Douglas, 4, 325
McClellan, Harriet, *see* Mencken, Harriet
McClellan
McCormick, Col. Robert, 323
McDavid, Raven I., Jr., 148
McKinley, William, 76
McPherson, Aimee Semple, 233
Mahler, Gustav, 260
"Mailed Fist and Its Prophet, The"
(Mencken), 123–24, 133, 230
Mailer, Norman, 313, 340
Main Street (Lewis), 11, 177–78, 189, 255
"Main Street's Been Paved" (Lewis), 202
Making a President (Mencken), 263–64
Making of a Lady, The (Sara Haardt
Mencken), 251
Malinowski, Bronislaw, 247
Malone, Dudley Field, 217
Manchester, William, 314, 317, 324,
325–26
Man Mencken, The (Goldberg), 32, 212
Mann, Col. William D'Alton, 84, 106
Mann, Thomas, 305–6, 341
Maple Heights Sanitorium, 208, 209
Marconi's, 20, 185–86, 207, 334
Maritzer, Lou, 158, 206
Marquand, John P., 188, 321n
Marshall, Beek & Gordon, 60
Maryland, in Civil War, 22–23
Maryland, University of, 284
"Maryland! My Maryland!," 23
Maugham, Somerset, 114, 179, 182
Mayfield, Sara, xi, 205, 206, 207, 222, 234,
241–42
Meekins, Lynn Roby, 63, 66, 68, 70–71
"Mellow Mencken" (editorial), 261
Melville, Herman, 29
Mencke, Johann Burkhard, 24
Mencken (Bode), xi
Mencken, Anna Margaret Abhau, 26–28,
34, 38, 39, 41, 43, 44, 121, 167, 169,
299
death of, 67, 222–24, 274, 297
family background of, 9, 24
HLM's devotion to, 67–68, 79

Mencken, August (HLM's brother), 32,
167, 208–9, 260, 273–74, 276, 282,
295, 301, 305, 306, 310, 328
HLM's death and, 333–35
HLM's health problems and, 316–17,
318, 320, 321
Mencken, August (HLM's father), 9, 19, 21,
24–28, 38, 39, 75, 98, 299
death of, 43–44, 79, 179, 223, 274, 297
as drinker, 25, 30
as father, 33, 34, 36, 37, 40, 41, 123
Mencken, Burkhardt L., 24–25
Mencken, Charles (HLM's brother), 31, 32,
43, 167, 305, 333
Mencken, Charles (HLM's uncle), 25
Mencken, Eva Gegner, 24
Mencken, Gertrude, 32, 167, 260, 276, 282,
305, 317, 333
Mencken, Harriet McClellan, 24
Mencken, H. L.:
adolescence of, 33, 37–42, 44, 45, 46, 61,
76, 87
anti–New Deal views of, 2, 4, 6, 276–81,
285, 341
appearance of, 4–5, 46, 59, 83, 235–36,
323–24
birth and early years of, 27, 30–37, 98,
280–81, 284
as book reviewer, 83, 88–91, 96, 97, 99,
101–6, 109, 135–38, 140, 159, 178,
179, 182–83, 191, 196, 250, 269, 270
death of, 12, 19, 325–38
diary of, *see Diary of H. L. Mencken, The*
as drama critic, 57, 62–63, 68, 70, 79, 80
as drinker, 66, 79, 106, 254, 260, 325
early writing of, 39, 41–43, 48–50
as editor of *Smart Set,* 10, 84, 108–18,
121–22, 143, 145, 151–52, 175,
179–83, 195
education of, 18, 26, 32–33, 37–38, 40,
42–43, 47, 48, 73, 75
European travels of, 123, 126, 130–31,
179, 250, 271, 274, 286
fame and notoriety of, 5, 73–74, 84, 96,
101–2, 107, 186, 233, 255, 321–24,
325
family background of, 9, 18, 24–27
German studies of, 33, 73, 81
at Gridiron Club dinner, 1–10, 273
health problems and hypochondria of, 9,
43, 44–45, 54–55, 122–23, 155–56,
178–79, 254, 259, 283, 284, 286, 291,
299, 301, 304–5, 309, 310, 316–21, 325
humor of, 15, 17, 42, 93, 307, 325
income of, 50, 54, 57, 67, 73, 74, 79, 83,
107, 108, 122, 162, 180, 262, 299, 307

Mencken, H. L. (*cont.*)
  as literary critic, 10, 13, 70–73, 115–18,
    343–44
  love interests of, 118–22, 138–43, 154–59,
    185, 205–9, 232–35, 237, 240–44,
    283–84
  as mama's boy, 67–68, 79
  marriage of, 19, 20, 67, 118, 242–44,
    250–54, 256, 282, 332
  memoirs of, 16; *see also Happy Days*;
    Heathen Days; *Newspaper Days*
  as music critic, 14, 46–47, 49–50
  novel attempted by, 56–57, 110
  obituaries of, 183, 269, 331–32, 333
  old age of, 78, 301–30
  paradox in life and work of, 13–18
  philistinism of, 75, 201, 202–3
  photography of, 39
  as pianist, 14, 33, 41, 169, 170–71, 319
  as police reporter, 52–54
  power and influence of, 10, 235, 341–49
  prejudices of, 13, 16, 137, 211–12, 249,
    268–70, 287–90, 296–97, 313, 323,
    335, 337–40
  printing press of, 34, 36–37
  pro-German views of, 9, 24, 123–27,
    130–32, 135, 136, 194, 322–23, 342–43
  prose style of, 12, 50, 70, 71, 79–80,
    92–93, 100, 117–18, 205, 347–48
  radicalism of, 1, 127, 238, 254
  reading of, 32, 34–36, 38, 39, 42, 51, 76,
    168
  as reviewer of ideas, 13, 116, 179
  revival of, 301–2, 337–41
  self-assessment of, 236, 345
  self-confidence and sense of superiority
    of, 16, 58, 69–70, 75, 124, 313, 348
  semiretirement of, 24, 131–33
  short story writing of, 42, 54, 56, 57,
    64–65, 73
  skepticism of, 14–15, 17, 53, 64, 213, 245,
    302
  social Darwinism of, 65, 97, 116
  in tobacco business, 40, 41, 44, 49, 50
  as translator, 10, 33, 97, 176
  verse of, 38, 39, 41–43, 54, 55–56, 59–63,
    75
  as Victorian, 17, 118, 125, 157, 208, 244,
    344
  will of, 334–35
  see also specific people and works
Mencken, Sara Powell Haardt, 20, 205–10,
  250–55, 259, 260, 264, 271–76, 282
  background of, 206, 208
  death of, 8, 17, 273–76, 283, 293, 297,
    301, 334

  health problems of, 3, 208, 209, 240–44,
    272–73
  HLM's correspondence with, xi, 206–9,
    217, 235, 250–51
  in Hollywood, 234–35
  marriage of, 242–44, 250–54, 256, 332
  writing of, 205, 206, 243, 244, 251, 274
Mencken, Virginia, 334
Mencken Chrestomathy, A (Mencken), 156,
  166–67, 219, 221, 228, 309, 311–12,
  319, 321, 329, 348
Menckeniana (journal), xi
Menckeniana: A Schimpflexicon, 11, 13, 237,
  340
Men Versus the Man (Mencken), 97
Methodism, 225–26, 247, 249
Metropolitan Opera Company, 268, 336
Mill, John Stuart, 126
Millay, Edna St. Vincent, 109
"Mind of the South, The" (Cash), 204
Minority Report (Mencken), 171, 312,
  327–30, 335, 338
"Miss Thompson" (Maugham), 179
modernism, 76, 200–201, 238
Moffett, W. Edwin, 170, 334
Moneychangers, The (Sinclair), 88
Monthly Story Magazine, 73
More, Paul Elmer, 258
Mozart, Wolfgang Amadeus, 171, 250, 318
"Mr. Mencken Sounds Off" (Mencken), 309
music, 131–32, 238, 313
  HLM's interest in, 14, 33–34, 46–47,
    49–50, 63, 169–72, 260, 306, 319,
    320–21, 325, 326, 327, 330
  jazz, 174, 201
  Saturday Night Club and, 63, 169–72,
    319
Mussolini, Benito, 9, 279, 337
My Ántonia (Cather), 152n
My Life as Author and Editor (Mencken), xi,
  xii, 96, 110, 137, 157, 179, 196, 206,
  212, 224, 303, 307, 309, 317, 337

Nabokov, Vladimir, 281
Nathan, George Jean, 83–88, 102, 172,
  177–82, 206, 319, 324
  American Mercury and, 180, 181,
    188–200, 203, 209, 254–55, 256
  background of, 85
  as editor of *Smart Set*, 84, 108–15, 117,
    122, 145, 151, 152, 179–82, 195, 200
  HLM's friendship with, 16, 86, 87–88,
    181, 195–200, 209, 224, 255, 305
  HLM's pen portrait of, 88
  as Jewish, 16, 87, 88
  self-sketch by, 86–87

Nation, 12, 72, 136, 202, 216
"National Letters, The" (Mencken), 184
National Review, 341
"Neglected Anniversary, A" (Mencken), 134
New Deal, 2, 4, 6, 10, 276–81, 285, 341
New Dictionary of Quotations on Historical Principles from Ancient and Modern Sources, A (Mencken), 106, 170, 302–3
New England Magazine, 61
New England Watch and Ward Society, 225
New Machiavelli, The (Wells), 84
New Orleans Times-Democrat, 102
"New Platinum Toning Bath for Silver Prints, A" (Mencken), 39
New Republic, 12, 84, 136, 175, 193, 261, 281–82
Newspaper Days (Mencken), 43–44, 46–49, 52–53, 59, 66–68, 121, 285, 292, 295–96, 312, 345
   George Bernard Shaw in, 70–71
   "Girl from Red Lion, P.A." in, 345
   great fire in, 66–67
   last chapter of, 74–75
   managerial authority in, 63–64
   Ventures into Verse in, 61, 62
Newspaper Guild, 294
New York City, 79, 88, 112, 180
   HLM's visits to, 67, 83, 93, 105–6, 108, 138, 139, 141, 152–53, 169–70, 172, 174, 177, 202, 210–11, 227, 233, 234
New Yorker, 84, 269, 278, 307
   American Mercury compared with, 204, 258
   HLM's writing for, 66, 148, 271, 276, 280–81, 283, 284, 291, 295, 303, 306, 311
   "Postscripts to the American Language" in, 311
New York Evening Mail, 133–34, 138, 140, 150
New York Evening Sun, 151
New York Evening World, 101
New York Herald, 74, 85
New York Herald Tribune, 225, 332
New York Newsday, 339
New York Post, 72, 105
New York Public Library, 301, 334, 337, 338n
New York Review of Books, 340
New York Society for the Suppression of Vice, 128
New York Sun, 51, 53
New York Times, 28, 72, 80–81, 252, 255, 321n, 332
New York Times Book Review, 324

New York World, 47, 201
Niebuhr, Reinhold, 245
Nietzsche, Friedrich, 28, 65, 90, 101, 106, 123–36, 194, 248, 266, 326
   HLM's translation of, 10, 33, 176
   see also Philosophy of Friedrich Nietzsche, The
Norris, Frank, 84, 117
North American Review, 290–91
Notes on Democracy (Mencken), 14, 228–31, 240, 245–46
"Novel of the First Rank, A" (Mencken), 102–6

"Ode to the Pennant on the Centerfield Pole" (Mencken), 41–42
Omaha World Herald, 261
"On Being an American" (Mencken), 184–85
O'Neill, Eugene, 10, 85, 114, 194, 195, 197
Only Yesterday (Allen), 17, 221
"Ordeal of a Philosopher" (Mencken), 280–81, 284
Order of the Star Spangled Banner, 22
"Orf'cer Boy, The" (Mencken), 62
Origin of Species, The (Darwin), 213
Orwell, George, 321n, 341
Owens, Hamilton, 16, 21, 252, 286, 331–34

Palestine, 271, 288, 308
Palmer, A. Mitchell, 153
Parisienne Monthly Magazine, 122, 128
Parke-Bernet, 158n
Parker, Dorothy, 204
Parmenter, James P., 227
Patterson, Paul, 3, 8, 162, 163, 232, 250, 251–52, 266, 284–85, 294, 304, 310, 317, 318
Penn, Irving, 305
Pennypacker, Isaac R., 192
Pepys, Samuel, 168
Philadelphia Evening Telegraph, 67
Philosophy of Friedrich Nietzsche, The (Mencken), 16, 73, 80–83, 96, 97, 122, 123, 175
Photoplay, 233
Piaget, Jean, 247
Picasso, Pablo, 29, 76
Pittsburgh Jewish Chronicle, 248–49
Plain Tales from the Hills (Kipling), 38
Platt, Thomas, 68
Poe, Edgar Allan, 10, 168, 336
Point of No Return (Marquand), 187–88, 321n
Portable Faulkner (Cowley, ed.), 312
Posner, Richard, 341
"Pother About Glands, The" (Hussey), 193

Pound, Ezra, 10, 107, 323, 341
Pratt, Enoch, 23
Pratt, Enoch, Free Library, 11, 20–21, 36,
    62, 73, 81, 158, 251, 301, 303, 334,
    335, 337–40
Prejudices (Mencken; Farrell, ed.), 338
Prejudices: First Series (Mencken), 159–62,
    345
Prejudices: Second Series (Mencken), 176,
    184, 345
Prejudices: Third Series (Mencken), 179,
    184–85, 345
Prejudices: Fourth Series (Mencken), 196,
    210, 345
Prejudices: Fifth Series (Mencken), 210, 232,
    345
Prejudices: Sixth Series (Mencken), 7–8, 210,
    228, 237, 345
Pride and Prejudice (Austen), 349
Pringle, Aileen, 232–35, 241, 242, 253, 305
printing, 34, 36–37, 40
Progressive Party, 311
"Progress of the Great Crusade"
    (Mencken), 294
Prohibition, 144–45, 152, 153, 169n, 194,
    265n, 332
Protestants, 22, 248
    fundamentalism and, 213, 214, 218–19,
    221, 222
Puritanism, 5, 10, 125, 127, 128, 175, 183,
    221, 225, 342, 346
"Puritanism as a Literary Force"
    (Mencken), 134–35

racism, 339, 340, 346
Rand, Ayn, 321n, 341
Randall, James R., 23
Ransom, John Crowe, 258
Rascoe, Burton, 161, 226, 343
Raulston, John T., 213, 219
Red Book, 73
"Reflection" (Bloom), 139–40
"Remember Us!" (Hecht), 301
Republicans, 25, 68, 144, 196, 262–63, 278,
    310, 332–33
Reviewer, 206
"Rhyme and Reason" column, 55
Rimsky-Korsakov, Nikolai, 191
Rinehart, Mary Roberts, 89
Ritchie, Albert, 4, 7
Roman Catholicism, 87, 240, 248, 314
Roosevelt, Alice, 85
Roosevelt, Franklin Delano, 1–9, 12, 29, 92,
    294, 297
    anti-Mencken speech of, 6–8
    death of, 9, 307

HLM's views on, 1–2, 4–9, 12, 13, 143,
    263–67, 273, 276–79, 283, 285, 287,
    289, 290, 293, 300, 306–7, 309, 311,
    312, 328
Roosevelt, Theodore, 68, 76, 85, 97–98,
    143–44, 263
Ross, Harold, 204, 258, 281, 291, 306
"Ruin of an Artist, The" (Mencken), 291
Rumely, Edward, 138
Runyon, Damon, 84, 180
Russia, 131, 279, 288

"Sahara of the Bozart, The" (Mencken), 134
San Francisco Bulletin, 72–73
Santayana, George, 191
Saroyan, William, 111
Saturday Evening Post, 63, 113, 177
Saturday Night Club, 63, 169–72, 208, 254,
    274, 299, 307, 309, 318, 319, 334, 336
Saturday Review, 272, 324
Saucy Stories, 122
Schaff, Harrison Hale, 70, 73, 80
Schattenstein, Nikol, 235–36, 332
Schellhase's, 170
Schilling, Mary, 37
Schubert, Franz, 127, 250, 326
Schuyler, George, 16, 203–4
Scopes, John T., 212–14, 216, 219, 221–22
Scopes trial, 204, 212–22
    HLM's contradictory views on, 216–18
    HLM's coverage of, 217–22, 225, 226,
    229, 238, 332
Sedgwick, Ellery, 57–58, 67, 73, 109, 197,
    274
Sedition Act, 144, 153
Senate, U.S., 4, 173
"Sex and the Co-Ed" (De Voto), 227
Shakespeare, William, 56–57, 202
Shaw, George Bernard, 10, 70–73, 118, 160,
    171, 193, 194, 320, 341
Sheppard, Sam, 326
Sherman, Stuart P., 136
Short Stories, 54
Sinclair, Upton, 88, 89, 117, 181, 194, 279
Sister Carrie (Dreiser), 91–96, 103, 128, 135
Sitting Bull, Chief, 260
Sixteenth Amendment, 144
Smart Set, 10, 83–85, 101–18, 128, 143,
    145–46, 150, 166, 175, 200, 203–6,
    224, 227, 274, 303, 312
    American Mercury compared with,
    189–91, 195, 203, 257
    Bloom's contributions to, 121–22, 138,
    139–40, 158n
    changes in ownership of, 106–8
    Cowley's views on, 115

HLM's book reviews for, 83, 88–91, 96,
97, 99, 101–6, 109, 135–38, 140, 159,
178, 179, 182–83
November 1914 issue of, 109–10
payment policy of, 113, 203
"Répétition Générale" in, 151–52, 157,
189, 242, 289
subscriber pamphlet for, 111–12
Smart Set Company, Inc., 108
social Darwinism, 65, 97, 116, 214, 215,
218
socialism, 97, 98, 256, 342, 346
Social Statics (Spencer), 215*n*
"Song of the Slapstick" (Mencken), 55*n*
Sousa, John Philip, 49–50
Southern Album (Sara Haardt Mencken),
274
Soviet Union, 173, 314
Spencer, Herbert, 215*n*
Spivak, Lawrence, 300–301
Splint, Fred, 96
Stalin, Joseph, 9, 92, 279–80, 314
Stearns, Harold, 188–89
Steinach, Eugen, 283
Stieff, Charles M., 33
Stopes, Marie, 194
Strauss, Richard, 76, 313
Stravinsky, Igor, 29, 238, 313
Strube, Gustav, 170
Suckow, Ruth, 192
"Suggestions to Our Visitors" (leaflet),
152–53
Sumner, John S., 128, 130
Sun Also Rises, The (Hemingway), 11
Sunpapers, see Baltimore *Evening Sun*;
Baltimore *Morning Sun*; Baltimore *Sun*
Sunpapers of Baltimore, The, 284, 320
Supreme Court, U.S., 214, 235
Swope, Herbert Bayard, 101

Tales of the Jazz Age (Fitzgerald), 11
Tate, Allen, 258
Teasdale, Sara, 109
television, 313, 320, 325
Tennessee v. *Scopes*, 212–14, 216–19, 221–22
Tennyson, Alfred, Lord, 38, 42, 202
Thackeray, William Makepeace, 38, 42
Thayer, John Adams, 106–8
Thirty-five Years of Newspaper Work
(Mencken), xii, 210, 217, 220, 259, 265,
269, 286, 303, 317, 337–38, 340
This Side of Paradise (Fitzgerald), 161, 162
"Three—Minus One" (Braley), 180–81
Thurber, James, 281
Thus Spake Zarathustra (Nietzsche), 124
Time, 191, 193, 271, 300, 302, 319

Times (London), 333
Times Literary Supplement, 291–92
"Time to Be Wary, A" (Mencken), 265
Titan, The (Dreiser), 127–28, 135
tobacco business, 24, 25, 38, 98
HLM in, 40, 41, 44, 49, 50
Tocqueville, Alexis de, 126*n*
Tolstoy, Leo, 104
"To R. K." (Mencken), 61
Town Topics, 84, 85, 106
Treatise on Right and Wrong (Mencken), 272,
312
Treatise on the Gods (Mencken), 240, 244–50,
259, 269, 272, 288, 307, 312
Trilling, Lionel, 189, 324
Triumph of the Egg, The (Anderson), 162
Truman, Harry, 4, 310–11, 313, 325
Tudor, Antony, 86
Twain, Mark, 25, 29, 34–36, 40, 48, 56, 90,
115, 118, 150, 281, 348

Ulysses (Joyce), 114
United Press, 252, 262
Untermeyer, Louis, 116, 185
"Untold Tales" column, 55, 56

Valentino, Rudolph, 233, 236
Van Doren, Carl, 181, 192
Van Doren, Mark, 322
Vanity Fair, 175, 255–56
Van Vechten, Carl, 204
Ventures into Verse (Mencken), 60–63, 81
Victoria, Queen of England, 27, 76
Views and Reviews (James), 89
Vintage Mencken, The (Cooke, ed.), 338

Wagner, Philip, 285, 286, 293, 325
Wagner, Richard, 29, 90, 330, 339
Wallace, Henry A., 311
Warner, Eltinge F., 108, 109, 122
Warren, Robert Penn, 258
Washington, D.C., 66–67, 98, 119, 120,
138, 140, 234
Gridiron Club dinner in, 1–10, 273
Washington Post, 66–67, 339
Waugh, Evelyn, 314–15, 317
Ways, Max, 44, 48–52, 63, 76
Wellek, René, 116, 343
Wells, H. G., 108, 140, 160, 341, 348
Wharton, Edith, 115–16, 178
"What Is Going On in the World"
(Mencken), 261–62
White, E. B., 12, 271
White, Katharine, 271, 284
White, William Allen, 129–30
Wilde, Oscar, 14, 84, 193, 281

Wilder, Thornton, 238
Wiley, Z. K., 43, 44
Wilhelm II, Kaiser of Germany, 80, 123,
    230–31, 251
Williams, William Carlos, 306
Willkie, Wendell, 293
Wilson, Edmund, 10, 13, 87, 117, 175, 181,
    220, 231, 261, 281–82, 307
Wilson, Woodrow, 123, 131, 143–44, 153,
    213, 221, 257
Woollcott, Alexander, 169
Woollcott, Willie, 169–70
World War I, 78, 107, 108, 112, 123,
    130–33, 140–45, 151, 153, 213, 220,
    268, 278, 286, 289
  effects of, 161, 173, 175, 227

World War II, 9, 15, 20, 78, 151, 171n,
    277–79, 288, 289, 294–300, 306–8,
    314, 342
Wright, Harold Bell, 89
Wright, James L., 2, 6
Wright, Richard, 346–47
Wright, Willard Huntington, 106–7, 109
Writer's Market, 256

Yeats, W. B., 107, 283
yellow journalism, 29, 47, 74
"Youth" (Conrad), 326

Zabel, Morton Dauwen, 302–3
Zionism, 271n
Zola, Émile, 89, 104

ABOUT THE AUTHOR

TERRY TEACHOUT is the music critic of *Commentary* and a contributor to *Time*, for which he covers classical music, jazz, and dance, and the *Washington Post*, for which he writes "Second City," a column about the arts in New York City. He writes about jazz and dance for the *New York Times*, books for *Book Magazine*, *National Review*, the *New York Times Book Review*, and the *Wall Street Journal*, and film for *Crisis*, and is a commentator on National Public Radio's *Performance Today*. He was previously a senior editor of *Harper's Magazine*, an editorial writer for the *New York Daily News*, and the *News'* classical music and dance critic.

Teachout is the author of *City Limits: Memories of a Small-Town Boy* (1991) and the editor of *Beyond the Boom: New Voices on American Life, Culture, and Politics* (1990) and *Ghosts on the Roof: Selected Journalism of Whittaker Chambers, 1931–1959* (1989). He also wrote the foreword to Paul Taylor's *Private Domain: An Autobiography* and contributed to *The Oxford Companion to Jazz*. In 1992 he rediscovered the manuscript of *A Second Mencken Chrestomathy* among H. L. Mencken's private papers, subsequently editing it for publication by Alfred A. Knopf (1995).

Born in Cape Girardeau, Missouri, in 1956, Teachout attended St. John's College, William Jewell College, and the University of Illinois at Urbana-Champaign. From 1975 to 1983 he lived in Kansas City, where he worked as a jazz bassist and as a music critic for the *Kansas City Star*. The oldest son of Evelyn Teachout of Sikeston, Missouri, he now lives in New York City.